# SECOND EDITION
# FIRE AND EMERGENCY SERVICES ADMINISTRATION

## Management and Leadership Practices

## L. Charles Smeby, Jr., MIFireE

JONES & BARTLETT
LEARNING

World Headquarters
Jones & Bartlett Learning
5 Wall Street
Burlington, MA 01803
978-443-5000
info@jblearning.com
www.jblearning.com

Jones & Bartlett Learning books and products are available through most bookstores and online booksellers. To contact Jones & Bartlett Learning directly, call 800-832-0034, fax 978-443-8000, or visit our website, www.jblearning.com.

Substantial discounts on bulk quantities of Jones & Bartlett Learning publications are available to corporations, professional associations, and other qualified organizations. For details and specific discount information, contact the special sales department at Jones & Bartlett Learning via the above contact information or send an email to specialsales@jblearning.com.

Copyright © 2014 by Jones & Bartlett Learning, LLC, an Ascend Learning Company

All rights reserved. No part of the material protected by this copyright may be reproduced or utilized in any form, electronic or mechanical, including photocopying, recording, or by any information storage and retrieval system, without written permission from the copyright owner.

The content, statements, views, and opinions herein are the sole expression of the respective authors and not that of Jones & Bartlett Learning, LLC. Reference herein to any specific commercial product, process, or service by trade name, trademark, manufacturer, or otherwise does not constitute or imply its endorsement or recommendation by Jones & Bartlett Learning, LLC and such reference shall not be used for advertising or product endorsement purposes. All trademarks displayed are the trademarks of the parties noted herein. *Fire and Emergency Services Administration: Management and Leadership Practices, Second Edition* is an independent publication and has not been authorized, sponsored, or otherwise approved by the owners of the trademarks or service marks referenced in this product.

There may be images in this book that feature models; these models do not necessarily endorse, represent, or participate in the activities represented in the images. Any screenshots in this product are for educational and instructive purposes only. Any individuals and scenarios featured in the case studies throughout this product may be real or fictitious, but are used for instructional purposes only.

**Production Credits**
Chief Executive Officer: Ty Field
President: James Homer
SVP, Editor-in-Chief: Michael Johnson
Executive Publisher: Kimberly Brophy
Executive Acquisitions Editor: William Larkin
Associate Editor: Carly Lavoie
Production Editor: Jessica deMartin
Associate Production Editor: Nora Menzi
VP Sales, Public Safety Group: Matthew Maniscalco
Director of Sales, Public Safety Group: Patricia Einstein
Senior Marketing Manager: Brian Rooney
VP, Manufacturing and Inventory Control: Therese Connell
Composition: Cenveo Publisher Services
Cover Design: Kristin E. Parker
Rights & Photo Research Associate: Lian Bruno
Cover Image: © Jones & Bartlett Learning. Photographed by Glen E. Ellman.
Printing and Binding: Courier Companies
Cover Printing: Courier Companies

**Library of Congress Cataloging-in-Publication Data**
Smeby, L. Charles.
 Fire and emergency services administration : management and leadership
practices — 2nd ed. / L. Charles Smeby, Jr.
     p. cm.
 Includes bibliographical references and index.
 ISBN 978-1-4496-0583-4
 1. Fire departments—Management.  2. Personnel management.  3. Leadership.  I. Title.
 TH9158.S64 2013
 363.37068'4—dc23
                                                                                            2012017924
6048

Printed in the United States of America
17  16  15  14  13    10 9 8 7 6 5 4 3 2 1

# Brief Contents

**CHAPTER 1** Historical Foundations of Fire and Emergency Services .......................................... 1

**CHAPTER 2** Introduction to Administration .......................................... 12

**CHAPTER 3** Management .......................................... 28

**CHAPTER 4** Leading Change .......................................... 45

**CHAPTER 5** Financial Management .......................................... 56

**CHAPTER 6** Human Resources Management .......................................... 79

**CHAPTER 7** Customer Service .......................................... 101

**CHAPTER 8** Training and Education .......................................... 113

**CHAPTER 9** Health and Safety .......................................... 126

**CHAPTER 10** Government Regulation, Laws, and the Courts .......................................... 141

**CHAPTER 11** Ethics .......................................... 155

**CHAPTER 12** Public Policy Analysis .......................................... 164

**CHAPTER 13** The Future .......................................... 183

    Appendix A .......................................... 189

    Index .......................................... 191

# Contents

| | | |
|---|---|---|
| Instructor Resources | | vii |
| About the Author | | ix |
| Reviewers | | x |
| Foreword | | xii |

**CHAPTER 1  Historical Foundations of Fire and Emergency Services** .......... 1
- Prologue to the Future .............. 1
- Local Beginnings ................... 2
- National Development Efforts ........ 2
  - Development of the Modern Fire Service ................... 2
  - Development of Modern Emergency Medical Services ...... 3
- Universal Professional Standards ..... 4
  - Fire Standards .................. 4
  - EMS Standards .................. 5
  - Federal Involvement ............. 5
- Melding of Fire Service and EMS ...... 5
- Unified Federal Emergency Response ...................... 6
- Current Problems Facing Fire Service and EMS ................ 8
  - Challenges of the Modern Fire Service ................... 8
  - Challenges of Modern EMS ........ 8
- Today's Fire and Emergency Services ........................ 9
- References ....................... 11

**CHAPTER 2  Introduction to Administration** ........ 12
- Combining Management and Leadership .................. 12
- Professional Qualifications for Administrators ............... 13
  - Staff–Line Distinctions .......... 14
- Selection of Administrators ......... 15
  - By Election .................... 15
  - By Appointment ................ 15
- Rules and Regulations .............. 16
  - National Consensus Standards .... 17
  - Standard Operating Procedures ... 18
  - Direct Supervision and Standardization ............... 18
- Bringing About Change ............ 19
  - Getting Feedback .............. 19
  - Sources of Power ............... 20
  - The Process of Negotiation ...... 25
- References ....................... 27

**CHAPTER 3  Management** ...................... 28
- What Is Management? .............. 29
- Decision-Making .................. 29
  - When to Use Group Decision-Making ............... 30
  - How to Select Group Members ...... 30
  - Optimal Group Size ............. 30
  - Techniques to Aid Group Discussions .................... 30
  - Disadvantages of Group Decision-Making ............... 33
- Organizing ....................... 33
- Directing ........................ 34
  - Creating Rules and Regulations ..... 34
  - Motivating Staff ................ 34
- Controlling ...................... 35
  - Measuring Performance .......... 36
  - Comparing Results to a Standard ... 36
  - Taking Corrective Action ......... 36
- Management Tactics ............... 37
  - Technology-based Programs ....... 37
  - Total Quality Management ........ 38
  - Leading by Example ............. 38
  - Broad-based Empowerment ....... 39
  - Managing by Walking Around ..... 39
- Assessing Managerial Performance .... 39
- References ....................... 44

**CHAPTER 4  Leading Change** ................... 45
- Embracing Change ................. 45
- Creating Change .................. 46
- Step 1: Identify the Need for Change and Create a Sense of Urgency ...... 46
- Step 2: Create a Guiding Coalition ..... 48
- Step 3: Develop a Vision ............ 49
- Step 4: Communicate the Vision ...... 49
- Step 5: Overcome Barriers .......... 50
  - Member Resistance ............. 51
- Step 6: Create Short-Term Goals ...... 52
- Step 7: Institutionalize the Change .... 53
- Leading Change at All Levels ........ 54
- References ....................... 55

**CHAPTER 5  Financial Management** ............. 56
- Introduction ..................... 57
- Budgets ......................... 57
  - Types of Budgets ............... 57

**Budget Process and Planning**......... 58
   Budgetary Cycle .................. 58
   Planning ........................ 58
   Submission...................... 58
   Review and Approval ............. 58
   Management ..................... 59
**Key Players in the Budgetary Process** ....................... 59
   Financial Manager ................ 59
   Bureaucrats and Other Agency Managers .................... 60
   The Public and Their Representatives ................ 60
**Revenues**........................... 61
   Taxes ........................... 61
   Fees ............................ 62
   Government Bonds and Short-Term Borrowing ........... 62
   Investments ..................... 63
   Other Forms of Borrowing ......... 64
**Techniques for Increasing Funding** .... 64
   Provide Evidence from National Standards ............. 64
   Perform a Cost-Benefit Analysis ..... 65
   Define Sources of Funding ......... 65
**Expenditures** ....................... 66
   Salaries and Benefits............... 66
   Tracking Expenditures ............. 66
   Purchasing Policies ................ 68
   Total Cost Purchasing ............. 68
   Cooperative Purchasing............ 68
   Legal Considerations .............. 68
**Accountability and Auditing** ......... 69
**Making Budget Cuts** ................ 70
   Adaptive Policies ................. 70
   Common Areas for Budget Cuts ..... 71
   Protecting the Budget ............. 71
**Tax Considerations**.................. 71
   Tax Incidence Analysis............. 72
   Interstate Considerations .......... 73
   Tax Avoidance and Reduction ....... 73
**Local Impact of a Global Economy** ..... 73
**References** ........................ 78

**CHAPTER 6 Human Resources Management** ...... 79
**The Most Valuable Resource** ......... 79
**The Function and Operation of Human Resources** ................. 80
**Diversity in the Department** ......... 80
   Affirmative Action Cases ........... 81
   Diversity Selection in Practice....... 81
   Diversity Sensitivity Training........ 82
   Recruitment and Selection of Firefighters and Volunteers..... 83
**Legal Issues** ....................... 83
   Hiring Issues ..................... 84
   Reference Checks................. 84
   First Amendment: Freedom of Speech ..................... 85
   Civil Rights ...................... 85
   The Americans with Disabilities Act .................. 85
   Injuries on the Job ................ 86
   Pregnancy Issues ................. 86
   Family and Medical Leave Act ....... 87
   Drug and Alcohol Testing........... 87
   Sexual Harassment ............... 87
   Fair Labor Standards Act ........... 88
   The Financial Impact of Lawsuits ..................... 89
   Recent Supreme Court Cases ....... 89
**Insubordination** .................... 90
   Silencing Complaints without Violating the Law ............... 90
   Public Sector Discipline............. 90
   Fair, Reasonable, and Evenly Enforced Discipline ............. 90
   Probationary Period............... 91
   Terminations.................... 91
   Constructive Discharge ............ 92
**State and Local Hiring Laws** ......... 92
   Validation ...................... 92
**Job Classification** ................... 94
   Emergency Medical Responder...... 94
   Emergency Medical Technician ...... 94
   Advanced EMT ................... 94
   Paramedic....................... 94
**Recruitment** ....................... 94
   The Selection Process ............. 95
**Unions**............................ 95
   Public Sector Unions .............. 95
   Private Sector Unions ............. 96
   Strikes and Job Actions ............ 96
   Bargaining Units.................. 96
   Local Government Representatives................. 96
   Grievances ...................... 96
   Progressive Labor Relations ........ 97
   Getting to Yes ................... 97
**Motivation**......................... 97
   Retention Within Volunteer Departments................... 97
   Retention Within Combination Departments................... 98
**References** ........................ 100

**CHAPTER 7 Customer Service**.................... 101
**Overview**.......................... 101
**What Is Private? What Is Public?** ...... 102
   Justification for Government Intervention ................... 102
**Keeping Pace with Community Demographic Change** ............. 103

Expanded Services. . . . . . . . . . . . . . . . . 103
   Fire Suppression . . . . . . . . . . . . . . . 103
   Emergency Medical Services. . . . . . . . 104
EMS Challenges. . . . . . . . . . . . . . . . . . . . 105
   Overcommitment of Resources . . . . . 105
   Reduction of the Volume of
      Medical Calls . . . . . . . . . . . . . . . . . 105
   Issues Surrounding Healthcare
      Reform. . . . . . . . . . . . . . . . . . . . . . 106
   Overloading of System . . . . . . . . . . . 106
   Integration with the Healthcare
      System . . . . . . . . . . . . . . . . . . . . . 106
What Is Fire and Emergency Service's
   Customer Service Duty? . . . . . . . . . . 106
   Fire Prevention . . . . . . . . . . . . . . . . . 107
   Home Fire Sprinkler Systems . . . . . . . 107
   Fire Safety Inspections . . . . . . . . . . . 108
   Enforcement of Fire
      Safety Codes. . . . . . . . . . . . . . . . . 109
   Preventive Medicine. . . . . . . . . . . . . . 110
Customer Service at Emergencies:
   The Extra Mile . . . . . . . . . . . . . . . . . 110
References . . . . . . . . . . . . . . . . . . . . . . 111

**CHAPTER 8** Training and Education. . . . . . . . . . . . 113
Introduction to Training
   and Education. . . . . . . . . . . . . . . . . . 113
Firefighting Training and
   Certification . . . . . . . . . . . . . . . . . . 114
   Basic Firefighter Training. . . . . . . . . . 115
   Live-Fire Training . . . . . . . . . . . . . . . 116
   EMS Training and Certification . . . . . . 116
   Technology-Based Training and
      Certifications . . . . . . . . . . . . . . . . 116
   Seminars and Training Sessions. . . . . 117
Recertification . . . . . . . . . . . . . . . . . . . 117
Training to Fit a Need: Standard
   Operating Procedures. . . . . . . . . . . . 117
   Consistency and Reliability . . . . . . . . 118
   Variances in SOPs. . . . . . . . . . . . . . . 118
   Critique of SOPs . . . . . . . . . . . . . . . . 118
Training to Fit a Need: Hazardous
   Materials. . . . . . . . . . . . . . . . . . . . . 118
Training to Fit a Need: NIMS. . . . . . . . . . 119
Training Goals: Initial Fire Attack. . . . . . 120
Higher Education . . . . . . . . . . . . . . . . . 121
Officer Education and Training . . . . . . . 122
National Professional
   Development Model . . . . . . . . . . . . . 123
References . . . . . . . . . . . . . . . . . . . . . . 125

**CHAPTER 9** Health and Safety . . . . . . . . . . . . . . . . 126
An Introduction to Health and
   Safety for Emergency Personnel . . . . 126
   Firefighter Safety . . . . . . . . . . . . . . . 126
   Emergency Medical Services
      Personnel Safety . . . . . . . . . . . . . 127

NFPA Safety and Health Standard . . . . . 127
   NFPA 1500 Major Topics. . . . . . . . . . . 127
   Federal Safety Regulations . . . . . . . . 128
Firefighter Injury and Fatality
   Prevention. . . . . . . . . . . . . . . . . . . . 129
EMS Injury and Fatality Prevention . . . . 130
Ambulance and Fire Apparatus
   Crash Prevention . . . . . . . . . . . . . . . 130
   Crashes Involving Ambulances . . . . . . 131
   Crashes Involving Fire
      Apparatus . . . . . . . . . . . . . . . . . . 132
Physical Fitness . . . . . . . . . . . . . . . . . . 132
   The Need to Stay in Shape . . . . . . . . 132
   Physical Fitness Program. . . . . . . . . . 134
   Physical Fitness Testing. . . . . . . . . . . 135
Other Occupational Hazards. . . . . . . . . 136
   Personal Protective Equipment. . . . . . 136
   Violence . . . . . . . . . . . . . . . . . . . . . 136
   Stress . . . . . . . . . . . . . . . . . . . . . . . 137
   Sleep Deprivation. . . . . . . . . . . . . . . 137
References . . . . . . . . . . . . . . . . . . . . . . 139

**CHAPTER 10** Government Regulation, Laws,
   and the Courts . . . . . . . . . . . . . . . . 141
Government Regulation. . . . . . . . . . . . . 141
   Role of Government to
      Protect Citizens. . . . . . . . . . . . . . . 142
   Market Failure Consequences. . . . . . . 142
   Monopolies. . . . . . . . . . . . . . . . . . . . 143
Federal Regulations. . . . . . . . . . . . . . . 143
   Offering Effective Input on Codes,
      Standards, and Regulations . . . . . . 144
State Regulations . . . . . . . . . . . . . . . . 144
   Building and Fire Codes. . . . . . . . . . . 144
   Zoning Regulations . . . . . . . . . . . . . . 145
Union Contracts . . . . . . . . . . . . . . . . . . 145
Tax Regulation . . . . . . . . . . . . . . . . . . . 146
Hiring and Personnel
   Regulations . . . . . . . . . . . . . . . . . . . 146
OSHA Regulations . . . . . . . . . . . . . . . . . 146
   OSHA 29 CFR 1910.146:
      Permit-Required Confined
      Spaces. . . . . . . . . . . . . . . . . . . . . 147
   OSHA 29 CFR 1910.134:
      Respiratory Protection. . . . . . . . . . 147
   OSHA 29 CFR 1910.1030:
      Occupational Exposure to
      Bloodborne Pathogens . . . . . . . . . 147
   OSHA 29 CFR 1910.120:
      Hazardous Waste Operations
      and Emergency Response . . . . . . . 147
   OSHA 29 CFR 1910.156:
      Fire Brigades. . . . . . . . . . . . . . . . . 148
   OSHA's General Duty Clause . . . . . . . 148

NFPA Codes and Standards.......... 148
   NFPA 1500, *Standard on Fire Department Occupational Safety and Health Program* ...... 148
   NFPA 1710, *Standard for the Organization and Deployment of Fire Suppression Operations, Emergency Medical Operations, and Special Operations to the Public by Career Fire Departments* ............... 149
National Highway Traffic Safety Administration .............. 150
Legal and Court Issues ............. 150
   The Court System................. 151
   Attorneys........................ 151
   Administrative Rule Making ........ 152
   Legal Aspects of FES .............. 152
References ....................... 153

**CHAPTER 11** Ethics.......................... 155
Ethical Behavior................... 155
   Professional Ethics................ 156
Why People Lie.................... 156
Justification for Lying............... 157
   "It's for the Public Good".......... 157
   "It's in Self-Defense" ............. 158
   "It's a Miscommunication" ........ 158
   Ethical Tests ..................... 158
Consequences of Lies .............. 159
   Denial of the Public's Free Choice .................... 159
   Damage to Professional Reputation .................... 159
Moral Obligations of Public Roles ..................... 159
   Duty to Obey.................... 159
   Fairness Among Employees......... 160
   Cost-Benefit Analyses ............. 160
References ....................... 163

**CHAPTER 12** Public Policy Analysis............... 164
Future Planning and Decision-Making ................. 165
   Strategic Planning ................ 165

The Role of Budgeting Within Public Policy Analysis ............. 166
Measuring Outcomes ................ 166
   Staffing Levels ................... 167
   Response Times .................. 168
   Professional Competency .......... 169
Consensus Building.................. 169
Public Policy Presentation ........... 170
   Consider the Audience............. 171
   Practice ........................ 171
   Be Positive and Cheerful .......... 171
   Maintain Eye Contact.............. 171
   Use Down-to-Earth Ideas........... 172
   Use Visual Aids................... 172
   Involve the Audience ............. 172
Using Statistics .................... 172
   Case Studies..................... 174
Program Budgeting ................. 174
   Cost-Benefit Analysis.............. 174
   Program Analysis ................ 175
Policy Analysis Reference Sources..... 175
   NFPA 1710 ...................... 175
   ISO Fire Suppression Grading Schedule............... 176
Changing Social Perspective ......... 177
   Empowering Employees ........... 177
References ....................... 181

**CHAPTER 13** The Future....................... 183
Persistent Change in the Past and Future..................... 183
Technology and Research ........... 184
Economic Impacts .................. 185
Higher Education and Training ........ 186
Risk Management .................. 186
Homeland Security ................. 187
Advice for the Future ............... 187
References ....................... 188

Appendix A ....................... 189

Index ............................ 191

# Instructor Resources

## Instructor's ToolKit

Preparing for class is easy with the resources on this CD. The CD includes the following resources:

- **Adaptable PowerPoint Presentations.** Provide instructors with a powerful way to create presentations that are educational and engaging to their students. These slides can be modified and edited to meet instructors' specific needs.
- **Detailed Lecture Outlines.** Keyed to the PowerPoint presentations, these complete, ready-to-use lecture outlines include all of the topics covered in the text. The lecture outlines can be modified and customized to fit any course.
- **Electronic Test Bank**: Contains multiple-choice questions and allows instructors to create tailor-made classroom tests and quizzes quickly and easily by selecting, editing, organizing, and printing a test along with an answer key. Includes page references to the text.
- **Image and Table Bank.** Offers a selection of the images and tables found in the text. Instructors can use these graphics to incorporate more images into the PowerPoint presentations, make handouts, or enlarge a specific image for further discussion.

## Navigate Course Manager

Combining our robust teaching and learning materials with an intuitive and customizable learning platform, Navigate Course Manager enables instructors to create hybrid/online courses quickly and easily. The system allows instructors to readily complete the following tasks:

- Customize preloaded content or easily import new content.
- Provide online testing.
- Offer discussion forums, real-time chat, group projects, and assignments.
- Organize course curricula and schedules.
- Track student progress, generate reports, and manage training and compliance activities.

# About the Author

CHIEF (Retired) SMEBY has been an adjunct professor at the University of Florida since 1998. He develops, creates, and presents courses and provides administrative support for the Fire and Emergency Service Bachelor's distance learning program.

Previously, he was a Senior Fire Service Specialist at the National Fire Protection Association (NFPA) in the Public Fire Protection Division. He coordinated several NFPA fire service technical committees and provided guidance to the implementation, understanding, and intent of NFPA fire service standards.

Before joining the NFPA, Mr. Smeby had a twenty-year career with the Prince George's County, Maryland, Fire Department, where he retired as a Battalion Chief. In addition, he served on the Fire Department Accreditation Task Force for the International Association of Fire Chiefs. Presently, he is a member of the NFPA 1710 Fire and Emergency Service Organization and Deployment-Career technical committee.

Prior to his salaried PGFD career, Mr. Smeby began his service as a volunteer with the Bethesda-Chevy Chase Rescue Squad in Bethesda, Maryland and the College Park Volunteer Fire Department in College Park, Maryland. He has also taught in the fire science programs at Prince George's Community College and the University of Maryland. His education includes a Bachelor's degree in Fire Protection Engineering and a Master's of Public Policy from the University of Maryland.

# Reviewers

**Don Beckering, M. Ed.**
Minnesota State Colleges and Universities
St. Paul, Minnesota

**John P. Binaski, MS, EFO**
College of the Seqoias
Hanford, California

**Billy Bradshaw, EMT-P**
College Station Fire Department
College Station, Texas

**Katherine Burton, MS**
Coastal Carolina Community College
Jacksonville, North Carolina

**Paul E. Calderwood, EFO**
Providence College
Providence, Rhode Island

**Dennis Childress**
California Fire Chief's Association
San Clemente, California

**Gary M. Courtney**
Lakes Region Community College
Laconia, New Hampshire

**Leonard C. Edge, Jr.**
Central Piedmont Community College
Charlotte, North Carolina

**Craig R. Farnsworth**
Westampton Township Emergency Services
Westampton, New Jersey

**Travis Ford**
Volunteer State College
Gallatin, Tennessee

**William T. Giannini, BS, CEOSH**
Providence Fire Department
Providence, Rhode Island

**Robert Halpin**
Guilford Technical Community College
Randleman, North Carolina

**Jamie Hirsch**
Mt. San Antonio College
Walnut, California

**Kevin M. Huben, MS**
El Camino College
Torrance, California

**Gary Kistner**
Southern Illinois University Carbondale
Carbondale, Illinois

**John Kubilewicz**
Mercer County Fire Academy (retired)
Pine Beach, New Jersey

**Paul D. Matheis, BS, EFO**
California Fire Chiefs Association
Irvine, California

**Byron Mathews**
Cheyenne Fire and Rescue
Cheyenne, Wyoming

**R. Jeffery Maxfield, Ed. D., MPA**
Utah Valley University
Orem, Utah

**Bonnie Maynard, EMT-P**
Nashville Fire Department
Nashville, Tennessee

**James P. Moore**
Crystal Lake Fire Rescue Department
Crystal Lake, Illinois

**John M. Moschella, Ed. D., EFO, MIFireE**
Anna Maria College
Paxton, Massachusetts

**Everett G. Pierce, MS, CFO, MIFireE**
Anna Maria College
Paxton, Massachusetts

**Adam D. Piskura**
Connecticut Fire Academy
Windsor Locks, Connecticut

**Tracy E. Rickman, MPA**
Rio Hondo Community College
Santa Fe Springs, California

**Randall J. Sellnow, CFOD, MIFireE**
Madison Area Technical College
Madison, Wisconsin

**Hugh Stott**
West Chicago Fire District
West Chicago, Illinois

**Robert Swiger**
City of Clinton Fire Department
Raleigh, North Carolina

**Thomas B. Sturtevant, CFPS**
Texas Engineering Extension Service
College Station, Texas

**William Trisler, M. Ed**
Commission on Fire Prevention and Control
Windsor Locks, Connecticut

**Ed Trigeiro**
College of the Redwoods
Fort Bragg, California

**Bobby B. Valles, BAAS, EFO, LP**
Odessa Fire/Rescue
Odessa College
Odessa, Texas

**Martin Walsh, CFPS**
San Diego Miramar College
San Diego, California

**William Wren, BA, SFI, FO-1, FSI-2**
New Hartford Fire Department
New Hartford, New York

# Foreword

This text is written for all levels of chief officers that administer, manage, or lead fire, emergency medical services (EMS), rescue, and/or emergency management organizations. After reading and studying this text, chief officers (who, when acting in their various roles, are also called "managers," "leaders," and "administrators") will have the knowledge to make policy decisions and the skills needed to lead the organization through progressive change. These same abilities will help anyone in a top position thrive—despite any political pressures that are present in the public government agency.

Many fire departments and EMS organizations were propelled into a new level of consciousness after the September 11, 2001, terrorist attack; these organizations are entering a new era of expanded and professional emergency services, including EMS, hazardous materials response, fire prevention, and disaster planning. In addition to these new services, however, traditional fires still occur. Although fewer fires occur now than in the past, these incidents remain a challenge, perhaps even more so now because firefighters have less real-world experience. In addition, today's fires can be hotter and faster spreading. Newer buildings use structural components that do not have the same fire resistance possessed by older construction, often resulting in earlier structural collapse during a fire. The contents of modern buildings have higher rates of heat release and smoke, resulting in shorter times to flashover. The future is full of ever-growing challenges.

The fire and emergency services profession has made many progressive changes in equipment, training, and funding, but the job of making improvements is never complete. This text provides the knowledge needed by chief officers and their staff to identify and implement progressive change that will help keep firefighters safe while providing the best public fire and emergency services.

## The Challenge of Gaining New Knowledge

Remember the firehouse pronouncement "200 years of tradition unhampered by progress"? (It is now actually closer to 270 years since the first volunteer fire department was created in 1736!) If you are reading this text, you are probably open-minded about the future; but remember, many people remain fearful of change. This fear can result in strong opposition to new ideas. Expect to hear "we tried that before and it did not work," "it is too expensive," "the union or management will never approve," etc. Keep in mind that this resistance is not personal.

The advice in this text will challenge accepted wisdom in fire and emergency services' custom, culture, and traditions. Be a risk taker as you read these words and consider new ideas, but remain cautious. Ask yourself if the idea is based on hard scientific data or if it is just someone's opinion. Seek out other open-minded individuals in your agency with whom you can speak privately. You may scare some members of your organization if you openly express your newfound ideas. Turn on the light of knowledge slowly for these members.

# CHAPTER 1

# Historical Foundations of Fire and Emergency Services

## Fire and Emergency Services Higher Education (FESHE) Course Objectives

**Module I: Leading and Managing Purposefully with a Community Approach**

The students will:

1. Describe the role of the fire/emergency medical services department as a part of the community government and comprehensive plan. (pp 1–10)
4. Identify local, state, and national organizations that will be beneficial to your department. (pp 2–7)

**Module IV: Leading Change**

The students will:

1. Describe the importance of accepting and managing change within the fire and emergency service department. (pp 1–10)

**Module V: CRM—A 21st Century FESA Responsibility**

The students will:

5. Identify direct and indirect costs associated with fire. (pp 2–3, 8, 9)
6. Analyze economic incentives that encourage and discourage fire prevention. (pp 2–3)

## Knowledge Objectives

After studying this chapter, the student will be able to:

1. Understand the history of fire and emergency services (FES) and its impact on contemporary organizations. (pp 1–10)
2. Examine the effect that the insurance industry has had on building and fire prevention codes. (pp 2–3)
3. Examine the influence of the National Fire Protection Association (NFPA) on the fire service. (pp 4–5, 7)
4. Explore federal involvement in regulation and funding of state and local FES. (p 5)
5. Examine challenges in FES. (pp 8–9)
6. Examine progressive trends in FES. (pp 9–10)

## Prologue to the Future

Leaders are always thinking about the future and looking ahead. However, it is important to begin with a basic understanding of how we arrived at where we are today. As the American philosopher George Santayana said, "Those who cannot remember the past are condemned to repeat it." By examining the history of FES, you can gain insight into the foundation of these organizations, avoid mistakes from the past, and identify what works in the present to serve as a guide toward a brighter future.

Your role in leadership, administration, and management can be enhanced with knowledge of the history and current status of the FES profession. Like any journey, you need to know where the trip began and where you are at present to plan the route to the future, including all the points along the way. The history of any profession is important because it explains many current policies and points to issues that remain unresolved.

## Local Beginnings

The first fire protection service on record in North America started in 1648 in the town of New Amsterdam, now part of New York City. Like other early fire departments that soon followed, it was an all-volunteer organization. However, it was not long before the public began to demand fire safety codes and paid fire departments to combat the massive fires that struck many large cities across the United States. In 1853, the nation's oldest fully paid fire department was organized in Cincinnati, Ohio **FIGURE 1-1**.

As demand for fire protection grew, other towns and cities developed their own independent fire departments. However, the need for some communication and standardization across departments soon became clear. When a large conflagration consumed Baltimore in 1904, the mayor called on fire departments in Washington, D.C., Philadelphia, and New York to assist. These agencies responded but were dismayed to find that their hose couplings would not fit the Baltimore fire hydrants, and the city continued to burn.

These local origins also explain why there are now more than 30,000 fire departments across the United States. Most of these departments protect a population of fewer than 2,500 citizens, and only 15 departments protect more than one million people each (Karter, 2010). Interestingly, the largest department—New York City Fire Department—is actually a consolidation of five counties.

Emergency medical response also started out on a small scale. The first prehospital medical response system on record served with Napoleon's army in 1797. This system was designed to triage and transport the injured from the field to aid stations. More modern field care and transport began after the first year of the Civil War. Civilian ambulance services in the United States began transporting patients to hospitals in Cincinnati in 1865 and in New York City in 1869. Later, funeral home hearses, which often doubled as ambulances, were eventually replaced by fire department, municipal, and private ambulances.

**FIGURE 1-1** Early American firefighters.

## National Development Efforts

### Development of the Modern Fire Service

Today, the United States has an estimated 30,170 fire departments. Because there are relatively few federal regulations for fire services in the United States, there could be 30,170 different ways to provide public fire protection. Volunteer firefighters staff most of these fire departments, slightly less than half (14,817) of which serve small populations of 2,500 or less. Presently, all-paid departments protect 47% of the US population, all-volunteer departments protect 19.5%, and 33.5% are protected by combination departments (Karter, 2010). Even with the lack of federal regulation, the equipment and procedures of these departments are relatively uniform because of past initiatives, such as the *America Burning Report* (US Fire Administration [USFA], 1973) and the Insurance Service Office (ISO) Grading Schedule.

**Insurance Service Office** Most of the uniformity among fire service practices and equipment that exists today is a direct result of the ISO Grading Schedule for Municipal Fire Protection. This document was created in the early 1900s by the National Board of Fire Underwriters, an organization whose members come from the insurance industry. The effort was undertaken after several insurance providers went bankrupt trying to pay claims from citywide fires, and a resultant financial panic in the insurance industry arose.

The insurance industry wisely realized that it needed to encourage fire departments to prepare for large-scale fires by giving them the ability to provide mutual aid to other fire departments unable to handle major fires on their own. The grading schedule surveyed public fire protection departments in urban centers throughout the United States and aimed to create uniformity. Municipalities that did not adopt the ISO's recommendations faced the threat of higher insurance costs.

Today, municipalities are graded with a numerical rating from Class 1 to 10, with 1 being the best rating; a community that has a Class 1 rating has lower insurance premiums than a community with a higher rating. Many fire chiefs (administrators) have justified budgetary increases for items recommended by ISO to improve a community's rating, thus lowering insurance costs. Unfortunately, this national standardization initiative has not been accepted by all departments.

Some fire officials and municipal administrators have been openly critical of the grading schedule, which was originally created to prevent large-scale conflagrations. To receive a high rating, a fire department has to invest heavily in many fire companies. However, as the probability of large-scale fires has decreased, public officials have questioned the need for large crews (up to six firefighters per company) and close spacing for fire stations (sometimes as close as three-quarters of a mile apart).

In addition, there have been complaints regarding the relevancy of the grading schedule. For example, in June 2007, a tragic fire in Charleston, South Carolina resulted in the death of nine firefighters. At the time, Charleston Fire Department was one of 38 departments in the country with an ISO Class 1 rating. In a December 2008 article in *Fire Chief* magazine, Charles Jennings, Ph.D., MIFireE, CFO commented: "Numerous studies have questioned whether compliance with the [ISO Grading] schedule is correlated with better fire services or lower losses. In the wake of the deaths in the Class 1 Charleston (S.C.) Fire Department, an embarrassed insurance industry trotted out proposed revisions to the schedule, which hasn't been updated since 1980" (Jennings, 2008). Some of the potential revisions include increased reference to NFPA standards, additional emphasis on firefighter safety and training, and credit for adoption and enforcement of model building and fire prevention code.

Still, although complaints regarding the validity of this grading scale may be well-founded, it has helped fire departments from different areas fight fires together, and many departments rely on the grading schedule to justify and plan improvements to their departments.

As an alternative to ISO for planning, the International Association of Fire Chiefs (IAFC) and the International City/County Management Association have created a new accreditation program for fire and emergency rescue agencies. The Commission on Fire Accreditation International self-assessment process can be used to measure fire department capabilities, response times, and compliance with nationally recognized standards. As of March 2013, 163 fire departments have been accredited. The program is a comprehensive self-analysis of each agency's ability to provide professional FES to the public. A committee of public fire protection experts appointed by the IAFC and International City/County Management Association evaluates the competencies using professional standards as guidelines. Although this effort is a step in the right direction, stricter requirements are still needed for accreditation and compliance with standards on firefighter safety, deployment, and staffing.

***America Burning Report*** In 1968, the Fire Research and Safety Act was created. This Act, which established the National Commission on Fire Prevention and Control, stated, "The [U.S.] Congress finds and declares that the growing problem of the loss of life and property from fire is a matter of grave national concern" (USFA, 1973). In 1973, the national effort to professionalize the fire service started with the Commission's *America Burning Report*.

The report influenced the development of several NFPA standards for training, safety, and professional qualifications of fire service members and hazardous materials response personnel. In total, the Commission formally listed 90 recommendations. A summary of some of these recommendations follows:

- Establish a national fire data system.
- Increase the research and medical facilities for burn treatment.
- Make fire prevention at least equal to suppression in planning priorities.
- Use women for fire service duties.
- Increase the use of automatic aid.
- Recognize advanced and specialized education in hiring and promotions.
- Provide federal financial help for planning, training, equipment, research, and fire safety education.
- Provide ambulance, paramedical, and rescue services where they are not provided by other agencies.
- Create the USFA and the National Fire Academy.
- Enforce adequate building and fire prevention codes.
- Require smoke alarms to protect sleeping areas and automatic extinguishing systems for high-rise buildings.
- Implement fire safety education in schools throughout the school year.
- Annually inspect homes for fire safety.
- Develop technology for automatic extinguishing systems for all kinds of dwellings.

Even though many of its recommendations have not been fully implemented, the report has been the basis for many of the changes in progressive FES organizations and continues to influence the modern fire service. Because of its continued relevancy, the USFA updated and reprinted the report in 1989.

## Development of Modern Emergency Medical Services

Modern emergency medical services (EMS) started with the 1966 publication of the paper "Accidental Death and Disability: The Neglected Disease of Modern Society," by the National Academy of Sciences' National Research Council (National Academy of Sciences, 1966). This document

---

### Facts and Figures

**Subzone Rating Factors**

When State Farm insurance company abandoned the ISO grading schedule in 2001, it was uncertain what budgetary planning tool would be used as a replacement. State Farm changed to a system that determines insurance rates based solely on the loss experience in *subzone rating factors*. These factors include fire, wind, hail, water damage, theft, and liability. For example, if the company pays out $20 million during a year, it wants to charge rates that recover the payouts along with a percentage profit. State Farm has stated that 70% of claims paid under its homeowner's program are nonfire losses, although it has not produced hard data to support this claim.

pointed out an injury epidemic that was the leading cause of death among persons between the ages of 1 and 37, and served as a blueprint for a national effort to improve emergency medical care.

The report revealed that many ambulances and emergency medical treatment facilities were inappropriately designed, ill-equipped, and staffed with inadequately trained personnel. When the Highway Safety Act of 1966 established the Emergency Medical Services Program in the Department of Transportation (DOT), the DOT was given the authority to improve EMS through program implementation and development of standards for provider training **FIGURE 1-2**. States were required to develop regional EMS systems, and some of the costs of these systems were funded by the National Highway Traffic Safety Administration (NHTSA).

Through this federal agency, funding became available for training, equipment, and planning. At about the same time, a number of military medics were returning from the Vietnam War. They had experience with professional emergency medical training, organization, and equipment that had never before been seen in any other military operation. These medics brought that experience, which included the use of medical air transport for rapid evacuation, to the field of prehospital care.

In addition, the war provided opportunity for trauma research, more effective tourniquets, chest seals, and bandages with clotting material. Even synthetic blood and intravenous solutions that can carry oxygen have now been field tested because of war experience. Prehospital care has benefitted from this improved equipment and the resulting procedures.

In 1971, another phenomenon changed the general public's attitude concerning the fire service and emergency medical care. When the television show *Emergency!* aired for the first time, there were only six regions in the country providing paramedic service: Seattle, Miami, Jacksonville, Los Angeles City and County, and Columbus. By 1978, however, there were more than 300 regions providing EMS. The show *Emergency!* generated an increased interest in being a paramedic, luring people nationwide to get involved with fire departments and ambulance services.

In 1999—more than 20 years after the show's debut—Project 51, a website dedicated to the show and its cast, stated, "Who would have guessed that the humble beginnings of the Mobile Intensive Care Unit would evolve to one of today's most recognized emergency based services, and one that is used most routinely by millions of people" (Project 51, 1999). The show had more to do with the fast adoption of advanced and basic EMS by the fire service, hospitals, and ambulance services than anything that had come before.

However, although *Emergency!* was visionary, it was not a completely accurate portrayal of the state of EMS at the time. For example, it suggested that paramedics existed everywhere, which was not true in the 1970s. It also portrayed paramedics as having frequent success saving lives, which may have led to unrealistic expectations by civilians. EMS television series that followed also created the misnomer that 911 calls are answered by personnel who are able to provide lifesaving instructions over the telephone. In reality, much of the country still cannot access EMS by calling 911, and prearrival instructions are not always provided. Even today, portrayals of EMS in television, movies, and other media may create unrealistic perceptions and expectations regarding EMS.

## Universal Professional Standards

To develop a professional service organization—one whose members are trained and qualified—fire and EMS professionals must rely on recognized national or international fire and emergency standards. In the fire service, this effort is led by the NFPA. The NFPA uses fire service experts and public input to create consensus standards for firefighter training and education, equipment, and emergency operations.

### Fire Standards

In the early 1900s, national and regional building and fire codes were developed in response to the loss of life and property occurring because of fires plaguing many US cities. The NFPA, which was formed in 1896 by insurance companies, was and still is the leader in national consensus fire, safety, and electrical codes. The first NFPA standard addressed uniform automatic fire sprinkler installation. Around 1913, in response to several fatal fires, the NFPA added a standard for Safety to Life in an effort to prevent fire-related deaths. To this day, NFPA continues to create, update, and adopt standards using a consensus method from committees that represent the fire protection community.

In 2001, an NFPA committee finished work on a new standard—1710, *Standard for the Organization and Deployment of Fire Suppression Operations, Emergency*

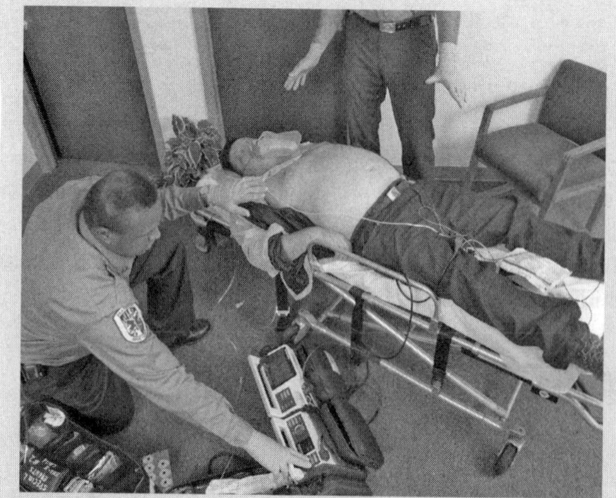

**FIGURE 1-2** Paramedics in action.

## Case Study

### The Need for National Standards

In a tragic example of not learning the lessons of history, 25 people died and nearly 3,500 homes were destroyed in a large wildfire in Oakland, California, in 1991. Surrounding jurisdictions that responded to the mutual aid call were unable to connect their hoses to the Oakland hydrants, severely handicapping firefighting capabilities. Even today, many FES are unable to communicate with one another and come to each other's aid because of different radio channels and systems of communication. Further efforts are needed to standardize fire service equipment, training, and communication. Adherence to national standards is necessary to keep fire departments up-to-date with personnel, equipment, and training.

*Medical Operations, and Special Operations to the Public by Career Fire Departments*—which measures fire and EMS deployment, response times, and staffing (NFPA, 2010). Chief Brown, president of the IAFC, recommends using NFPA 1710 as the future benchmark for FES (IAFC, 2001).

Expert consultants can also be hired to study and make recommendations on any aspect of a FES organization; this practice is discussed in more detail in the chapter *Public Policy Analysis*.

## EMS Standards

Within the field of EMS, the path toward national unifying standards began with the nationally recognized EMT-A curriculum created by the NHTSA in 1969. The field of EMS quickly grew with the Statewide EMS Technical Assessment program initiated by the NHTSA in 1988, along with the Trauma Care Systems Planning and Development Act of 1990. These efforts provided guidelines for a trauma system to address the needs of injured patients and match them to available resources. Local EMS authorities were invested with the responsibility of establishing trauma systems and designating trauma centers in their areas, and states were encouraged and given funds to develop inclusive trauma systems.

In 1991, the Commission on Accreditation of Ambulance Service began administering an EMS accreditation program that is still used today. This program is a comprehensive self-evaluation analysis with the main objective being to provide quality and timely EMS to the community. In 1992, the National Association of EMS Physicians and the National Association of State EMS Directors created the *EMS Agenda for the Future* to gather input from EMS stakeholders. This document created a plan to standardize the delivery of EMS care throughout the country (e.g., by standardizing the levels of EMS education). In 1994, the National Standard Curriculum, a national training model developed by the DOT, was developed and included examinations, skills sheets, and evaluation tools from the National Registry of EMTs, the entity that conducts testing for national EMS certification.

The *EMS Agenda for the Future* led to the *EMS Education Agenda for the Future: A Systems Approach*, created in 2000 to establish standardization, ensuring the competency of EMS professionals and working toward creating shared EMS resources for large-scale disasters. In 2009, the NHTSA created the National EMS Standards, which replaced the National Standard Curriculum, and which are another step toward realizing the vision of the *EMS Agenda for the Future*. The Standards defined the minimal entry-level educational competencies for EMS personnel as identified in the National EMS Scope of Practice Model.

In addition, a 2007 white paper entitled "Prehospital 9-1-1 Emergency Medical Response: The Role of the United States Fire Service in Delivery and Coordination" identified another possible source of guidance for EMS chief officers: the Federal Interagency Committee on Emergency Medical Services established in 2005 to coordinate federal agencies involved with state, local, tribal, and regional EMS and 911 systems.

## Federal Involvement

Federal agencies, such as the Federal Emergency Management Agency (FEMA), along with the Department of Defense and Federal Bureau of Investigation, provide training, equipment, and resources to prepare local FES organizations for terrorist activities, weapons of mass destruction, and other extremist violence. The Environmental Protection Agency, a major regulator of hazardous substances, has created regulations for storage, use, and transportation of hazardous materials. The Occupational Safety and Health Administration has developed rules to protect fire and EMS personnel against the hazardous materials that they come into contact with in their jobs. The USFA and its National Fire Academy also provide valuable training and education for the fire service, although their impact has been limited as a result of insufficient federal budget support.

However, the federal effort continues to expand as a result of the September 11, 2001, terrorist attacks on the United States, which led to increased national attention and prestige for FES. These federal programs continue to play an important role in creating a path to national uniformity by stressing curriculum, ideas, management, and leadership practices that prove to be successful.

## Melding of Fire Service and EMS

Over time, the role of the fire service has expanded greatly beyond fire suppression to include medical response, fire prevention, and public education. Citizens now depend on the fire department to protect them against the dangers of fire, entrapment, and explosion, and to come to their aid in any emergency event that might occur in the community. Domestic acts of terrorism, such as those in Oklahoma City, Washington, DC, and New York City, have added a new mission to the fire service. As such, the job of the chief officer is becoming more complicated and challenging every day.

With the recognition that firefighters are dedicated to saving lives and are strategically positioned to deliver timely response, fire service–based EMS have gained popularity. Currently, four levels of EMS providers exist: (1) emergency medical responder, (2) emergency medical technician, (3) advanced emergency medical technician, and (4) paramedic. Firefighters may or may not be trained to an EMS level of certification, and fire departments may or may not provide ambulance transport.

On a national level, nearly 60% of fire departments provide some EMS. Those that do not are mostly all volunteer departments serving areas with small populations who do not have sufficient resources to provide EMS. In many areas, volunteer fire departments have trouble recruiting and retaining members, so there is concern that the additional demands of a volunteer's time to train for and provide EMS might discourage membership.

## Unified Federal Emergency Response

After a series of devastating fires in New Hampshire in 1803, the US Congress passed a measure that provided financial relief to the affected New Hampshire residents. This measure is considered the first piece of legislation passed by the federal government that provided relief after a disaster. Over the years, the federal government created various other programs to fund recovery from disasters, including flood control by the US Army Corps of Engineers.

During the 1960s and 1970s, such major disasters as hurricanes and earthquakes brought focus to this fragmented approach to federal disaster assistance FIGURE 1-3. In 1979, FEMA was established to consolidate the federal disaster response, which at the time included the USFA and the Civil Preparedness Agency in the Defense Department. With the end of the Cold War, resources once allocated to civil defense were redirected to support disaster relief, recovery, and mitigation programs. FEMA started programs to prevent and reduce the risk before a disaster struck. Currently, the agency is focused on preparing and delivering training to mitigate and recover from terrorist incidents.

The US Forest Service, under the Department of Agriculture, also played a crucial role in developing the ability to provide a unified emergency response through the help of two key initiatives: the development of the National Interagency Incident Management System—now called the National Incident Management System (NIMS)—and the red card system. The Forest Service developed the National Interagency Incident Management System in the early 1970s to ensure that fire protection agencies were prepared and organized to be able to fight destructive wildfires in southern California. In the red

**FIGURE 1-3** Hurricane damage.
© Dustie/ShutterStock, Inc.

card program, firefighters who are assigned to a wildland fire managed by a federal agency (US Forest Service, Bureau of Land Management, National Park Service, Bureau of Indian Affairs, or US Fish and Wildlife) or by many state agencies are required to have a red card, which documents the current wildfire qualifications of an individual. In a sense, it is similar to a driver's license. The credentials specify levels of competency, including firefighting and incident command, achieved by successfully completing training and testing to a national standard.

This credentialing system is remarkably effective and prepares firefighters from all over the country to work together, efficiently and collaboratively, on large wildfire firefighting efforts. For example, a wildland firefighter from North Carolina can travel to Wyoming and be assigned to a team made up of individuals from all over the country because there is a national system of competency credentialing.

On November 22, 2002, the federal government took the effort for emergency response a step further with the establishment of the US Department of Homeland Security (DHS). At the time, threats against the United States were growing in magnitude, and the DHS was created to protect American citizens from terrorist attacks. In 2003, FEMA became part of the DHS and continues to be the nation's incident manager for all hazards and major disasters. This command system provides a direct line of authority and communications from the President of the United States—by way of the Secretary of Homeland Security—down to the local level.

The number one priority of the DHS is to prevent the occurrence of a terrorist attack. The same goal is also very important to fire and EMS providers, because preventing problems has always been more effective than responding to an emergency in progress. Traditionally, however, prevention has not been a major priority with most emergency services. In December 2008, the DHS released its NIMS. This system supersedes all previous emergency management systems when the federal government is involved and is the default system in the United States. In responding to future disasters, a plan that details mutual aid across jurisdictions and state lines using the incident management command system will be a top priority.

In addition, there are now more mutual aid agreements being signed to prepare for major emergency incidents that could result from terrorist actions and catastrophic natural disasters. National credentialing efforts for structural firefighters and incident commanders are being developed with the encouragement of the DHS. The National Fire Academy has formed a committee to design a national credentialing system that includes identification demonstrating that a firefighter has qualifications to contribute to a national response.

As successful as these efforts have been, greater standardization among fire departments is still necessary to ensure departments and individual responders are able to work together safely and effectively. The NFPA and DHS are continuing to develop standards and initiatives to improve the fire service's ability to present a unified emergency response. These include NFPA 1500, *Standard on Fire Department Occupational Safety and Health Program* (NFPA, 2013) and NFPA 1710, *Standard for the Organization and Deployment of Fire Suppression Operations, Emergency Medical Operations, and Special Operations to the Public by Career Fire Departments* (NFPA, 2010).

When many different emergency services departments come together to respond to national events, effective management is critically important. According to NFPA 1561, *Standard on Emergency Services Incident Management System*, "The absence of incident command at an incident scene puts firefighters at great risk and is one of five leading contributing factors of firefighter fatalities, as reported by the National Institute of Safety and Health" (NFPA, 2003). Therefore, incident scene management is critical to the successful mitigation of large- and small-scale emergencies and the health and safety of emergency responders.

## Case Study

**An Example of Progress in Fire and Emergency Services**
In 1994, a state fire marshal was in the process of reviewing plans for a new casino ship that would dock in his state and sail on the Mississippi River when he decided to make inquiries to find out if there was a fire safety standard that covered casino ships. He found that NFPA did not have a standard at that time (subsequently, at the request of the US Coast Guard, NFPA did create a fire safety standard).

The fire marshal then noticed that the permit applicant was the previous owner of the Beverly Hills Supper Club in Southgate, Kentucky. This nightclub was the location of a tragic fire in 1977 that took the lives of 165 innocent people.

The marshal asked himself how this person, who was at least partially responsible for many fire deaths, could be allowed to own and operate a ship that needed to be safe enough to protect the lives of thousands.

As he began to think about this situation, he remembered a more recent tragic fire that occurred. In 1991, a fire broke out at a meat processing plant in Hamlet, North Carolina; the fire claimed the lives of 25 workers. The owner was convicted of 25 counts of manslaughter, lost all his wealth in civil court action, and spent many years in prison. This sentence was in sharp contrast to the consequences encountered by the owner of the Beverly Hills Supper Club some 14 years prior.

*(continues)*

## Case Study (Continued)

As this case exemplifies, there has been a great deal of progress in the FES, but because it is relatively slow and not always obvious to the casual observer, some might believe that these tradition-based services are unaffected by change. This is not true; the following examples demonstrate relatively recent progressive changes in the FES:

- Smoke alarms
- Hurricane- and earthquake-resistant construction
- Paramedic services
- Professional firefighter and officers competency standards
- Federal grants for FES
- Compressed air foam
- National standards for emergency medical training and certification
- Higher education opportunities
- National Fire Academy
- Incident management system
- Residential fire sprinklers
- Safety standards for firefighters
- Defibrillators
- Federal support for disaster mitigation
- NFPA standard for the deployment and staffing of fire companies
- Thermal imagers
- Hydraulic rescue tools
- National focus of emergency response to major disaster events
- Chem-bio detectors
- FEMA and the USFA
- Seat belts and air bags

## Current Problems Facing Fire Service and EMS

Although the fire service in the United States has come a long way from its local beginnings to create a system of national standards and unified emergency response, there are still many areas where further professional progress is needed. The modern fire service is continuing to work toward measurable outcomes for firefighter and fire officer training, education, and physical fitness nationwide.

There are still many challenges to be faced, including breaking down traditional ways of operating and getting officials to help set emergency management priorities. A few specific challenges are mentioned in the following sections.

### Challenges of the Modern Fire Service

In 2009, NFPA conducted research on fire-related trends and statistics, and noted a number of issues relating to smoke alarms. To begin with, 40% of all home fire deaths reported occurred in households without smoke alarms. What is even more concerning is that, although most households in the United States have at least one smoke alarm, these alarms were found to operate correctly in only about half (47%) of reported home fires (NFPA, 2011b). This is a trend that the fire service will have to continue to battle.

A more recent NFPA study conducted in 2010 found the following additional concerns regarding the US Fire Service's needs and response capabilities:

- There is a great need to refurbish or replace many fire stations; 38% of fire stations are more than 40 years old.
- Except for cities protecting populations of greater than 250,000, most cities do not assign at least four firefighters to an engine or ladder company and are therefore not in compliance with NFPA 1710, which requires a minimum of four firefighters on an engine or pumper.
- Both the largest and the smallest communities have seen increases in the percentage of departments needing additional stations to meet minimum response time criteria. Some of these needs might be met through automatic mutual aid or consolidations.
- Many departments are not able to provide all of the fire prevention services that would help reduce fire deaths and property loss.
- When grouped by the size of population protected, most departments said they would be unable to provide technical rescue with EMS at a structural collapse of a building with 50 occupants with local specialized equipment. This finding indicates that all departments would need mutual aid to handle such a collapse.
- Most of the revenues for all-volunteer or mostly volunteer fire departments come from taxes.
- Most firefighters serve in fire departments with no program to maintain basic firefighter fitness and health.
- Most of the departments with no formal structural fire training are in rural areas served by volunteer firefighters.
- Thirty-three percent of the US population is protected by fire departments that do not have a program for free distribution of home smoke alarms.

### Challenges of Modern EMS

A common challenge of modern EMS is that much of the country still cannot access EMS by calling 911, and people place many unnecessary 911 calls. Fees charged for transport are paid by Medicare, Medicaid, and private health insurance, or not paid at all. Just like anything that is free, emergency transport is demanded more often than it is actually needed. As stated by Pratt, "Prehospital 9-1-1 emergency patient medical care is a major part of the safety net for the American healthcare system. They might be the provider of last resort for the needy, yet they can be one more mechanism for overloading the health care system" (Pratt, Pepe, Katz, & Persse, 2007, p. 14). The easy access of ambulance transport that is common in most densely populated areas results in high numbers of noncritical patients requesting medical care. Because emergency rooms cannot turn anyone away from medical care, they are overtaxed and may have to turn real emergencies away.

At the present time, fees can be charged only for transportation of a patient and not for emergency medical care provided by first responders. As a result, there are many private ambulance companies providing EMS along with transport. The Patient Protection and Affordable Care Act of 2010, which is reforming the nation's healthcare system, does not have any specific requirements for EMS but might contain regulatory authority for a federal agency to write rules affecting EMS providers. Many of the specifics of this legislation have not been implemented or are not fully understood. This Act is evolving and might be changed or repealed in the future.

**Financial Challenges** The financial pain of the recession that started in December 2007 will continue for years. The full impact on FES organizations has yet to be felt as a result of federal aid and the nature of tax revenues.

However, most federal financial help dried up by 2011, and a wave of political action with a distinct anti-tax and antigovernment sentiment is sweeping the country. One leading example of this backlash occurred in Wisconsin in 2011. The state passed legislation that limited bargaining rights for public employee unions. The legislation also increased state employees' contributions to retirement and health insurance. Numerous other states are in economic peril and will have to take similar actions in the near future.

There are many other sources of concern for the US economy, such as unprecedented deficits and high unemployment rates. Although the unemployment rate is falling, the federal deficit keeps rising, with the total cost per taxpayer of more than $146,000 at the time of publication (to access the latest numbers, go to http://www.usdebtclock.org/). The danger that inflation and increasing oil prices might cripple the economy is always present.

Every week sees a new story about a fire union giving back some of its members' salaries or benefits to help a local jurisdiction meet a budget shortfall and save firefighter jobs. This sacrifice is both good and bad; bad in that members have to give back benefits, but good in that there is mutual cooperation that generally saves firefighter jobs. More unions will face this dilemma in the future. Issues that are part of this discussion include:

- Union bargaining rights
- Employees' salaries and benefits
- Revenue from EMS transport service
- Consolidations
- Reducing either staffing or stations (or both)

The only thing that is for certain is that change will happen.

## Today's Fire and Emergency Services

What is the status of today's FES? Often, there is a large gap between the public's perception and the reality that many FES leaders encounter. Veteran chief fire officer Dr. Harry Carter states, "The fact is these leaders do not have the staffing, equipment, or most importantly, the financial support to back up their claims. Sadly, far too many of our folks in positions of leadership use their education and training to weave an intricate web of falsehoods and deceit. These folks are unwilling to face reality and provide a true assessment of their agency's capabilities to their citizens and their local governments" (Carter, 2005). Dr. Carter points to many departments that run fire companies with one, two, or three firefighters. This does not meet the minimum requirement of four firefighters as listed in several national standards, a requirement that was validated by a 2010 scientific-based study at the National Institute of Science and Technology (NIST) that showed that the size of firefighting companies largely affects the company's ability to protect lives and property (NIST, 2010). Carter also cites the example of Hurricane Katrina's aftermath, which exposed numerous issues involving mutual aid response, including failures of local, state, and national plans for catastrophic incidents.

As is true in any profession, there is always room for improvement; however, there are also a number of positive trends that should be recognized. For example, although careless smoking habits are a leading cause of fire deaths, these deaths are decreasing thanks to the increased effectiveness of smoke alarms and a decrease in the number of smokers (from 42.4% in 1965 to 20.6% in 2008). In addition, the Coalition for Fire-Safe Cigarettes has worked to save lives and prevent injuries and devastation from cigarette-ignited fires through state passage of fire-safe cigarette legislation. A report released by the NFPA on October 21, 2010, stated that "the long-term trend in smoking-material fires has been down, by 73% from 1980 to 2010, helped by the decline in smoking, the effect of standards and regulations that have made mattresses and upholstered furniture more resistant to cigarette ignition, and more recently, the adoption of fire-safe cigarette requirements throughout the country" (Hall, 2012, p. i).

Furthermore, fire departments are now placing more emphasis on fire prevention, specifically through the use of modern building and fire codes. Automatic sprinklers, long used for property protection, have shown to be very effective tools for reducing or preventing loss of life or property (Karter, 2010). In addition, fire safety education, which was started in the 1970s, has now become part of many schools and fire departments. Public safety education by fire departments includes information about vehicle accidents, swimming pool and water safety, and other common accidents that cause injuries and deaths (Karter, 2010). There is also the trend of departments adding new services, such as rescue, hazardous material response, emergency preparedness, and EMS.

Still, the chief officer should never underestimate the difficulty of making changes. Preparation is the key to success. Useful solutions to common FES issues, and techniques to implement change, can be found in the following chapters. The first step in preparation is being informed and educated.

## CHAPTER ACTIVITY #1: An Example of Progress—Maybe?

When the DHS promulgated the NIMS, there were those in the FES community who opposed it. Some noted that it was not designed to address the multialarm fire command system that was typically used in parts of the country.

Another problem identified was that no common terminology for fire and EMS resources existed. One simple example is the definition of the basic unit of a fire department: the engine company. The name of this unit can vary by department, such as engine, pumper, wagon, or unit. One state plan has the following definition of a Type I Engine: 1,000 GPM pump, 750 gallons of water, 1,200 feet of supply hose, 200 feet of handlines, and four firefighters. Each firefighter on a Type I Engine must be state-certified as a Firefighter II.

The NIMS currently does not specify minimum qualifications and certifications for personnel, and no consensus or consistency exists in the fire service for determining competency of fire personnel. Although a few states have mandatory firefighter certification systems, most fire departments operate independently, and each decides how to train its members.

Before answering the discussion questions, please download and review the following documents: *National Incident Management System* (www.fema.gov) and *America Burning* (www.usfa.fema.gov).

### Discussion Questions

1. Do all FES organizations have to comply with the federal emergency management plan?

2. Would your department's engine companies measure up to the Type I Engine requirements listed previously? If not, would your department change, or do you think that the requirements are not realistic and should be changed? Please provide a complete justification for your answer.

3. Are there any major differences between the plan outlined in the DHS document and your department's standard operating procedures for incident management? If so, please list the differences.

4. After reading the FEMA document, what challenges do you see in your fire or emergency services agency's ability to respond to a large catastrophic event? Please provide a detailed point-by-point analysis.

## CHAPTER ACTIVITY #2: A Second Example of Progress

In 1976, the US General Services Administration identified minimum requirements for new EMS ambulances, and medical equipment companies began catering to the prehospital EMS market. As a result, patient care devices became smaller, lighter, and more efficient. Local governments standardized patient care protocols within their jurisdictions based on state and national requirements and recommendations. Original patient care protocols were based on what little research had been done at the time. Data collection from paper patient care reports was difficult to understand, and the information acquired was incomplete and often inaccurate.

In the mid-1990s, electronic data collection systems, made possible by advances in medical equipment, came into use. With the ability to collect good data and the emergence of quality improvement programs to evaluate data, EMS providers learned that some accepted patient care practices were not only ineffective, but potentially harmful. Today, decisions concerning diagnostic and therapeutic equipment and patient care protocols are based on research showing accurate, definitive conclusions about the effectiveness of a product, drug, or treatment protocol. This practice is known as "evidence-based medicine."

### Discussion Questions

1. Evidence-based medicine has changed long-held beliefs in patient care treatment. How has evidence-based medicine changed EMS protocols in your community?

2. Are the EMS protocols the same in adjoining cities, the county, or the state? If no, why not?

## References

Carter, H. (2005). *Perception and reality*. Retrieved from http://www.firehouse.com/topic/leadership-and-command/perception-and-reality

Hall, J. R. (2012). *Smoking materials fire problem*. Retrieved from http://www.nfpa.org/assets/files/PDF/OS.Smoking.pdf

International Association of Fire Chiefs. (2001, July 12). Testimony of Chief Brown. *IAFC OnScene*, 15:1–3.

Jennings, C. (2008). *ISO system obsolete*. Retrieved from firechief.com/biog/iso-system-obsolete

Karter, M. J., Jr. (2010). *U.S. fire department profile through 2009*. Quincy, MA: National Fire Protection Association.

National Academy of Sciences. (1966). *Accidental death and disability: The neglected disease of modern society*. Washington, DC: National Academies Press.

National Fire Protection Association. (2003). *NFPA fire protection handbook* (19th ed.). Quincy, MA: National Fire Protection Association.

National Fire Protection Association. (2010). 1710, *Standard for the organization and deployment of fire suppression operations, emergency medical operations, and special operations to the public by career fire departments*. Quincy, MA: National Fire Protection Association.

National Fire Protection Association. (2013). 1500, *Standard on Fire Department Occupational Safety and Health Program*. Quincy, MA: National Fire Protection Association.

National Fire Protection Association. (2011a). *A third needs assessment of the U.S. Fire Service 2010*. Retrieved from http://www.nfpa.org

National Fire Protection Association. (2011b). *Smoke alarms in U.S. home fires*. Retrieved from: http://www.nfpa.org

National Institute of Science and Technology (NIST). (2010). *Report on residential fireground field experiments*. Retrieved from: http://www.nist.gov

Pratt, F. D., Pepe, P. E., Katz, S., & Persse, D. (2007) *Prehospital 9-1-1 emergency medical response: The role of the United States Fire Service in delivery and coordination*. Retrieved from http://www.iaff.org/Tech/PDF/FB%20EMS%20Whitepaper%20FINAL%20July%205%202007%20.pdf

Project 51. (1999). *History of the TV show Emergency*. Retrieved from http://www.squad51.org

United States Fire Administration. (1973). *America burning—The report of the National Commission on Fire Prevention and Control*. Retrieved from http://www.usfa.fema.gov/downloads/pdf/publications/fa-264.pdf

# CHAPTER 2
# Introduction to Administration

## Fire and Emergency Services Higher Education (FESHE) Course Objectives

### Module I: Leading and Managing Purposefully with a Community Approach

The students will:

2. Explain the importance of a good working relationship with public officials and the community as a whole. (pp 19, 21–22, 23–24)
3. Assess ways to develop a good working relationship with public officials and the community. (pp 21–22, 23–24)
6. Identify effective skills for developing a cooperative relationship with fire and emergency services personnel as well as public officials and the general public. (pp 19–24)

### Module II: Core Administrative Skills

The students will:

1. Identify the core skills essential to administrative success. (pp 12–14)
6. Recognize the formal and informal dynamics of public organizations and describe strategies to ensure success. (pp 19–20)
7. Discuss the components and styles of leadership. (p 13)

## Knowledge Objectives

After studying this chapter, the student will be able to:

1. Recognize how both management and leadership are integral to effective administration. (pp 12–13)
2. Outline professional qualifications for fire and emergency medical services (EMS) administrators. (pp 13–14)
3. Comprehend how problems caused by staff–line distinctions can be overcome. (pp 14–15)
4. Describe the pros and cons to selection of administrators through the process of election or appointment. (pp 15–16)
5. Examine how national consensus standards and standard operating procedures (SOPs) affect the consistency and effectiveness of emergency services. (pp 17–18)
6. Know the value of direct supervision and standardization in firefighting and EMS. (p 18)
7. Explain how administrators can work to bring about change and what common challenges they face. (pp 19–25)
8. Describe how to gather feedback from staff members and peers as part of the decision-making process. (pp 19–20)
9. Recognize how to engage influential sources of power including tradition, political groups and individuals, unions, and the public. (pp 20–24)
10. Explain how to gain influence over the political process. (p 20)
11. Discuss the process and challenges of negotiation. (p 25)

## Combining Management and Leadership

Effective administration requires two skills: management and leadership. Management ensures that the organization is prepared and able to accomplish its goals by establishing that sufficient personnel and equipment are available for the organization to perform its duties. Leadership can be observed when trained personnel safely and efficiently complete their mission using the resources provided by the FES organization. At the organization level, leadership is required to "sell" the need for new programs, additional resources, or progressive changes. After the resources become available or the new change is implemented, effective management is needed to use them to support the mission.

A person who excels in management skills will be very good at making the existing organization work efficiently. Many US companies have trained great managers who propelled the country to the top of the industrialized world. For example, Disney—a great visionary company—had a period of time when the management was focused on increasing profits and dividends for stock owners in lieu of truly visionary media products. Although this sequence of events does not occur in every organization, it has been common in many industries. For instance, management-oriented people generally represent most chief officers in FES organizations. Management skills are necessary for this role, because the chief officer often needs to be a good caretaker of public funds, staff, and resources.

However, managers may not be proficient at leadership skills. A person with strong leadership skills has a clear vision of where the organization needs to go and the courage to attempt the journey. There is no guarantee that the leader will arrive at the visionary goal, but because good leaders are inherently risk takers, they try anyway. Conversely, good leaders may not be good managers.

Administrators should possess management and leadership skills, even if they feel more comfortable in one of the roles than the other. A leadership-oriented administrator may focus on a visionary goal 5–10 years down the road but forget to plan for tomorrow. In contrast, a management-oriented person generally is not comfortable with risk taking and would rather not get involved in visionary goals that have no guaranteed outcome. Managers may try to implement change slowly or make it voluntary; be aware that a voluntary plan can foster inconsistencies because members and supervisors set their own goals. When the voluntary phase results in a commitment of 10–15% or more members, this is a good time to phase in a mandatory policy.

A leadership-oriented person is more of a risk taker when it comes to new ideas and change. Many people drawn to the FES have a reduced sense of fear and are risk takers in physical activities. There is an old saying that highlights the heroic behavior of firefighters: when firefighters are entering a burning building, everyone else is leaving it. However, when it comes to changes in the organization, fear of the unknown may make many FES and EMS chief officers hesitant.

Although these descriptions of leadership and management may be oversimplifications, it is important for administrators to recognize the value of management and leadership skills and work to develop them through formal education, self-study, seminars, and experience. The National Commission on Fire Prevention and Control's *America Burning Report* of 1973 further comments on the duality faced by the chief officer:

> Presiding over this tenuous alliance [fire department] is the fire chief, who wears two hats—one, the administrative hat required to run the organization; the other, the helmet he dons when the alarm is sounded to lead his firefighters in the suppression of a fire. Since the fire chief usually has come up through the ranks, the second hat probably fits comfortably. (United States Fire Administration, 1973, p. 20)

The *America Burning Report* uses the term *leadership* when discussing the ability to lead firefighters at the scene of an emergency incident. In this text, the term *leader* is used at the organizational or administrative level, which calls for a special set of skills and knowledge. Leaders are problem solvers; therefore, this text discusses the many aspects of problem solving. More precisely, this text is about the very special challenges to problem solving that FES and EMS chief officers encounter in the public sector.

## Professional Qualifications for Administrators

Career and volunteer organizations might have difficulty selecting the best person for the chief officer's position. Whether the person is elected or appointed, the result may be the same: the administrator (chief) may be ineffective. FES is evolving as a profession, and the selection of chiefs should be similar to other professions, such as doctors, engineers, and lawyers.

Professional qualifications for fire and EMS administration include three areas of competency: (1) training, (2) experience, and (3) education. In all three areas, fire and EMS administrators should meet minimum levels of proficiency; these minimal qualifications should be the same throughout the country, just as they are for medical, engineering, and legal professionals. For example, a medical doctor must complete college-level education and medical school, and then an internship/residency (training and experience). Only after completing these requirements and passing a comprehensive examination is the doctor granted a license to practice.

Many FES organizations can improve their departments' reputation and ability by adopting minimum professional requirements for officers. Although this is a complex and controversial subject, the following are some examples of recommended requirements for fire administrators:

- Minimum of 3 years of experience at each prerequisite level—firefighter and fire company officer
- Certification as a Fire Officer III or IV
- Training and certification as a Firefighter II and Fire Instructor I

- A degree from a regionally accredited college or university that is recognized by the US Department of Education (an associate's degree for smaller departments—those with less than six stations—or a bachelor's degree for larger departments)

EMS administrator qualifications are not as standardized as those of fire departments and vary across companies or positions. The three areas of competency—training, experience, and education—may be less defined in the EMS world. Fire departments have a hierarchical structure based on standardized training, but this is not the case with many private ambulance companies, because EMS supervisors have no national oversight organization to define those supervisory standards.

Ambulance companies may offer supervisor or management positions to paramedics with field experience, but they may not require management experience or higher education. A paramedic who may be an excellent clinician does not necessarily have the skill set to be a competent manager. The extent to which a new supervisor, manager, or administrator is offered training opportunities also varies greatly from company to company and may be considered the responsibility of the hired individual.

As EMS professionalism continues to increase, qualifications for management positions are becoming better defined. Common and preferred requirements for EMS administrative positions include:

- High school diploma or GED equivalent (college degrees may be preferred)
- Advanced Cardiac Life Support and professional-level cardiopulmonary resuscitation certification, with specialized training (e.g., HAZMAT, emergency vehicle operator, National Incident Management System, International Trauma Life Support, Pediatric Advanced Life Support) preferred
- EMT-P license and state driver's license
- Three to five years active prehospital experience, with supervisory experience preferred

## Staff–Line Distinctions

Most FES organizations have a distinct dividing line between staff and line functions. This classification is especially apparent in larger departments where it is common to have specialized divisions, such as fire prevention, training, dispatch, logistics, or human resources.

There are several reasons why this traditional organization can cause conflict, especially when those on the line are asked to move into staff positions. First, firefighters and EMS providers often enjoy working at emergency incidents, so they may resist a transfer to staff positions if doing so takes away their ability to respond to emergency calls. Second, the crew can almost become a second family, and individuals may be hesitant to leave that family to move into staff positions. In addition, the daytime work schedule often required in staff positions may not be appealing for those who have second jobs or enjoy having extended time off. In the common 24-hour shift, an employee works for 24 hours straight, and then is off for 2 or even 3 days. Firefighters and paramedics might become comfortable working this shift and look forward to having time for their second job, family, or social commitments.

To address these concerns, some organizations have combined the staff–line jobs at the mid-supervisory level. For example, a fire department may have a battalion chief

---

### Chief Officer Tip

**Tips for an Outstanding Reputation**

The following is a list of suggestions for how a chief officer can build an outstanding professional reputation:

- Build a solid experience base, starting at the lowest level and moving up through the ranks. In small organizations the chief may have to help out and, therefore, should still be skilled in basic emergency tasks.
- Take advantage of every opportunity to learn the skills and knowledge of the job, through hands-on experience and training.
- Keep in the best physical shape possible; this includes strength, aerobic fitness, and body weight within recommended limits. Even if the chief officer no longer has to perform physical skills, fitness is a great idea for leading by example.
- Acquire a formal education, especially through programs that lead to an associate's or bachelor's degree in fire science, administration, management, and other job-related areas.
- Stay up-to-date by attending seminars and conferences, and by reading FES, government, and management periodicals.
- Read local newspapers and publications from influential organizations. Be familiar with all controversial issues in the community, not just those that relate to FES.
- Visit other FES organizations and learn from their ideas.
- Have high personal and professional standards for ethics and morality.
- If appointed to a top position, choose your initial proposed changes carefully. Choose only those changes that are certain to succeed. It is very important to start out as a winner.
- Last, but perhaps most important, lead by example.

in charge of each shift. This same battalion chief would also be assigned to supervise certain staff functions. During the weekday, the battalion chief is the administrative head of prevention but is available to respond to major emergencies when needed. Another alternative organizational strategy is to rotate command officers every few years to supervisory positions of both staff and line divisions. Serving in different commands ensures that the next chief officer has a well-rounded knowledge of the entire organization.

A similar model may be found in EMS administration. Paramedic or emergency medical technician supervisors are much like battalion chiefs in terms of job function and span of control. During a 24-hour shift, they might be responsible for administrative tasks, meetings, and field oversight and also might be expected to respond to certain types of calls or function as a provider when staffing levels are low.

Some companies have the controversial rule of requiring supervisors to staff an ambulance while supervising some staff functions. Some EMS providers believe that, although supervisors should be required to maintain field skills and certifications, there should be a clear distinction between staff and line roles. In addition, union affiliation may create some conflict, because in many companies field providers are union members but supervisors are not members. In companies where union field employees work as part-time supervisors, they may be restricted in their ability to discipline other field employees, handle sensitive company information, or investigate complaints.

### Case Study

**Resolving Staff Divisions**
In 1996, the Madison, Wisconsin, Fire Department made a decision to reduce its budget by decreasing the number of division chiefs and combining staff and line functions. Madison's Assistant Fire Chief Phillip Vorlander recognized that the distinction between line and staff functions had created divisiveness in the department, alienating shift commanders from other administrators. The department changed its organizational structure so that each battalion chief worked two 24-hour shifts and six 8-hour shifts per 2-week period. Between the 8- and 24-hour shifts, each battalion chief spent 4 days a week serving as a staff supervisor.

With this new organization, staff officers were more able to participate in the administrative decision-making. This increased the quality of the decisions and their acceptance, and the commitment to the management team. This structure also had a positive effect on the cohesiveness of the department because chief officers no longer had allegiance to a particular battalion, division, or shift (Vorlander, 1996).

## Selection of Administrators

Most fire and EMS administrators are selected by election or appointment. Both methods present unique challenges.

### By Election

In many volunteer FES organizations, administrators are elected. These organizations are often independent and self-regulated private corporations that are operated without formal scrutiny by the municipality or public (customers) they serve. Without public accountability or oversight other than the members named as its commissioners or board of directors, they may be more likely to make unintentional faulty decisions. To counteract this lack of supervision and accountability to the public, elected officials for the area served may appoint the chief officer from a list of qualified applicants.

In addition to issues of accountability, these organizations may also have less stability and consistency than those municipal FES organizations. This is because appointed officials typically have more latitude than elected officials in enforcing, interpreting, and changing the SOPs, training requirements, and equipment and apparatus needs and specifications.

In practice, volunteer chiefs may initiate change each time a new chief is elected. The resulting inconsistency is not caused by bad or incompetent people; it is the result of a system in which officers are elected by a democratic process.

Furthermore, because chiefs are typically elected for a 1-year term, they must start planning on reelection as soon as they are elected, tempering their ability to be strong leaders. A similar situation exists in the US House of Representatives, in which representatives are elected for 2-year terms. Almost as soon as representatives are elected and take office, they must start planning for their reelection. Thus, they spend a considerable amount of time and effort on reelection strategies, taking away from their key mission, which is to represent the people in their district and the interests of the United States. In most cases, these are honorable people who are placed in a system that has limitations.

### By Appointment

Just as there are limitations to volunteer organizations that elect their officers, there are problems when chief officers are appointed by government officials or administrators. When appointed, the FES administrator might have conflicted loyalty to those who appointed him or her, the labor group, and the public. It is not uncommon to find FES or EMS administrators who feel alienated from their organizations. The chief officer may believe that the labor organization, the elected officials, and other appointed administrators all want to convince or force their particular ideas on the organization. Government officials and labor organizations have powerful self-interest, but the public may only have a weak influence because of its lack of knowledge about emergency services. Therefore, the chief should be an advocate for the public interest.

For example, an EMS service may propose incentive pay for employees who have been certified as paramedics. The union supports the change, but the city officials oppose it. The public has no opinion because the change was not publicized by the local press. The municipal administration and the labor organization agree that funds for this incentive pay should be taken from a new, successful Risk Watch program that the chief officer championed last year and that had been successfully funded. The National Fire Protection Association (NFPA) Risk Watch is an educational program for young children intended to reduce and eliminate loss of life and property damage from accidents. Although administrators have a responsibility to their organization to enforce its rules, support its mission, and serve its members, in this situation it is the responsibility of the chief officer to speak up and advocate for the public interest.

## Rules and Regulations

One of the main duties of a fire or EMS service administrator is dealing with rules and regulations. This includes analyzing, reviewing, creating, determining, and enforcing rules and regulations for the entire organization. In some cases, such as state or federal regulations, rules may be mandated by a higher authority in government. These rules may also be the result of a task force of members empowered by the administrator to create SOPs for safety and efficiency during emergency operations. It is important to note that, despite what some critics say, if SOPs are promulgated and enforced rigorously and consistently, there should be no concern for liability issues. Vigorous enforcement actually helps protect the organization from lawsuits.

Other than requesting additional funding from elected officials, enforcing rules and regulations is one of the most difficult duties of the chief officer. There are always some members and supervisors in fire departments and EMS organizations who do not like rules and regulations or who disagree with the rules and regulations. However, rules are absolutely essential for accountability and consistency, especially during emergency operations. Too often, support for rules and regulations comes only after a major tragedy. For example, after a devastating hurricane in Florida, the construction of new buildings required the use of specific techniques to reduce or eliminate the damage from these powerful storms **FIGURE 2-1**. Building and fire codes are rules and regulations that chief officers can use to accomplish their mission of fire safety, and it is often easier to find justification and public support for strict enforcement and changes after a major disaster.

In a fire department or EMS organization, there are always people or special interest groups that want more rules, and some that do not want any at all. As a general guideline, rules and regulations can fall into two major categories: management and emergency operations. For emergency operations, basic rules and regulations are absolutely necessary to provide consistent, safe, and efficient FES, especially in the first few minutes.

Additionally, employees who are unionized want to have a written contract that contains rules on personnel management. Such items as when overtime is to be paid and many other conditions of employment are typically included in a labor agreement. Most of these agreements are necessary to prevent inconsistent and unfair actions in personnel management. However, some agreements also contain items that can restrict effective administration.

### Words of Wisdom

"Every fire department should have a set of rules and regulations that outline performance expectations for its members, the SOPs for the department, and disciplinary action that may be taken for failure to follow the regulations...Both the rules and regulations and subsequent orders from the chief should be written and distributed in such a manner as to ensure all persons are made aware of them."

—NFPA

*Source*: National Fire Protection Association (2003). *Fire Protection Handbook*, 19th ed. Quincy, MA: National Fire Protection Association.

### Case Study

#### Administrator Responsibilities

A supervisor receives a complaint from his superiors that a crew "abandoned" a patient. The supervisor is told that the crew was treating an abused child when the abuser returned and threatened them with bodily harm if they did not leave. The crew left, parked around the corner, and then called the police. The local news has a very negative spin on the story, and the public is outraged because the patient was an injured baby.

The company is in the middle of negotiating a new transportation contract and has a lot of competition. Therefore, upper administration is planning to suspend the crew over this incident to appease the media and the public.

Does the supervisor advocate for the crew based on the threat to their safety at the scene, or does he act on the need to mitigate the political firestorm before it has a negative impact on the potential for winning the new contract? Administrators have a responsibility to the organization to enforce the rules and the mission of the company, but they also need to be an advocate for the people doing the work that makes the organization viable.

FIGURE 2-1 Rules and regulations regarding building construction are created to prevent tragedies.
© ZUMA Wire Service/Alamy Images

## National Consensus Standards

National consensus standards, such as NFPA standards and Occupational Safety and Health Administration regulations, are relatively new to the fire service. Although the necessity of such standards is well accepted, FES organizations are still in the process of implementing changes to comply with many of these standards.

For example, an April 2001 article in the magazine *Fire Chief* asked four safety officers if their departments had completed a safety audit as described in the appendix to NFPA 1500 *Standard on Fire Department Occupational Safety and Health Program*. None of the safety officers interviewed reported that they were in full compliance with this safety standard, which was first adopted by NFPA in 1987, 14 years before the interviews. Instead, these safety officers estimated anywhere from 60% to 90% compliance (*Fire Chief*, 2001, p. 4). Although there currently may be higher percentages, seldom does a department comply with 100% of the safety requirements.

Other NFPA standards are designed to reduce inconsistency in operations, staffing, training, and communications. These inconsistencies in the FES result in differences from agency to agency, town to town, and state to state. These differences add to the public's difficulty in judging the service provided and determining if the community is truly receiving professional-quality emergency service. At this time, a sign on the side of an engine that proclaims FIRE DEPARTMENT does not guarantee the same level of service for all similarly marked fire trucks throughout the country. Although the fire service has made a lot of advances, the pursuit of quality and consistency is still a work in progress.

Some fire and emergency officials, along with elected and appointed municipal officials, outwardly resist complying with standards and regulations that they personally disagree with or that have fiscal impact. In some areas, local FES officials want the freedom to provide the level of public fire protection that is determined necessary by the local community. In many cases, the citizens of these communities have not done enough research to determine the needed local level of protection or quality of service (e.g., number of stations, training, companies, and staffing). Whatever resources and types of organizations exist at the local level are the result of a very complex and independent evolutionary process, not a systematic planning effort.

However, FES organizations are entering a new era. In the EMS field, responders are now certified to national standards for emergency medical technicians and paramedics. In the fire services arena, national standards for training and safety are now available.

One example of a national standard for the fire service is NFPA 1021, *Standard for Fire Officer Professional Qualifications*. The chief may propose a new program to certify all officers to this standard. This type of program may require training, incentive pay, and overtime pay to cover classroom attendance (especially if education is a requirement for promotion). The selling point for this proposal is that, when completed, the city will have a more professional, competent fire department—a goal on which the municipal administration, the public, and elected officials can easily agree.

## Standard Operating Procedures

Within an FES organization, the level of service can vary by shift or battalion. Freelancing or independent goal-setting can be deadly on the scene of a major fire; the public deserves an emergency services organization that can operate with a high degree of consistency and conformity. SOPs further this goal by achieving a high degree of conformity and consistency throughout the organization.

SOPs also ensure emergency scene safety and encourage situational awareness, which is sometimes a problem for a younger workforce that has little life experience and feels invulnerable. In particular, emergency response driving is a high-liability issue, and fatal accidents can occur. For that reason, many public agencies and private companies have developed SOPs regarding emergency response, such as those requiring units to stop at intersections with red lights even in areas where emergency vehicles have exemptions from this traffic law. Infractions of this safety SOP are taken seriously for the safety of the crew, the patients, and the public.

Enforcement of SOPs that guide the decisions made in emergency situations—when first responders have only 5–15 seconds to consider all the alternatives—is critical for the safety and success of emergency operations. Although there should be some room for exceptions based on the judgment of the officer, an appropriate system to critique these emergency field decisions ensures that this latitude is not used without justification.

For example, at a fire in a large vacant house, the first due engine stretched its hose line to the front door and began to enter. The department's SOPs required the first due engine to attack the fire from a position on Side 1 (the front), where the front door is found in most buildings. On entering, the firefighters noticed that the floor had burned through, leaving a large hole into the basement, which was initially blocked from view by heavy smoke and fire. The firefighters quickly retreated and radioed this information to the Battalion Chief. Based on the Battalion Chief's judgment, a revised plan of attack was implemented, resulting in successful fire suppression.

## Direct Supervision and Standardization

Many modern entrepreneurial businesses are organized around the principle of employee empowerment. Business administrators may argue that empowering employees is the best way for individuals to learn new skills and deal with changing environments. Employee empowerment can improve a business's service and competitiveness in the world where revenue generation is the goal. However, although the empowerment theory may work very well in a corporate or industrial organization, it does not lend itself to the FES organization at the company level.

FES organizations could not operate effectively if each company officer was empowered to determine what actions the company would take at an emergency incident. Employees or companies acting independently (freelancing) at the scene of an emergency cannot be tolerated. Standardization and SOPs are needed to coordinate efforts, and direct supervision is required to ensure that each member works as part of the overall team **FIGURE 2-2**.

Direct supervision and standardization are especially critical for the FES organization during the first few minutes after arriving on the scene. This is the time before the incident commander has arrived and has been able to evaluate and formulate a comprehensive strategy for the emergency. Consistent and reliable emergency operations at all times are very important, if not critical, to the successful and safe handling of emergency incidents.

**FIGURE 2-2** Safety committee meeting (Alachua County Fire Rescue, Florida).
Courtesy of L. Charles Smeby, Jr./University of Florida

## Bringing About Change

Do not underestimate the chief administrator's power to make change. In most cases, the administrator's power goes beyond simply giving orders that may or may not be followed. When chief administrators speak and initiate a new project or change, the attention of the organization and its members is immediately focused on the proposed change.

The chief officer also has the power to set the agenda for change in the organization. Generally, changes come about when the chief officer initiates the process, with the exception being changes that are forced from the outside by a court order, a new state or federal regulation, union pressure, or elected officials. In most cases, initiating change takes real courage and detailed preparation.

The chief officer's ability to get the administrative job done, identify problems, and, when necessary, make changes in the organization makes up a dynamic process. To comprehend these dynamics it is critical to understand the chief's power, the interdependence on professional capabilities and agency reputation, and the chief's ability to influence others. The following questions help determine if a chief officer is ready to manage and lead the march to initiate change:

- Is the officer knowledgeable about all aspects of emergency operations?
- Does the officer have the appropriate formal education?
- Can the officer lead by example?
- Can the officer gather all the facts needed to make an informed decision?
- Does the officer really know what is going on in their department and community?
- Can the officer schedule and set an agenda for change that is reasonable and evenly paced?
- Will the officer's proposed change improve the service to the public (customer)?
- Will the officer take risks for the sake of making real progress in improving service to the public?

## Getting Feedback

The chief officer must attempt to maintain a professional, independent position, which, at times can make the job a very lonely one. As Chief Ron Coleman noted, "There are people who jump at every opportunity to criticize the management of an organization...You can call that the loneliness of command, because that is what leadership is when we put it on the line" (Coleman, 2004, p. 5). Although this loneliness may make it difficult for an administrator to get unbiased advice and feedback on controversial administrative decisions, it is important to continue to solicit this feedback from staff members and peers.

Before seeking feedback, administrators should gather information about new policies or equipment by reading and studying the issue. Care must be taken that the informational sources are nonbiased and based on expert opinion and facts. Doing so may be a challenge in the FES, where many periodicals are not peer-reviewed for accuracy and legitimacy. Effective fire chiefs will find they do more listening and reading than talking.

**Feedback from Staff Members** To be a part of the decision-making process, it is important that staff members demonstrate loyalty to the agency. Knowing that staff is loyal allows administrators to communicate honestly, trusting that any sensitive or potentially controversial information remains private.

Administrators should solicit opinions in an open format. Preliminary discussions with staff may be held in meetings behind closed doors to facilitate an open, honest dialogue. However, after the administrator has chosen a particular policy, staff should be expected to support the change. Never equate a privately expressed negative or opposing view as a sign of a person's lack of loyalty. In fact, this person may be giving the chief officer the advice needed to stay away from a bad decision.

The chief officer should be aware of this, realizing that there are situations for which the decision-maker does not know all the facts or circumstances. These types of situations are sometimes referred to as imperfect information. Imperfect information may occur, for example, when the top staff members are reluctant to make an argument that opposes what the chief officer would like to do, or when the officer is facing pressure from elected or appointed high-level officials to act before knowing all the facts. Behind a desk, it is easy for administrators to forget the field perspective. Making decisions without an accurate understanding of the current situation is not effective leadership. Decisions need to be balanced between administrative goals, public needs, political realities, and the needs and concerns of those in the field.

When the subject is one of major importance to the organization, it is best to empower a *task force* to study the issue and make recommendations. The task force should include members that represent diverse opinions and all levels of the organization, including experienced

---

**Words of Wisdom**

"Command is lonely."

— Colin Powell

Source: Powell, C & Persico, J. (1995). *My American Journey.* New York: Random House Ballantine Publishing Group, P. 308.

firefighters, field providers, and officers. A diversified task force opens communication and prevents information obstructions that can occur between different levels of management.

Chief officers should encourage open and honest dialogue and explain that it helps them to make fair policy decisions. Staff members should feel safe bringing their concerns to the chief officer; angry reactions from the administrator can discourage staff from raising negative information in the future.

A management style called "management by walking around" is also useful in fire and EMS administration. In some cases, although the chief officer has the authority to order a change, members may make this change only in name, without actually changing their actions. Departments may report to their peers and community that they are progressive, but chief officers can use "management by walking around"—along with accountability—to find out what is really going on in their organizations.

Administrators should visit stations often and float ideas for informal discussion. It is important to remember that there may be some members who are unable to differentiate a brainstorming session from a formal announcement of a new policy, so administrators need to develop a keen listening ability. During actual implementation of a decision, an administrator should be their own intelligence officer, confirming that all the details to fully implement the decisions are complete.

**Feedback from Peers** Often other chief officers or professional acquaintances may be able to help in evaluating and discussing changes, especially if these discussions are better kept confidential in the preliminary assessment. While attending conferences and professional development courses, administrators can meet with professional peers who may be able to provide advice and insight into difficult decisions.

Although some administrators may feel uncomfortable asking for help, it is important to work to overcome this reluctance. Administrators must be able to gather and analyze information to present a full picture to their peers. It is also important to be able to ask more than one person for advice to gain different perspectives on the situation. To illustrate this process, it is helpful to look at how a patient would deal with a life-threatening medical problem. The decision to undergo major medical surgery should not rest on the opinion of one surgeon. Just as in the field of medicine, consulting with two or more experts provides a second opinion and ensures that individual biases are weighed against one another. Furthermore, speaking to administrators in other parts of the state or country can provide a wider perspective and help prevent the conversation from being repeated locally.

Although feedback from peers is important, talking freely on a day-to-day basis about potentially controversial changes is discouraged. The casual nonstop conversation around the dining table in the firehouse kitchen is a great socializing time for firefighters and company officers, but the minute the officer puts on the five bugles, everything they say takes on new meaning. As a firefighter, talk about controversial change may have ended with a good-natured argument; but, as the fire chief, this kind of talk may scare members to the point of damaging moral.

To help this effort, NFPA sponsors an annual meeting of Metro Fire Chiefs. It sends invitations to fire chiefs in some of the largest fire departments in the United States and several international departments. Only fire chiefs—not deputy assistants or other chief officers—can attend. This exclusion is intentional to allow fire chiefs to feel safe that all conversations and discussions at the conference are confidential. NFPA realizes that the fire chief position is special and that those underneath and those above the chief's position can become adversaries when the chief attempts to initiate change. Real change always has opponents. It is not a good idea to give opponents a heads-up on the change, because this gives them more time to fight the change.

For those chiefs who cannot attend or who do not command a metropolitan department, networking with fire chiefs in other departments can be very helpful for honest discussion of controversial subjects. A good place to meet other chiefs is at the National Fire Academy or national conferences. An exclusive environment of peers gives the fire chief a safe venue to discuss potential or controversial changes.

Finding a similar environment may be more of a challenge in the field of EMS, where no national organization provides a safe environment for sharing sensitive information. Administrators from private EMS companies must also recognize that competition for 911 contracts makes peer discussion of decision-making more challenging. EMS administrators may know each other well but share little proprietary information. EMS administrators may learn from others by examining errors other companies have made; doing research; being honest in determining if a choice is cost effective; and making a decision based on the best interests of the company, the community, and the work force they serve.

## Sources of Power

One of the challenges that FES chief officers encounter in bringing about change in the public sector is related to other sources of power that affect decision-making. This includes the power of tradition, political entities, employee unions, and the community.

**Power of Tradition** The power of existing traditions within fire and EMS services can provide a barrier to implementing change, often with tragic consequences. For example, a career fire department in Pittsburgh, Pennsylvania, customarily staffed its engine companies with two or three members. At a fire, the officer was often in charge of operating the nozzle. This custom was caused by the lack of adequate staffing and the privilege of the position. However, when a newly appointed chief increased staffing levels, all engine and truck companies increased to four members, but the traditional engine company operations did not change. This failure turned out to be a contributing factor in the deaths of two firefighters and their lieutenant.

On February 14, 1995, with the lieutenant on the nozzle and the two firefighters behind, the officer was unable to supervise the operation. When the hose line burned through, there was no longer any water to control the fire. The hose line had extended down one flight of stairs into a basement, and the fire had burned through their only known avenue of escape—the stairs leading up to the ground-level floor. This tragedy may have been prevented had the officer been supervising the operation.

Too often, organizations are driven not by their missions, but by their rules, traditions, or budgets. Although rules are important, too many administrative rules can restrict an organization from being mission- or customer-focused.

**Political Power** To some, the term *politics* may bring up negative feelings of uncaring elected officials only taking care of special interest groups. In some communities it may seem that the big money contributors have more influence than ordinary citizens. Although some individual elected officials might abuse the power entrusted to them, this does not mean that the entire political system is not legitimate. Remember that politics is the foundation of the United States' representative form of government.

Although elected officials have many goals, one goal that is fairly common is to get reelected. In most cases, this means taking care of the voters' or the campaign contributors' needs. For this reason, many elected officials are heavily influenced by the local voters. Because many voters are concerned about the amount of taxes they pay, elected officials almost always question any request for additional funding for the department. These politicians are acting as the voice of the individual taxpayer who may believe that taxes are already too high. This is especially common in jurisdictions with large numbers of taxpayers who are on fixed incomes, such as Social Security. This group of voters continues to increase in numbers with the retiring "baby boomer" generation.

Furthermore, the view that fire and EMS services are monopolies has given elected officials justification to impose regulations, just as they do with local law enforcement agencies and public utility companies. A monopoly occurs when only one provider or company has complete control of the market and can charge prices that are above the market value—that is, what they would be worth in a competitive market. If a house catches fire, for example, the occupants do not go to the yellow pages to choose a fire department to call. FES is provided to anybody who calls, but only one primary provider is responsible for the service provided. In the United States, monopolies are illegal unless they are a government-regulated industry or service. In this way, the government can meet the public's expectations that it will maintain high-quality emergency services. Public FES can be viewed as monopolies either provided by a government or allowed to operate without control by the local government (e.g., independent volunteer departments). Although municipal departments may have no trouble understanding the issue of government oversight, independent volunteer companies might strongly resist any oversight by elected representatives.

In addition to formal political authorities, such as elected or appointed officials, fire and EMS service administrators must be aware of informal sources of political power. This includes external groups of influential citizens such as:

- Employee or member associations (e.g., unions and volunteer fire companies)
- Media
- Special interest groups
- Professional organizations
- Businesses and industries
- Homeowners' groups
- Social clubs
- Religious organizations

Fire and EMS administrators should be in touch with administrators or staff in these organizations and work to maintain positive relationships, because these groups can influence political decisions that affect the service's budget. The general public may affect approval of any proposed changes—yet another reason why public opinion is so important. Generally, the public trusts that fire and EMS services are providing quality emergency services. By demonstrating actions that show fairness, honesty, responsiveness, openness, and accountability, the FES agency builds an even stronger support base with the general public, elected and appointed officials, and influential citizens in the jurisdiction.

Some administrators may find it challenging to achieve the changes they desire in the face of the many opposing goals of these political groups. Before attempting a policy change, the fire and EMS administrator should consider the following:

- What political party is now in control?
- Are the elected officials short-term thinkers, long-term thinkers, or both?
- Are there any elected officials who are expected to lose or not run for reelection in the upcoming elections?
- Has a single-issue candidate recently been elected?
- Are elected officials elected at-large or from districts?
- Are there any signs of taxpayer disapproval?
- Is there any expectation of what is politically correct?
- Are there any elected officials who would have a reason to not support the FES organization?

- Are there any officials who would like to be a champion or strong supporter of the fire or EMS service?
- Do any of these elected representatives owe a special interest group that may oppose your changes?

An administrator must be willing to show courage, sacrifice, and determination to achieve their goals. The difference between caution and courage can be the difference between the status quo and successful progressive change. Always taking the safe path and insisting on having all the answers before making a decision prevents the chief officer from leading change in any organization. Seeking new frontiers for one' vision and thriving on the unpredictability of the future helps achieve real change. Courage and vision reduce the unexpected changes that tend to push many organizations into crisis management mode. It is relatively easy to be a hero at the emergency incident; it takes a lot more determination to be a courageous leader who ventures out to make significant change in the organization.

Remember, there are always some people who are against change simply because it is change. As an administrator, one needs to be prepared to accept some inevitable criticism.

The following abilities are helpful for administrators to overcome political opposition:

- Accepting the legitimacy of politics and elected officials
- Understanding the structure and process of politics and government
- Building political alliances
- Obtaining the support of special interest groups
- Acquiring public support through effective marketing
- Practicing open and honest communications with employees, citizens, media, and appointed and elected officials
- Using conflict resolution, negotiation, and bargaining techniques
- Identifying the various stakeholders and any benefits they may receive or lose as a result of a change
- Developing trust between members in the organization, the public, the media, and government representatives

### Words of Wisdom

"Being responsible sometimes means pissing people off."

– Colin Powell

Source: Powell, C & Persico, J. (1995). *My American Journey*. New York: Random House Ballantine Publishing Group, p. 35.

**The Power of Unions** In FES organizations, unions work to protect the interest of their members and can have substantial influence over administration decisions. For example, the International Association of Firefighters (IAFF) is a labor union representing nearly 300,000 firefighters and paramedics who protect 85% of the nation's population. Since 1918, the IAFF has worked to unite firefighters and achieve better wages, improved safety, and greater service for the public. Heading this organization is the general president, who is elected to a 4-year term by the membership. It is typical for the general president to serve for many years because there is no term limit. The lack of a term limit is advantageous because it emphasizes long-term goals over short-term success.

The IAFF actively lobbies in Washington through its political action committee FIREPAC, which has influenced Occupational Safety and Health Administration regulations, NFPA Standards, and federal legislation. IAFF contributes to congressional election campaigns for individuals who support its agenda and goals. In addition, IAFF provides research and advisors to assist government officials in legislative decision-making. At the local municipal level, IAFF's influence varies greatly throughout the United States, having less influence in the 23 "Right to Work" states **TABLE 2-1**. In a "Right to Work" state, an employee cannot be required to join a union or pay union dues. This reduces the number of members and their influence on wages and benefits.

For many FES workers, the union becomes a social fraternal organization along with a platform for employee benefits and safety. However, in "Right to Work" states, the influence of the union varies depending on whether the administration formally bargains with the union. In

### TABLE 2-1 Right-to-Work States

| | |
|---|---|
| Alabama | Nevada |
| Arizona | North Carolina |
| Arkansas | North Dakota |
| Florida | Oklahoma |
| Georgia | South Carolina |
| Idaho | South Dakota |
| Indiana | Tennessee |
| Iowa | Texas |
| Kansas | Utah |
| Louisiana | Virginia |
| Mississippi | Wyoming |
| Nebraska | |

Courtesy of L. Charles Smeby, Jr.

any case, the real influence is primarily derived from the union president's ability to impact public policy. This is strongly affected by the same items that affect the chief administrator's power.

There are also national, state, and local efforts that aim to reduce the influence of public-sector unions. For example, in March 2011, Wisconsin's state government passed legislation restricting union bargaining rights for public workers (except for police and firefighters). These efforts are often driven by economic considerations and the desire to reduce expenses in the forms of worker wages and benefits, which the unions aim to protect. The union's job is to represent the workers in gaining better wages and benefits and safer working environments.

Although in difficult economic times unions have agreed to take a reduction in wages or benefits to save the jobs of union members, it is not the union's duty to save money for the community or government by declining wage or benefit offers. It is also important to understand that in other industries, employees who do not believe their wages, benefits, or working conditions are fair or safe may change jobs relatively easily. In FES, this is rare, because the system is designed to offer rewards and benefits based on seniority, discouraging employees from seeking other opportunities.

Many disagreements between the government and unions stem from each group not understanding the duties of the other. The job of chief officer, appointed official, and elected representative is to provide the best services to the public, given a specific tax revenue base. This is a very complex task. Offers of future retirement and health insurance benefits that once seemed realistic may now be problematic. Economic issues at the national level, such as Social Security and Medicare, more than likely require change to fix the problem of solvency.

Although there are many issues on which the government and unions may agree, there are also disagreements. These should be settled by principled negotiations, in which all parties should act in good faith toward their stated purpose. FES administrators should avoid being in a position where they feel forced to accede to union demands. Unions like to use a process of binding arbitration to settle disagreements. In the process of binding arbitration, both parties choose an impartial arbitrator or third party to represent their position and reach a compromise, which is binding on both parties. Binding arbitration should be considered for all disagreements, especially safety issues; however, elected and appointed officials often do not want wages and benefits to be considered for binding arbitration.

For administrators involved in negotiations, the book *Getting to Yes, Negotiating Agreement Without Giving In* (Fisher & Ury, 1991) provides additional guidance and recommendations.

**The Power of the Community** Administrators should identify and become acquainted with the community's power elite at the earliest opportunity. The power elite are a small number of highly influential citizens rarely seen at public meetings. These are local people who have easy access to elected officials as a result of longstanding friendships, wealth, or influence at the local, state, or federal level of government. The powerful elite can either help with the adoption of a new policy or, with active opposition, doom the policy to failure. In most cases, it is unusual for the power elite to get directly involved in FES policy decisions. However, for major policy proposals, such as the construction of a new fire station or a request for funding that requires an increase in taxes, the chief officer should consider approaching these individuals privately and explaining the request personally. Although the help of the powerful elite may not be needed in all cases, they may have the potential to influence policy changes.

The following are examples of potential conflicts with the power elite:

- A new fire station on a vacant piece of property is proposed. The information about the neighboring properties contains the name of one of the power elite. However, the chief fails to make a personal contact with this person before announcing publicly that a site has been selected.
- A new physical fitness program is proposed that contains reasonable goals along with an implementation plan for those existing members who cannot comply immediately. However, the firefighter's union was not consulted. The president of the union is a long-time resident of the jurisdiction and knows many of the power elite by first name.
- The chief has just finished a high-rise fire seminar at a national conference. On returning to the city, a reporter asks for any revelations on fire safety discussed at the conference. The chief announces to the reporter that he is now convinced that all existing high-rise buildings should be equipped with automatic sprinklers, and that a proposed local ordinance will soon be submitted to require it. What the chief is not aware of at that time is that the power elite own many of the existing high-rise buildings.

Private ambulance companies are not in the same situation as fire departments in terms of community power elite having direct impact on internal policy decisions. However, EMS management is wise to seek out and court the power elite who may have influence with those who have political clout. When contracts are up for bid, community stakeholders can have a big influence. Incumbent companies may lose contracts even if they have a great response time record or few patient care complaints. The winner is usually the company that comes in with the lowest bid. Recently, companies with a history of 40 or 50 years of good service have been ousted by others that offer a lower bottom line.

When faced with a conflict with the powerful elite, the chief officer might not always be fighting a losing battle. Approach the potential change by slowly building a consensus among those people who will have an influence over the final approval. In general, arguments that are backed up by professional judgment and solid research have the best chance of approval. When national consensus standards support the request, it is easier to gain the support of the power elite and others who shy away from public debates.

Remember, the power elite and special interest groups (e.g., firefighters' unions and volunteer fire associations) can have a big influence over elected officials. Approach these groups and their officers at the earliest opportunity to gain their support. Although it may not always be necessary to gain their support to be victorious, in many cases their support can make the process smoother, resulting in a better chance of success.

**Administrative Power** In 1952, Dwight D. Eisenhower, who had been a great army general during World War II, was elected president. President Truman, the outgoing president at the time, had this to say about Eisenhower: "He'll sit here…and he'll say, 'Do this! Do that!' And nothing will happen. Poor Ike—it won't be a bit like the Army. He'll find it very frustrating" (Neustadt, 1990). As noted by President Truman, power is never guaranteed. Being promoted to chief officer does not ensure that all members of a company will follow all of one's orders. A FES administrator, when managing and leading change, would do well to study how the most powerful administrators in the world use position and power to make changes.

The FES or EMS administrator is familiar with the power of the direct order or command. The following criteria are necessary for a command to be followed without question:

- The reason for involvement is instantly recognizable.
- The orders are widely publicized.
- The words are clearly understood.
- The supervisors and members who receive the order have control of everything needed to comply.
- There is no doubt about the authority to issue the command.

It is rare that orders on the emergency scene are not carried out verbatim. Because the FES are quasi-military organizations, it is well understood by all members that they must obey the orders of a higher-ranking officer. Officers are trained and educated to follow orders even if the justification is not provided or not understood. Individual officers and their company assignment fit into an overall plan that may not be clear to a specific unit, but fits into an overall strategy to mitigate the emergency. Therefore, orders must be obeyed without question, unless it is clearly an unsafe directive.

Private ambulance companies are less likely to have the same level of quasi-military supervision, but field crews know there is a chain of command they must follow (e.g., medical and trauma care is approved by online or offline medical control). Offline medical control is accomplished by protocols, policies, and standing orders approved by the local medical director. Online medical control occurs when a provider calls a base station and talks to a Mobile Intensive Care nurse or emergency department physician to ask a question, get an opinion, or seek approval for a situation that might not be covered under written protocols. Seeking approval from a higher-level provider provides legal protection for the paramedic.

Complications sometimes arise when fire departments are working alongside private providers who may not have the same training, expertise, or understanding of Unified Command or Incident Command System protocols. This is especially true when the fire department provides basic life support and the private company provides advanced life support. There might also be the issue of private providers not feeling a responsibility to follow the fire department's chain of command in a larger incident. If a breakdown in command were to happen in a multiple patient incident, patient care could be greatly compromised and both agencies could be at risk for litigation.

However, in nonemergency situations, the ability to effectively administer the FES organization ultimately depends on whether the chief officer can persuade or convince others of what ought to be done for their own good or for the betterment of their service to the public. In addition, follow-up deadlines and delegation all play a very important role in everyday managing and change projects.

The ability to recognize a problem and take action to make sure it does not happen again is a great attribute for an administrator. The chief officer should establish an informal information-gathering system made up of friends and confidantes throughout the organization. In addition, it is critical for fire and EMS administrators to maintain their professional reputation. If the chief has the reputation of not following-up on assignments, then when a request for members to comply with one of the department's orders is issued—either verbally or in writing—many will not comply. Noncompliance is much more likely if members do not agree with the change or request.

## Case Study

**Employee Noncompliance**

A good example of how employees may have their own agendas that keep them from complying with an order occurred during President John F. Kennedy's administration. In 1962, the President became aware that the Soviet Union was constructing nuclear missile facilities in Cuba. Because Cuba was so close to the mainland of the United States, these missiles could be launched and hit targets in the United States in an alarmingly short time.

As it became clear that his negotiations with the Soviets were not making any progress, President Kennedy ordered a blockade around Cuba. A blockade can be construed as an act of war, and the crisis brought this country to the brink of nuclear war with the Soviet Union.

During the negotiations that followed, an offer was received from the Soviet Union—if the United States would remove nuclear missiles from Turkey, the Soviet Union would remove its missiles from Cuba. At the time, Kennedy remembered that he had previously given a presidential order for the US military to remove its missiles from Turkey. Kennedy had received expert advice that the missiles were not necessary for the defense of the United States and could safely be removed.

The military officials—who were trained to obey all orders from superiors—had deliberately failed to carry out an order by the most powerful person in the United States, the commander-in-chief. The military officers believed that these missiles were critical to the defense of the country. Fortunately, the military's disobedience provided Kennedy with a bargaining token to make a deal with the Soviet Union. The United States still had the nuclear missiles in Turkey that Kennedy then used in negotiations to trade for the removal of the Soviet missiles from Cuba.

However, after this experience, Kennedy set up a system to follow-up and collect information about what was really going on in the many levels of the government. Not able to manage such a large organization by walking around, he established direct contacts with friends and trusted confidantes throughout the government.

Accurate feedback can be hampered either by subordinates following their own agendas or by high-ranking government officials not wanting to give the boss any bad news. This type of behavior by subordinates can be prevalent in departments where the administrator has a lot of influence over who is promoted. Many subordinates see their chances of being promoted as being based on their ability to do a great job and keep the boss happy. Colloquially, this is called surrounding yourself with "yes" people. Kennedy did not want to be blindsided again and therefore made extensive changes to provide impartial avenues by which information could reach him.

## The Process of Negotiation

In most cases, when a new vision or policy is proposed, the chief officer must negotiate or convince others that the change is good. Most administrators would think that they have the advantage in bargaining because they are the recognized leader. However, bargaining power only starts with the formal position; it also includes professional reputation and knowledge and skill in debating.

There may be opponents to the new vision because some members may lose privileges or simply because they resist any change. Member opposition does not mean that it is impossible to make policy changes, but the chief officer must be prepared, informed, and patient. Officers must have a plan to compensate losers (if possible) and be able to prove that the change is based on solid research, logic, and professional judgment.

Chief officers who have a trustworthy and justifiable proposal may be able to negotiate an agreement with the members to implement the policy. For example, faced with irrefutable justification for the betterment of the public and the individual, many labor organizations cooperate with a negotiated agreement. If administrators have the facts and can prove them, they have the bargaining and influence power.

Administrators in government agencies should stay in contact with a number of individuals outside their formal organization, such as staff members of legislative committees, researchers from nearby think tanks, and representatives of lobbying firms and public interest advocacy groups. In addition, other government agency administrators and their staff need to be contacted on a regular basis to facilitate information flow and, when needed, support for policy changes.

Informal organizations can be used to help facilitate the negotiations and data gathering that must go along with the decision-making and consensus building revolving around a proposed public policy objective. Informal organizations can help overcome barriers to change and facilitate communication between different groups; however, if not approached correctly, they can also sometimes initiate and facilitate resistance to change. An effective administrator must take full advantage of the informal organization to gather facts and acquire support for any policy changes.

## Chapter Activity #1: An Example of Policy Analysis

A voluntary physical fitness program increases the average physical fitness of firefighters. It is well recognized that firefighting requires a high level of strength and endurance. Therefore, the assumption is that the public would be better served by firefighters who can perform at more advanced physical levels.

Firefighting is a team activity, so the weakest link limits the entire company or team. When a two-person team enters a structural fire and one individual is in good physical fitness and the other is not, the total team effort is limited by the less fit person. For example, a person in good aerobic condition can have a useful work time with a self-contained breathing apparatus of around 20–30 minutes, whereas the unfit person may not be able to stay more than 10 minutes before the low air alarm sounds. Therefore, the team of firefighters is limited to 10 minutes because when one firefighter leaves, the other must also leave for safety reasons.

Conclusion: A mandatory physical fitness program is needed.

### Discussion Questions

1. Make a list of management and leadership goals that have to be accomplished to implement the conclusion from the above case study.
2. Do you have a management or leadership preference? How would this preference affect your actions necessary to complete the change process? Give several examples.
3. If you are stronger in one preference, describe how you would select and incorporate people with the other preference into your implementation process.

## Chapter Activity #2: Chain of Command

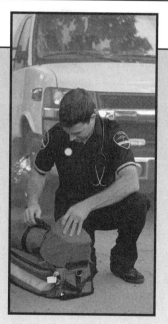

There is a shooting incident in your district. After police secure the scene, your BLS first responder engine arrives and ICS is initiated. As your company begins assessment, an EMT/paramedic unit from the private transport company arrives. With the help of firefighters, the paramedic assesses the patient with a bullet wound to the abdomen. The EMT assesses the patient with an apparent extremity injury. They determine that the first patient meets trauma center criteria.

The company officer calls for a medical helicopter, as is county policy because ground transport to the trauma center would exceed 30 minutes. He is given a 15 minute ETA. In 10 minutes, he calls the fire dispatch center to confirm the ETA of the helicopter and is advised that the request was canceled by the ambulance provider through his dispatch center.

When the second engine company arrives, they assist the ambulance EMT with what is described as a "through and through" gunshot wound to the thigh. The captain isn't comfortable with the patient presentation and insists on seeing the injury. There is no exit wound and the patient complains of abdominal pain. The captain also calls for a helicopter for the patient and is also advised that the helicopter was canceled by the paramedic who didn't agree that the patient met trauma criteria.

### Discussion Questions

1. As the FES administrator, how would you handle the paramedic canceling the helicopter without consulting IC, thereby failing to use the chain of command?
2. Who was ultimately responsible for the delay in patient care?
3. What can be done to assure this type of problem doesn't occur in the future?

# References

Coleman, R. (2004). For leaders, behaviors should outweigh looks. *Fire Chief*. New York: Penton Publications. Volume 48, page 42.

*Fire Chief* staff. (2001). What's new on the safety scene? *Fire Chief*. New York: Penton Publications, Volume 45, April 2001. Retrieved from: http://firechief.com/mag/firefighting_whats_new_safety.

Fisher, R., & Ury, W. (1991). *Getting to yes*. New York: Penguin Books.

National Fire Protection Association. (2003). *Fire protection handbook*, 19th ed. Quincy, MA: National Fire Protection Association.

Neustadt, R. E. (1990). *Presidential power: The politics of leadership*. New York: John Wiley & Sons.

Powell, C. & Persico, J. (1995). *My American Journey*. New York: Ballantine Books.

United States Fire Administration. (1973). *America burning: The report of the National Commission of Fire Prevention and Control*. Retrieved from http://www.usfa.fema.gov/downloads/pdf/publications/fa-264

Vorlander, P. (1996). An innovative approach to fire department command staffing. *Fire Engineering, 149*(8), page 131.

# CHAPTER 3

# Management

## Fire and Emergency Services Higher Education (FESHE) Course Objectives

**Module II: Core Administrative Skills**

The students will:

2. Describe the integrated management of financial, human, facilities, and equipment and information resources. (pp 29–40)
4. Describe the key elements of successful communication. (pp 30–32)
5. Recognize the basic management theory in use in your agency. (pp 37–39)
8. Identify and discuss a practical agency evaluation process. (pp 35–37)

**Module III: Planning and Implementation**

The students will:

1. Describe the process of consensus-building. (pp 30–32)
6. Analyze the importance of an organizational culture and mission in the development of a strategic plan. (pp 30–32)

**Module IV: Leading Change**

The students will:

1. Describe the importance of accepting and managing change within the fire and emergency service department. (pp 29–40)
3. Summarize the steps of the change management process. (pp 29–37)
4. Assess ways to create a positive climate for change and introduce new ideas within the organization. (pp 30–32, 34–35, 37–39)
6. Explain the benefits of employee involvement in departmental decisions. (p 30)
8. Describe ways to increase and reward professional development efforts. (pp 34–35)

## Knowledge Objectives

After studying this chapter, the student will be able to:

1. Understand the role of management in fire services and emergency medical services (EMS). (p 29)
2. Know the four key managerial tasks of planning, organizing, directing, and controlling. (pp 29–37)
3. Discuss the value of experience and intuition in decision-making. (pp 29–30)
4. Describe techniques managers can use to increase staff compliance with rules and regulations. (pp 30–32)
5. Explain techniques for motivating staff members. (pp 34–35)
6. Apply the three-step control process to accomplish organizational goals by measuring performance, comparing against a standard, and taking corrective action. (pp 35–37)
7. Discuss the management strategies of technology-based management, total quality management, leading by example, broad-based empowerment, and managing by walking around (MBWA). (pp 37–39)
8. Comprehend managerial performance standards and methods of performance assessment. (pp 39–40)

## What Is Management?

Management is the glue that holds an organization together on a day-to-day basis. Management is what enables an organization to achieve its goals, efficiently and effectively, with the necessary staff, equipment, and resources. As explained in the chapter *Introduction to Administration*, management and leadership skills are two very different components of fire and EMS administration. Leaders provide the vision and are change agents for an organization, whereas managers supervise and facilitate the work of others within the organization.

As Winston Churchill said, "The price of greatness is responsibility." Although many people may gravitate toward management for its status, being in a supervisory position comes with a great deal of responsibility. There are four main responsibilities of management: (1) planning, (2) organizing, (3) directing, and (4) controlling.

Planning includes creating and defining goals, and making important decisions and establishing strategic plans to achieve these goals. Without proper planning and implementation, the change required to meet new goals can be painful for the organization, its members, and its customers. Planning is discussed in greater detail in the chapter *Public Policy Analysis*.

Organizing is the process of determining what tasks need to be done and who is to perform these tasks. The manager determines the organizational structure (who reports to whom) and which parts of the organization respond to day-to-day problems. A good organizational chart indicates the major groups and who they report to for direction.

Directing the organization involves many tasks such as motivating staff, supervising staff, communicating with members, spotting noncompliance, administering appropriate disciplinary action, and resolving conflicts. These actions are what most people think of when they picture a manager.

Controlling refers to monitoring performance, measuring outcomes and progress toward goals, and making any adjustments necessary to keep the agency moving toward those goals. As part of this process, managers also provide feedback for future planning.

These four responsibilities are discussed in greater detail in subsequent sections.

## Decision-Making

One of the key responsibilities of a manager is to make day-to-day decisions to respond to emergencies and situations that are not expected or planned for in the strategic plan. This section explores one key aspect of the planning process—decision making.

Standard operating procedures (SOPs) cannot cover every situation at the scene of an emergency, so managers must be able to make quick, critical decisions. One of the qualities of an effective manager is the ability to make the right decision in the face of unfamiliar circumstances. These successful officers are often described as having good instincts, intuitively sensing when something is dangerous or wrong, or listening to a "gut feeling." Although intuition should not replace consideration of the facts, accepted knowledge, or prior experience, it can along with common sense contribute to decision-making.

Many experts believe that a person's subconscious mind holds all of the acquired knowledge, ethical values, past experiences, training, and skills that are used for complex problem solving. That is why it is always better when making an important decision to "sleep on it" if possible and allow the subconscious mind to process the information.

---

### Facts and Figures

**Incident Command System**

The Incident Command System is the initial model of command used at local emergency incidents. It contains all four main responsibilities of management discussed previously. With the exception of large disasters, all emergencies start as local emergencies. When the situation gets more complex and starts to involve multiple local, state, or federal jurisdictions, then an expanded model of management is necessary. National Incident Management System (NIMS) compliance is required of a wide variety of first responders, including EMS.

NIMS establishes a uniform set of processes, protocols, and procedures for all responders, and works with the National Response Framework. NIMS provides the template for incident management, whereas the National Response Framework provides structure and mechanisms for national-level policies. Key incident command functions include:

- Incident command
- Operations
- Planning
- Logistics
- Finance and administration

A NIMS-compliant Incident Command System should be used on every incident involving more than one resource. Initial command is initiated by the first arriving unit. This could be a police vehicle, private ambulance, or fire engine. As other agencies appear on scene, command should be passed to the most appropriate and qualified person and agency. Which agency takes that role depends on the type of incident and also on local structure and resources.

In part, intuition comes from experiences and acquired knowledge. In his bestselling book *Outliers: The Story of Success* (2008), author Malcolm Gladwell claims that the key to success in any field is, to a large extent, a matter of training, education, and about 10,000 hours of professional experience. Although a lot of experience is needed, it is not necessary to get all the experience oneself; the experience can be acquired from others through the review of case studies and writings of those with years of experience. Learning from others allows managers to gain knowledge of difficult situations without having to put themselves or their staff in danger. To gain knowledge of emergency incident experiences, managers can learn from textbooks, reports of work at emergency incidents, visits to other fire and emergency services (FES) agencies, investigations of emergency responder deaths and injuries, articles about incidents, interviews with other emergency responders, professional websites, and professional qualifications and safety standards.

Managers may wish to make decisions on their own as individuals, or they may make decisions based on group input. Group decision-making occurs when individuals, typically those that are affected by the decision, collectively study and make a choice from the potential realistic options. In most cases, the chief manager provides a definition of the problem, some background information, and any limitations on the group's decisions.

## When to Use Group Decision-Making

When the final decision directly impacts organizational activities, members' benefits, or performance benchmarks, group decision-making may make acceptance easier, particularly when sacrifice is required. Groups may find unique solutions to problems; for example, unions have voted to accept pay decreases instead of layoffs. With group decision-making, managers are less likely to feel blindsided by a lack of information about possible solutions. One of the major reasons noted for bad decisions made by management is lack of information. Groups ensure that many diverse opinions are discussed. This process also results in the collection, verification, and analysis of a lot of background information regarding the problem, thus dramatically reducing the chance of a poor decision being made.

## How to Select Group Members

Effective group decision-making should include a group of individuals from all levels of the organization. Outside experts such as elected officials or other agency representatives can also contribute to the process. Group members need to be selected carefully based on their openness to discussion, absence of bias, and ability to come up with rational solutions. Managers must be cautious not to be predisposed in selecting group members who are in favor of the preferred administration solution. The advantage of increased acceptance is lost if members believe the decision-making process was manipulated. Group members should have a minimum amount of experience, because there are some idiosyncrasies of the profession that simply take time to learn; however, they should not be near retirement. Members nearing retirement may not have a stake or interest in solving the problem. One common policy is to allow groups, such as the union, to appoint one or more members to the decision-making group. These selections should be accepted by the administration even if these members might be biased and would not have otherwise been selected.

## Optimal Group Size

Although there is no unanimously accepted number, groups of 5–15 members seem to be the most effective size for timely group decision-making. Some organizations that use group decision-making to determine consensus standards (e.g., the National Fire Protection Association [NFPA]) may have committees with up to 30 members. Larger groups may have greater difficulty arranging in-person meetings, although electronic audio-video conferencing can help. Ideally, at least half of all meetings should be in person. Whatever number is chosen, it should be an odd number to enable the group to reach a final decision among a majority of members.

## Techniques to Aid Group Discussions

Several different techniques may be useful in the process of group decision-making.

**Understanding Limitations** Managers are responsible for communicating any limitations of resources or other relevant limitations (e.g., specific allocations or regulations) to the group members. This may include

---

### Chief Officer Tip

**Emergency Incident Research**
When conducting emergency incident research, consider using the following internet resources:
- *Firehouse Magazine,* www.firehouse.com
- *Fire Engineering Magazine,* www.fireengineering.com
- *The Fire Chief,* www.firechief.com
- *JEMS: Journal of Emergency Medical Services,* www.jems.com
- *EMS World Magazine,* www.emsworld.com/magazine
- Firefighter Close Calls, www.firefighterclosecalls.com

mandates from elected officials, legal restraints, and state or federal regulations. In addition, the manager may wish to outline possible solutions and describe any that are not accepted by the administration.

**Gathering Information** Managers may wish to encourage group members to consider ideas for solutions that have demonstrated success in other departments. Group members may wish to consult research articles or visit departments that have had similar problems. This allows group members to gain insight into what has been tried elsewhere and what has succeeded or failed. Managers should encourage members to have an open mind; just because a solution failed somewhere else does not mean it will not be successful in their department.

**Reviewing National Consensus Standards** Fire services and EMS may be required to meet standards defined by national organizations, state governments, or local contracts to ensure quality and accountability. The guidelines provided by federal agencies, state authorities, professional organizations (e.g., NFPA, National Highway Traffic Safety Administration), and union agreements should be considered when determining company policies. For example, it can be argued that if a department complies with NFPA 1500, *Standard for Fire Department Occupational Safety and Health Program*, it will be enabled to deliver outstanding emergency service to the customer.

**Brainstorming** In the process of brainstorming, members may suggest any solutions without initial judgments; this is essentially a technique to generate ideas **FIGURE 3-1**. Anonymous input should be arranged, if possible, to encourage those members who may not want to publically advocate their ideas for fear of criticism. Members may wish to have a method to provide input anonymously. However, this may not be necessary because only ideas—not criticism—should be allowed during brainstorming sessions. After all of the suggestions have been documented, the group should be given time to consider the available options before discussing them.

**Discussion** At the next meeting, members may discuss the options and offer criticisms of the proposals. Members or staff should be selected to research the ideas, including all previous successful or unsuccessful experiences. Pros and cons should have solid data or reasoning behind them and should not be dismissed just because they have not been done before. Going through all the available options may take several meetings, but it is time well spent

**FIGURE 3-1** Brainstorming allows any and all solutions to be recommended without any initial judgments.

to ensure that group members have examined all possible solutions.

Only through questioning the status quo (e.g., the culture or tradition) does real change occur in organizations. For this reason, group members may choose to take on the role of "devil's advocate" during the decision-making process. This role is not for everyone; the devil's advocate is someone who thinks outside the box, is a risk taker, and is not afraid of change. In many respects, the devil's advocate is a true leader who may lead the department through substantial, controversial, or progressive changes. He or she should aim to present input without angering the group, because this could lead to ostracism and being excluded from future group decision-making processes.

The desire to be liked by others and not take an unpopular stance is one of the most challenging issues a person may face in his or her private and professional life. To express one's concerns without angering others, it is best to keep the focus on how the customer, member, or department will benefit from the new policy or decision and how alternate solutions may be harmful for the safety of the customer or efficiency of the department. Playing devil's advocate is a valuable exercise to make other group members justify their support of a new or old idea.

**Voting** At the end of the discussion, members may determine if there is sufficient support for any of the ideas. It is possible that a combination of proposals may be the best solution. Managers should encourage group members to compromise; sometimes the best compromise includes some form of future implementation (e.g., allowing several months or years to achieve the required change). Group members may determine how they will reach a consensus at their first meeting, or this process may be designated by the administration. Consensus may be reached by unanimous decision-making, a simple majority, or a specified majority (eg, the NFPA requires a two-thirds majority and the US Senate requires a three-fifths majority for most important laws). Although not necessary, having a majority consensus is preferable. It is easier to obtain as opposed to unanimous agreement, and it indicates that most members agree with the group's decision.

**Delayed Implementation** A delay in initiating the new decision may be necessary because of funding or some other practical limitation. Delayed implementation is sometimes combined with a "phased-in implementation" to smooth out the blow of some projects that create a lot of change.

**Forward Thinking** This technique is used to ensure that the decision takes into consideration what may happen in the future. Many decisions do not consider the future, and, therefore, it is common to have unintended consequences.

### Case Study

**The Devil's Advocate at Work**
A fire prevention program manager would like to increase the number of home fire inspections performed by station personnel. The station personnel have reported that they do not have any time available as a result of emergency calls, in-service training, physical fitness, and station and apparatus maintenance. Firefighters work a 24-hour shift during which the required nonemergency duties are performed during daylight hours. The time after dinner until the morning is reserved for personal time or call response.

However, the prevention manager would like to suggest that the time after dinner (and during weekends) be used to do these inspections. The manager believes that this would be a good time because many residents who work would be at home during the evening hours and weekends. Because these inspections are voluntary, the requesting residents would be advised of the time of their appointment with the understanding that an emergency call would necessitate rescheduling.

When reviewing this issue, the devil's advocate might ask the following questions: How popular would this idea be in most departments? Would this idea be controversial or resisted by the members and their labor unions? The devil's advocate would then broach this subject and, hopefully, be able to lead the group to a progressive change.

### Case Study

**Examples of Group Decision-making Techniques**
The following case studies provide examples of the previously described group decision-making techniques.

GATHERING INFORMATION
One major metropolitan fire department appointed a committee to design a new fire station. It had been many years since they had the funding for a station, and all of the existing stations were built by volunteer companies before the county consolidated. Because there was no uniformity in these designs and they were outdated, the committee decided to visit newly built stations in adjoining departments. Eventually, they found a station that was almost perfect for their needs. They obtained permission from the other department to use the same architect, which saved a great deal of time and money by eliminating the need for new engineering and architectural plans.

DELAYED IMPLEMENTATION
A department adopted a physical fitness program and testing system; they used delayed implementation
*(continues)*

### Case Study (Continued)

to allow incumbents (who may not have exercised for years) extended time to comply. For example, in the first year, physical fitness equipment would be purchased for each station, and medical evaluations and physical fitness testing would be completed. A plan for improvement would be given to each member. During the second year, members would start exercising at work using their individual goals as a guide. Between years 2 and 10, the existing members would be expected to achieve a percentage of the stated goal starting at 20% and increasing at 10% per year. At the end of 10 years, all members would meet the same fitness requirements, achieving the desired goal.

#### Forward Thinking

A large ambulance company purchased hydraulic gurneys for all its ambulances across the country. It was a monumental expense on the front end, but the company realized the gurneys would all but eliminate career-ending back injuries secondary to incorrect lifting. The future goal was that hugely expensive worker's compensation, and possibly patient lawsuits, would be significantly reduced. Field personnel were happy with the decision, which would help protect their backs and maintain their careers.

do not always reach ideal decisions. Poor results can be caused by failure to focus on the customer or department's goals, or by a lack of leadership or preparation.

Even if the results of the group process are valid, they may be unnecessarily complex and have internal inconsistency. For example, if a department's procedure for the first due engine on a structural fire contains more instructions and decision points than an officer could realistically complete in the typical 7 to 10 seconds before action must be taken, this may be the result of poor group decision-making. Because meetings occur in a nonemergency setting, the group has plenty of time to discuss and analyze all possible actions, often resulting in the tendency to overthink the problem and make it more complicated than necessary.

There is also a danger of minority domination in group processes. A charismatic group member or several members representing a particular group may dominate decision-making. Other members may feel a need to go along with the crowd to avoid conflict. This desire to go along with the group can be very strong in emergency response agencies where strong personal bonds often exist between members. Managers can help avoid this situation by providing clear direction and limitations.

## Organizing

In addition to planning and decision-making, another key task of most managers is determining the organizational structure. Departmental organization in FES is usually very similar in all departments, making it less likely to be an issue. Budgetary considerations have led a few organizations to create alternative departmental structures, but this is rare. Most departments are organized the same, although there may be some minor variation in job titles based on tradition or department size FIGURE 3-2.

The department is organized by a principle called *unity of command*—the idea that each individual reports to only one manager. This clear hierarchical structure is

### Disadvantages of Group Decision-Making

Group decision-making offers many benefits; however, it does have some disadvantages. For example, groups almost always take substantially longer than an individual to reach a decision. Also, as demonstrated by the saying "a camel is a racehorse put together by a committee," groups

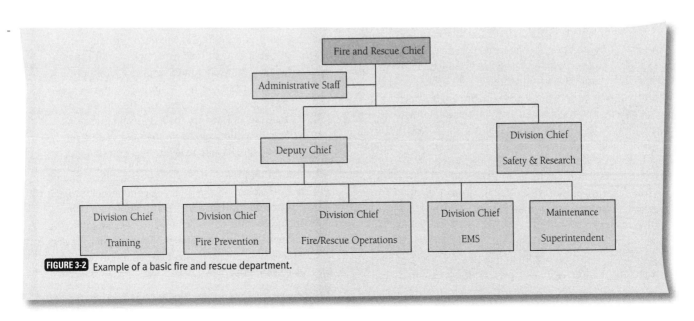

**FIGURE 3-2** Example of a basic fire and rescue department.

especially necessary to ensure order at the emergency scene. Unity of command becomes even more critical at large incidents where there are several different departments providing mutual aid. In national emergency incidents, the NIMS, discussed more thoroughly in the chapter *Introduction to Administration*, is used to ensure unity of command at catastrophic emergency incidents. FES organizations that use this system at everyday incidents will be comfortable with it and better prepared when large-scale incidents occur. Making anything a routine is more likely to result in it being done and done correctly.

At the top of the department, managers may report to individuals outside of the organization. In large departments, managers report to a mayor, other elected official, or appointed public safety director. In special taxing districts, a committee of elected officials is formed to oversee the FES department and represent the public interest. However, these elected officials may not have sufficient knowledge or expertise to judge the service, which can present a challenge for department managers. Unlike large departments, many volunteer departments do not report to anyone. This lack of accountability is usually offset by the high degree of commitment and motivation volunteers have to provide high-quality emergency service. However, managers must keep in mind the following quote from historian and politician John Emerich Edward Dalberg Acton: "Power tends to corrupt, and absolute power corrupts absolutely."

## Directing

The third major responsibility of managers—directing—can take many forms, such as orders, controls, and instructions. Directing is the process of a manager instructing the members under his or her authority to accomplish an organizational objective. Instruction may occur from person to person or take the form of written procedure or policy. At its simplest level, it is the verbal order of an officer to his or her crew on an emergency scene.

Directing can occur through direct supervision or voluntary compliance and varies with the objective to be accomplished. For example, fire inspectors normally operate under voluntary compliance (self-direction) as a result of the nature of the role. Generally, they are specially trained and educated members who operate individually when performing a fire inspection. When members are motivated and informed, enforcement methods are seldom needed. However, to ensure total compliance, mandatory methods of direction, including follow-up and discipline, may be necessary.

In addition, managers can achieve compliance through the use of rewards, positive role modeling, supervision, and a general culture of compliance. Ideally, managers would not have to closely supervise every member, because this opens the very real possibility that members are noncompliant when they are not being watched. Self-discipline is preferable for functional and safety issues. Close supervision may be needed when a new change is implemented, but this level of supervision is not needed after the culture and individual behaviors are changed.

Managers can use several of the following techniques together to achieve compliance.

### Creating Rules and Regulations

Rules and regulations are the backbone of directing. They achieve consistency by ensuring that all managers are following the same guidelines. Important organizational rules and regulations should be in line with the goals of the organization, measurable, and written down. In creating an agency's SOPs, a manager should strive to limit the number of SOPs required and to ensure that they are simple and easy to understand. Creating SOPs is often a complex process that identifies relevant state or federal regulations, local laws, traditions, and emergency practices affecting the department. Other departments' SOPs can be a great resource to help check the accuracy and completeness of new or existing SOPs.

### Motivating Staff

One of the goals of management is to motivate members to achieve certain tasks or goals. Ideally, members should be motivated to the point that they no longer require direct supervision. Motivated members are more inclined to cooperate, complete tasks willingly, and follow departmental rules than dissatisfied or unmotivated members. Members who are self-motivated can also add to the

---

**Chief Officer Tip**

The chief officer may find some of the characteristics of younger generations challenging to manage. Page (1999) explains this generation gap in the article *Your 21st Century Firefighter*:

> Our future firefighter will have precious little experience with teamwork, self-sacrifice, personal organization and respect for authority. It's not necessarily his fault. Young people have grown up in the soft, affluent society that was created for them. We sold our farms to corporations and won the Cold War so our kids didn't have to bale hay or get drafted.

Although a disparity does exist, blame should not be placed on the younger generation. As Page goes on to acknowledge, the ultimate question is, should the fire service change to accommodate the differences between generations, or should the new generation change to meet the traditional fire service standards?

motivation of those around them. These members might be good candidates for management positions because they demonstrate management skills and leadership traits.

Fire and EMS providers can be motivated by many considerations, including a desire to gain security in their membership status, obtain the respect and trust of supervisors, or maintain safety. They may also be motivated by the administration's commitment to their well-being, by salary and benefits, or by the pride they have in the organization. Members may be self-motivated by their job duties, a sense of achievement, or a desire for recognition. For example, motivation can take the form of public recognition or ceremonies that recognize heroic actions (managers should be careful when considering rewarding heroic acts; unsafe actions should not be rewarded because receiving a reward or recognition elevates the act to be an aspirational goal) **FIGURE 3-3**.

Cash awards or incentives can also be a great tool to encourage members to sacrifice. For example, it is common to offer incentive pay for advanced certifications that require many hours of studying to achieve. Furthermore, if a potential policy would be met with a great deal of opposition, incentives may help smooth its acceptance.

However, according to David Javitch from Entrepreneur.com, managers often make the mistake of assuming that workers are motivated by money alone. Studies show that recognition and status have a more lasting impact on output than raises or bonuses. Furthermore, although employers may not worry about motivating their smartest employees, it is often these employees who are the most vulnerable to boredom and frustration (Javitch, 2009). Money may not be an effective motivator for these employees or professionals, who typically already have good pay and benefits; instead they may be looking for increased responsibilities, challenging work, or professional growth opportunities.

One system that encourages professional growth and responsibility at the firefighter level is a tiered promotion ladder (e.g., recruit firefighter, firefighter, and senior firefighter). This system is based on professional standards for training, education, experience, and salary increases. Because managing is about getting the most and best out of each member regardless of rank, not every member should be encouraged to seek promotion. Promotions should be evaluated based on clearly defined requirements and responsibilities of the new position. Promoting the wrong person can easily lower morale and upset other members of the organization. Furthermore, not everyone can be an officer, and many members are more comfortable not being an officer. However, every department needs well-trained professional firefighters; and there should be a system that recognizes and rewards these vital members. This system can lead to improvement in the quality of the service provided to the public.

## Controlling

Controlling is the management function that ensures completion of all steps to accomplish the organization's goals and provide feedback on any unintended consequences. Honest feedback and accountability are very difficult to obtain for the boss. Many goals and objectives may not be able to be measured directly, which is the preferred accountability tool because it is not subject to interpretation. To be an effective manager, helpful criticism and suggestions for improvement are necessary.

If managers do not control, there would be no way to know if policies, plans, and procedures are being achieved correctly and on time. The control process is made up of three steps: (1) measuring performance, (2) comparing the performance results to a standard, and (3) taking corrective action to ensure compliance to the standard. For example, if a fire department had an everyday goal of staffing each fire company with four firefighters, the number of firefighters present at the start of each shift would be recorded; this data would be compared to the standard of four firefighters per shift; and if the number of firefighters was not realized, corrective action would be initiated, such as detailing or calling back firefighters on overtime and/or taking disciplinary action for firefighters who were absent.

Another example is if an EMS company required that all patient care reports be completed within 24 hours of the call or before the end of the member's shift. The supervisor would research the time that patient care reports are completed and the time of the corresponding incidents. These results would be compared to the standard of 100% compliance, although the company might allow time variances for certain complications such as technical difficulties, multiple late calls, vehicle breakdowns, Computer Aided Dispatch issues, and so forth. Corrective action would occur when the manager identified a crew member not in compliance with the policy; the manager would determine the reason why the patient

**FIGURE 3-3** Awards are opportunities for excellent public relations, but be careful that the acts rewarded are not unsafe actions.
© Elise Amendola/AP Photos

care reports were not completed and provide an opportunity for the provider to correct the problem.

## Measuring Performance

The first step in the control process—measuring actual performance—seems relatively straightforward. However, managers must determine the correct outcome to measure to avoid unintended consequences. For example, if a manager monitors and measures the time it takes for turnout, members will likely show improvement in this area but may resort to unsafe methods to shorten turnout times. Some outcomes can be measured quantitatively (e.g., a specific number); others may be measured categorically (e.g., yes or no); and others may be measured qualitatively (i.e., no absolute answer but a general measure of several different criteria). However, most "outcomes" should be measurable. What managers measure can determine what members will focus on to improve their performance.

## Comparing Results to a Standard

If comparing the results of the case study presented below to a standard, data would need to be collected. This data would need to note the total number of fire incidents and the property lost per incident over a calendar year for storage room fires in garden-style apartments. This data would have to be collected for several years after full implementation and be compared to the historical data before a change was made.

Furthermore, in that particular example, there was also anecdotal evidence of success because many firefighters openly complained about the lack of multi-alarm fires. The data supported these observations.

## Taking Corrective Action

When a manager finds it is necessary to take corrective action, the first step may be a feedback session with the employee. Feedback is an important part of a manager's responsibilities because it can help correct the performance of an individual or identify problems with a policy. Regular feedback meetings can be used for managers to get complete and honest information. Feedback meetings are like a form of counseling and should be scheduled in advance to discuss specific concerns in private at a time with minimal interruptions. In these meetings, the manager notifies the employee member of the issue and sites an example of when the issue occurred. The goal of these meetings is to discuss the concern and possible solutions. The manager must determine if the employee's noncompliance is the member's responsibility or indicates a problem with the rule, policy, or SOP. Although disciplinary action may be necessary later, the manager should aim to put the member at ease during these sessions and establish a supportive atmosphere in which positive feedback is given to help correct noncompliance.

During feedback sessions, a manager should focus on the facts of the issue and not generalized or subjective comments. The purpose of the meeting is to correct job-related behavior; personal judgments or criticisms have no place in these sessions. Any negative feedback should be directed toward the job-related behavior and non-compliance, backed by examples, facts, and verifiable observations of the inappropriate actions. The member should be encouraged to present his or her side of the story and explain any extenuating circumstances. The manager should listen to what the member has to say even if it is not factual or accurate. Excuses should be acknowledged, because they may point to issues that the member needs to correct to ensure that the infraction does not happen again; however, the facts should override excuses. After presenting the issue, the manager should explain in clear terms what the member can do to correct the negative job-related behavior. If the member is unable to correct the problem, the manager may need to discuss disciplinary action. Although rare, there are members who are not able to meet job performance requirements; these members should be quickly separated from the department, because this is a safety issue for the member and the team. The feedback session should end with a clear understanding of the issue and any solutions necessary. Some discipline policies require a written report of the conclusion.

> ### Case Study
>
> **Fire Prevention**
>
> In the early 1980s, the Prince Georges County, Maryland, Fire Department was experiencing numerous large-scale fires in garden-style apartments. Many of these buildings had two uninhabitable basement areas that were used as shared laundry and storage rooms. These areas had no windows and no fire detectors, and were commonly used to store combustible items, such as renters' personal items, furniture, and car tires.
>
> The fire department had computerized fire data for many years, and one of the items captured was whether there was a delayed or immediate notification of the fire. After analyzing the data, a strong correlation appeared between the ultimate property loss and the detection stage. As one might suspect, delayed detection (caused by the lack of fire detectors) resulted in higher property losses.
>
> By measuring this data, the department was able to identify a problem and implement special fire code requirements to address the problem. The owners were either required to install a sprinkler system or vacate and lock the storage rooms and install a smoke detector. In this case, the data identified a measurable problem that was analyzed and used to ultimately improve the department's performance in the area of fire safety.

When feedback sessions do not provide the desired result, disciplinary action may be necessary. For example, consider the use of the three-step control process to address the concern of firefighter safety; the manager would first collect data on firefighter death and injury. This information is well documented by the US Fire Administration and the NFPA. Next, the manager would work to determine a standard of comparison for safety goals (NFPA 1500, *Standard on Fire Department Occupational Safety and Health Program* and the National Fallen Firefighters Foundation sponsored "Everyone Goes Home" 16 Firefighter Life Safety Initiatives may provide useful guidelines to develop these goals.). After the data was compared to the goal, the manager would take corrective action to work toward the goal. The manager could start by creating a policy or SOP in consultation with a group of appointed experts. After the policy was created and approved by the administration, then compliance with this policy could be measured. For example, if the policy required an annual physical examination, members might be required to submit verification that they had completed the examination or face disciplinary action.

Unfortunately, many FES and EMS disciplinary systems are ineffective because managers do not take appropriate disciplinary actions, employees are not aware of what constitutes rule infractions, or low-level officers take disciplinary actions that are outside the scope of their knowledge or abilities. For a disciplinary system to be effective, a manager should be appointed as the point person on all disciplinary issues; selection should be based on an exceptional reputation of ethical behavior in the manager's personal and professional life. Having one person ultimately responsible for all discipline negates any unfair actions by subordinate officers and ensures consistency. This manager must be trained on issues regarding discrimination, sexual harassment, and employee rights.

A list of common rule violations and appropriate consequences should be created and given to all members. The consequences should be developed in consultation with legal and labor representatives. It may also be useful for the manager to compare the policy to those of neighboring departments. Most disciplinary policies involve progressive consequences, requiring more severe disciplinary actions for repeat violations or serious offenses. These steps should be followed precisely without deviation. A formal training session on the disciplinary system should be presented to all new employees and, when first implemented, to all existing members. Chief officers should watch for violations of the rules and regulations by walking around to visit stations and emergency incidents. Any employee who violates a rule should be reported to the manager for disciplinary action. The manager should aim to be fair and consistent in applying disciplinary action. If it is obvious that a company officer is deliberately not paying attention to violations by members of his or her crew,

## Case Study

### The Control Process

In Florida, an applicant for a firefighter position must sign an affidavit that states that he or she has not used tobacco for 1 year before the date of application. In the state of Florida, a diagnosis of heart or lung disease can be considered a job-related illness. Requiring applicants to sign this affidavit reduces the possibility that heart or lung disease was caused by smoking tobacco rather than job-related factors. The goal behind this initiative was to reduce the number of firefighters who were smokers with the hope that firefighters would remain smoke-free for the length of their careers.

However, in one fire department, the manager observed several new firefighters smoking. The manager wondered if they had started smoking after joining the department or had not been honest in signing the affidavit. The process of control (measuring outcomes, comparing to standards, and taking corrective action) could be used to solve this issue. The manager would measure new firefighters' compliance with this condition of employment by an initial nicotine test, followed by subsequent random tests. The results of the testing would be compared to the tobacco-free standard. Any new firefighters found in violation of the non-tobacco use policy would be counseled about the policy on the first occurrence. If still noncompliant when tested again, they could face job separation.

the officer should be disciplined. This situation can be a special problem in the fire service, where officers may be close friends with members.

## Management Tactics

Several different strategies may be used by managers to help their organization achieve its goals.

## Technology-based Programs

There are many technology-specific programs that are helpful for the manager. The use of technology is one of the most effective approaches to providing safety. Although not inclusive, the following are some examples:

- Some of the best examples of the use of technology in FES have been implemented in the area of fire prevention. For example, residential occupancies are attributed with the greatest number of fire deaths. Although humans are the most sensitive detector of smoke, they can be unreliable at times, such as when sleeping. That is why the simple and inexpensive smoke alarm (a measuring device) has been so effective. It is always ready to perform its life-saving function. Therefore, if the management goal is fire safety for residential occupants, this electronic device significantly and reliably makes progress toward this goal.

- It is not uncommon for FES departments to have rules and SOPs that need to be enforced to be effective, because this is the only method to guarantee 100% compliance. For example, one popular mandate is to require all firefighters to be seated and belted before responding to a call. This is not easy to measure because the officer is always looking forward with firefighters riding in a back compartment. Recently the NFPA standard for fire apparatus was updated to require a Seat Belt Monitoring System. This system provides an audible and visual warning when a seat is occupied without a fastened seat belt.
- Driving is the most dangerous part of an emergency response. Because some driving laws are suspended or altered for emergency response driving, the process of "due regard" must be honored. An intersection accident where the agency is held liable can cost them not only in money, but in lives, reputation, and rising insurance rates. However, it is impossible for managers to monitor emergency vehicles every time they are running "hot." Therefore, some agencies have outfitted vehicles with "black boxes"; the information the devices provide has saved companies from losing lawsuits and provides a measure of the driver's adherence to driving regulations.

## Total Quality Management

Total quality management is a quality improvement tactic that was first extensively used in Japan's auto industry. It was created under the assumption that customers are more likely to buy products that are high in quality. Therefore, this tactic strives for the goal of 100% perfection.

Jeff Dewar, quality consultant of QCI International in Red Bluff, California, further argues the need for eliminating defects altogether. To make his point, Dewar came up with various cases of what life would be like if things were done right 99.9% of the time. For example, being just 0.1% less than 100% would result in 500 incorrect surgical operations per week in the United States (Dewar, 1989)! This approach emphasizes the importance of striving for zero defects and is a good safety goal.

## Leading by Example

Leading, or rather managing, by example is one of the most powerful methods of encouragement a manager can offer. For example, a manager who always uses a seat belt when responding to calls encourages others to do the same. If a manager does not lead by example, management efforts can be sabotaged and, in many cases, change will not be institutionalized, creating the possibility for efforts to be overturned in the future.

As mentioned in the chapter *Introduction to Administration*, a manager is not necessarily a leader. A leader embodies special behavioral traits that help achieve new and innovative changes. However, managers can strive to be leaders. Successful leaders are courageous risk takers who do not fear the criticism that often accompanies significant change. Leadership means speaking up before there is consensus on an issue or problem. Also, a leader has the courage to admit mistakes and adjust plans accordingly.

For management, job-relevant knowledge—one of the key traits of leadership—can be gained through higher education. Continued learning increases a person's confidence and intuition. In selling new ideas to others, confidence is extremely important. Another trait that could be added to this list is the ability to listen. An important skill for all leaders, listening helps managers gain vital knowledge about the support or opposition to their proposals. Listening also helps define any problems, which is the first step of solving any management difficulties.

---

**Chief Officer Tip**

**Traits Associated with Leadership**

1. **Drive.** Leaders are self-starters with high energy. They are persistent and strive for success.
2. **Desire to lead.** Leaders enthusiastically take on new responsibilities and enjoy influencing others to accomplish enlightened goals.
3. **Honesty and integrity.** The personal and professional ethics of leaders are above reproach.
4. **Self-confidence.** Leaders believe in themselves and have faith in their own beliefs.
5. **Intelligence.** Leaders must be intelligent enough to analyze large quantities of data and ideas. Perseverance and hard work make up for any lack of knowledge.
6. **Job-relevant knowledge.** Leaders have extensive knowledge about the agency and the profession from practical experience, training, and formal education.
7. **Extraversion.** Leaders are generally happy, optimistic, and energetic. When needed, they can be socially outgoing; but they also take time to be alone for deep thought.

*Source*: Judge, T. A., Bono, J. E., Ilies, R., & Gerhardt, M. W. "Personality and leadership: A qualitative and quantitative review." *Journal of Applied Psychology*, August 2002, p. 767.

Managers can be leaders by setting a positive example for others to follow. By serving as a role model, managers can gain voluntary compliance of members. Leading by example can also help managers create a positive reputation and gain the trust of their staff. Managers should always strive to:

- Be honest.
- Display competency in professional knowledge and skills.
- Be loyal to the organization and to subordinates.
- Be fair and consistent.
- Show a commitment to sharing ideas and information.
- Keep promises.
- Keep confidences of superiors and subordinates.

## Broad-based Empowerment

In some industries, empowering low-level supervisors is necessary to help the organization run smoothly and achieve its goals. This management technique may work best in dynamic fields such as technology firms. However, broad-based empowerment is not appropriate for FES organizations, which require direct supervision and adherence to rules and regulations for efficiency and consistency of the overall emergency response.

The only time when broad-based empowerment can be used effectively in FES or EMS is when employees are given a voice in decisions that might benefit the company, such as purchasing equipment, organizing training sessions, or scheduling shifts. In these situations, acceptance of change increases with employee empowerment. People who believe they are a part of the decision-making process tend to be more productive and content. Allowing members to offer opinions on decisions that affect them makes them feel valued and inspires them to be more company oriented. Their involvement may also lead to creative suggestions for ways to solve problems specific to their jobs.

However, besides these types of decisions, direct hierarchical decision-making is necessary. Independent decisions at the company level can reduce the quality of response provided and create chaos at large-scale incidents where there are specific expectations of the capabilities and operations of all companies involved. The individuals within a fire or emergency service often do not have a sufficient understanding of the larger picture to be able to make appropriate decisions. For example, in one major metropolitan fire department, a manager empowered company officers to develop their own in-service training program after making a local assessment of the hazards in their first due areas. One officer in a rural area devised a yearly schedule of drills that did not include the use of fire department standpipes used in high-rise firefighting. In the officer's mind, this decision made sense because the first due area did not have any high-rise buildings. However, what the officer overlooked was that he, or any of the members of his company, could be asked to fill in at another station that did respond to calls in high-rise buildings. By empowering company officers to make the decisions, this manager reduced the quality of the department's service and put firefighters into a situation where they were not adequately prepared.

Broad-based empowerment can be equally dangerous in EMS situations. Processes and protocols are required for ambulance maintenance, patient care, and operational safety; there can be no flexibility. This is especially true when EMS, fire, law enforcement, and other agencies are working together at an incident.

## Managing by Walking Around

MBWA is a management strategy in which managers physically visit all areas of the organization. This can be extremely helpful for the typical FES organization whose stations are physically separate from headquarters. The chain of command can still be used to send and receive information from the different units, but MBWA allows for direct supervision and increased communication, because managers can spend up to 60% of their time out in the field. Information can become distorted or purposely suppressed as it travels from bottom to top and top to bottom through the lines of a large and dispersed organization. Using MBWA, FES and EMS managers can visit their stations to speak with personnel about issues they are having, make firsthand observations at emergency incidents, and create a formal critique procedure. Managers can more accurately assess emergency operations from these firsthand observations.

Many MBWA managers consider themselves facilitators who strive to listen carefully during their out-of-office visits for any signs of things that may not be going well. Managers can use their senior position to remove any roadblocks to enable members to accomplish their goals. In this way, MBWA sends a powerful message to members that the manager believes that the work that members are doing is important to the mission of the organization.

MBWA can also be effective in dealing with supervisors, local elected officials, neighboring officials, and other department heads in government. Using MBWA, the manager becomes personally and professionally acquainted with these influential officials and can increase support and cooperation from outside agencies.

## Assessing Managerial Performance

A manager's performance should be measured by the two key criteria of efficiency and effectiveness. This control process is used by senior administrators to judge the proficiency of an individual manager. Efficiency is measured by the ability to perform well using limited funds or resources, whereas effectiveness is measured by the manager's success in attaining goals.

Management effectiveness is directly related to the power, expertise, knowledge, ethical reputation, and personnel and professional relationships of the manager. Formal rank is the most obvious sign of power; however, power and relationships are codependent and rely on the manager's ability to create an effective power base. Creating this support is one of the most important goals for a successful career in management. To do so, individuals can undertake the steps listed below:

- Show respect to everyone you meet and treat others like you would want to be treated. Sometimes just saying "I need help" is enough to get cooperation.
- Give people the benefit of the doubt; assume they are really trying their best.
- Get to know other influential managers or associations at local, regional, state, and national levels of your profession.
- Create good relationships with your neighboring departments and unions.
- Develop skills and information sources that make you a unique member, such as the go-to person to research issues, generate policy papers to support departmental goals, or know the right person that can help solve a problem. Reading and studying this text are information-gathering tasks that add to expert knowledge. One word of caution, however: tone down your level of pride and enthusiasm for your own self-worth; others may be resentful.
- Seek assignments in all segments of the department to gain a broad perspective of the entire organization.
- Never think you know everything and never stop learning.
- Seek out a mentor; mentors are invaluable for helping you prepare for career advancements, new challenges, and other opportunities.

Building a power base is a long-term project. Managers should seek out opportunities as they become available and always keep their eyes open for possibilities to learn something new, meet other managers, or embark on a new challenge.

For managers to increase their efficiency, they can analyze basic tasks, seek to eliminate unnecessary motions or tasks, and aim to hire the best-qualified applicants for the work. One of the most important management functions is selecting new members who meet or exceed minimum job performance levels. One of the most valuable assets a manager possesses is his or her staff. A manager cannot be expected to perform every task that makes an organization run. Delegation is a key to increasing efficiency. To delegate effectively, you must clearly define the task for your subordinates, state the time frame in which the task should be completed, and follow up to ensure that the task was completed in a satisfactory manner.

For example, a chief officer sends the supervisors of each operations area a memo instructing them to check their snow chains. In the memo, the officer states a time for completion and requests a written reply that the task has been completed. The officer must record this deadline and set up a reminder system so that he or she can verify completion of the task within the appropriate time frame. Follow-up is absolutely necessary in all cases, and it should be done as inconspicuously as possible so that the members do not think they are viewed as untrustworthy. If the delegated task is a long-term or ongoing project, other management techniques, such as MBWA or leading by example, can be used to supplement a written report.

Many managers believe that multitasking allows them to increase their efficiency. It would seem to make sense that the ability to perform multiple management functions at the same time would result in increased productivity. However, this is not necessarily true. A Stanford study compared two groups: one that engaged in extensive electronic multitasking and another that did not. Several experiments were conducted to test the capability of both groups to pay attention, remember information, and switch from task to task. The results showed that multitaskers do much worse in all three areas than those who focus on one thing at a time. In fact, the more the participants multitasked, the worse they did compared to the focused group (Gorlick, 2009). Multitaskers may be easily distracted and unable to perform well.

Multitasking can also negatively affect communication. If you want to make someone feel important, give him or her your complete attention. When a manager is texting, emailing, or answering the telephone during a face-to-face conversation, it may be insulting to the other person in that conversation. The only interruption that should be allowed is an emergency call or the fire alarm.

To be effective, managers must recognize that there is always room for improvement and should seek out opportunities for improving their managerial skills, such as self-study through books, courses, or seminars. They should also be open to receiving helpful criticism and suggestions for improvement. Nobody is perfect; everyone has room for improvement. A *360-degree performance review* may be a useful tool to obtain this feedback from employees. This type of evaluation uses surveys sent to employees in all directions: those below the manager (who report directly to him or her), other managers on the same level, and the supervisor directly above the manager. Where this is not possible, feedback can be provided by other staff members or supervisors who have direct knowledge of the manager's work. Through this review process, managers may gain information they can use for self-improvement and uncover criticisms that they need to address. The results of this evaluation may be part of the annual performance evaluations, but should also be substantiated by the supervisor's own observations.

## CHAPTER ACTIVITY #1: MBWA

A safety chief for a large metropolitan fire department was in a staff vehicle traveling to a meeting when he heard a dispatch for a townhouse fire over the radio. At this time, the department had just started a comprehensive safety program. General orders covering safety policies and SOPs had recently been signed and issued by the fire chief. According to the SOP, the safety chief was dispatched only to second-alarm and higher emergency incidents, but he was in the neighborhood and decided to proceed in a nonemergency mode to observe the companies arrive and operate at this fire. About two blocks from the scene, the safety chief was passed by the first due engine. After it passed, he noticed a firefighter standing on the back step—a violation of the department's safety general order for emergency vehicle operations.

After arriving on the scene, the safety chief continued to watch the firefighters work at the townhouse, which had heavy smoke pouring from the front door and second floor windows. He observed one firefighter place a ladder to the front of the structure, climb the ladder, and start breaking out the windows from outside. This is a common ventilation procedure to remove heat and smoke from the structure. However, this firefighter, who was the driver for the first due engine, had on a helmet and no other protective clothing—another violation of the safety rules.

The fire was quickly extinguished by the first alarm assignment. After the fire was extinguished, the safety chief continued his trip to the meeting without speaking to anyone on the scene. The battalion chief in charge of this incident was a personal friend, so when the battalion chief returned to his office, the safety chief called to advise him of the observations. Formal disciplinary actions could have been initiated, but that would have breached the normal chain of command. As the two spoke, the battalion chief immediately volunteered that the firefighters had already admitted to the violations of the safety rules. The firefighters had recognized the safety chief and knew that they had been observed. As a result of this exchange, these individuals became committed to following the safety rules in the future. This commitment is a preferred result of MBWA—to gain compliance with rules and regulations while demonstrating support for the worker's safety.

### Discussion Questions

1. If this behavior reoccurred, what actions should be outlined in the department's disciplinary procedures? Be specific.

2. Would a comprehensive education safety program help to prevent these types of SOP violations, or is strong discipline the only effective method?

## CHAPTER ACTIVITY #2: Group Effectiveness

In most FES agencies, groups are commonly assigned to make decisions regarding controversial or change-based issues. However, some management experts doubt the effectiveness of groups to make good management decisions. For example, Robbins states, "Research comparing participatively set goals with assigned goals has not shown any strong or consistent relationship to performance…Groups almost always stack up as a poor second in efficiency to the individual decision maker" (Robbins, 2011). Still others continue to argue the value of group decision-making.

Review the following resources, keeping in mind the comparison of the decision-making process of groups compared to that of the unilateral decision-making of the fire chief or other agency head. In 400–800 words, discuss the best way to determine the goals of your organization. A minimum of two reference sources is required.

- "Strategic Planning for the Bloomington, Indiana Fire Department" by Jeff Barlow, http://www.usfa.dhs.gov
- "Group Decision-making" by Andrew E. Schwartz, http://www.nysscpa.org
- "How to Help Groups Make Meaningful Decisions," http://www.managementhelp.org
- "Group Decision Making," http://en.wikipedia.org

## CHAPTER ACTIVITY #3: Control System

One goal of the FES is to provide a professional level of fire suppression services to the public. As a management function, this service can typically be measured by looking at outcomes. However, as explained in the chapter *Public Policy Analysis*, outcomes are hard to measure. Many of the measurements that are kept by the traditional FES organization are measurements only of workload, not of outcomes. For example, FES organizations keep accurate statistics on the number of emergency incidents to which they respond, but that does not show how many lives and how much property was saved, which is an outcome.

Assessing the number of lives and properties saved is very difficult, if not impossible. For example, if the department responds to a small trash can fire in a high-rise building and extinguishes the fire quickly, could the department claim to have saved the entire dollar value of the high-rise building?

Outcomes for a FES organization must be measured using other items that are easily and verifiably measured. For example, if it can be demonstrated that response times can be reduced by either changing existing procedures (e.g., streamlining the call-taking process) or adding new resources (e.g., increasing the budget for new staff or new facilities), the argument can be made that improved service to the public has been accomplished. Even in volunteer organizations, implementing a system to have members on duty in the station substantially reduces total response time, resulting in an excellent outcome.

Use the following resources to research measurement control systems:

- http://management.about.com
- http://www.nfpa.org

Then, devise a measurement control program that ensures that firefighters are proficient in the skills and knowledge needed to provide outstanding fire-suppression service.

## CHAPTER ACTIVITY #4: Power Base Contacts

Management effectiveness can be directly related to power, expert knowledge, and relationships. Identify and develop a "power base" that increases your influence and effectiveness as a manager.

Use references of your choice to create a list of at least three people and/or organizations that a chief officer should network with for management effectiveness at the local, regional, state, and national levels.

## CHAPTER ACTIVITY #5: Trust

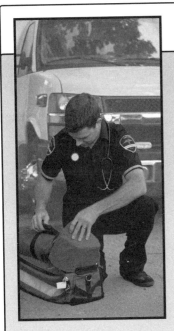

Trust plays a large part in the ability to be an effective manager and leader. One of the best ways to understand this concept is to look at examples. Review the following articles for examples and descriptions of trust:

- "The Supervising EMS Officer" by Tray Hagen, www.emsworld.com
- "Mentoring Tomorrow's EMS Leaders" by Jay Fitch, www.jems.com

After reviewing the articles, provide a short story exemplifying four of the following eight items:

1. Practice openness
2. Be fair
3. Speak you feelings
4. Tell the truth
5. Be consistent
6. Fulfill your promises
7. Maintain confidences
8. Demonstrate confidence

These stories may be from your personal or work life and can be either positive or negative examples.

# References

Dewar, J. (1989). *Almost perfect isn't good enough.* Retrieved from http://www.shsu.edu/~mgt_ves/mgt481/lesson9/tsld029.htm

Gladwell, M. (2008). *Outliers: The story of success.* New York, NY: Little, Brown.

Gorlick, A. (2009). *Media multitaskers pay mental price, Stanford study shows.* Retrieved from http://news.stanford.edu/news/2009/august24/multitask-research-study-082409.html

Javitch, D. (2009). 5 employee motivation myths debunked. *Entrepreneur.* Retrieved from http://www.entrepreneur.com/article/202352

Judge, T. A., Bono, J. E. Ilies, R., & Gerhardt, M. W. (2002, August). Personality and leadership: A qualitative and quantitative review. *Journal of Applied Psychology, 87*(4), 765–780.

Page, J. (1999). Your 21st-century firefighter. *FireRescue, 17*(11), 10.

Robbins, S. P. (2011). *Fundamentals of Management: Essential Concepts and Applications* (9th ed.). Upper Saddle River, NJ: Prentice Hall.

# CHAPTER 4

# Leading Change

## Fire and Emergency Services Higher Education (FESHE) Course Objectives

**Module I: Leading and Managing Purposefully with a Community Approach**

The students will:

1. Describe the role of the fire/emergency medical services department as a part of the community government and comprehensive plan. (pp 47, 49–51)

**Module IV: Leading Change**

The students will:

1. Describe the importance of accepting and managing change within the fire and emergency service department. (pp 45–46)
2. Identify models of change commonly used in organizations. (p 53)
3. Summarize the steps of the change management process. (pp 46–53)
4. Assess ways to create a positive climate for change and introduce new ideas within the organization. (pp 49–50)
5. Describe how an organization can respond to current or emerging events or trends. (pp 46–53)
6. Explain the benefits of employee involvement in departmental decisions. (pp 48–49)
7. Demonstrate innovative ways to address traditional problems within the organization. (pp 50–52)
8. Describe ways to increase and reward professional development efforts. (pp 51–52)

## Knowledge Objectives

After studying this chapter, the student will be able to:

1. Understand the importance of change in fire and emergency services organizations. (pp 45–46)
2. Recognize the influence the chief administrator has on the likelihood of change success. (p 46)
3. Describe the steps to create change. (pp 46–53)
4. Explain the need for a sense of urgency in the change process. (pp 46–48)
5. Describe the ideal characteristics of a guiding coalition. (pp 48–49)
6. Describe techniques that can be used to communicate a vision of change. (pp 49–50)
7. Know the roadblocks that can defeat efforts to accomplish change. (pp 50–52)
8. Explain the importance of short-term goals within a long-term project. (p 52)
9. Summarize the steps to institutionalize change within an organization. (p 53)

## Embracing Change

Change is inevitable; it is also a necessary condition for a business to survive and prosper. To stay competitive, businesses must constantly change to meet customers' expectations. Businesses change to reduce costs, improve the quality of products and services, locate new opportunities for growth, and increase productivity. Change has many forms including restructuring, reengineering, transforming, acquiring, merging, innovating, modernizing, downsizing, rightsizing, quality programs, and cultural renewal.

For public service agencies, change is driven by the need to improve service. In this sense, change may take the form of additional staff, expanded services, improved equipment, higher standards of training, and education for personnel. A good example of this in the fire and emergency services is the growing number of fire departments providing emergency medical services (EMS). In many cases, change is only temporary; one of the challenges of true leadership is to institutionalize change.

Some members may be resistant to change, so leaders must educate members and convey the importance of change. Leaders understand that not to change is to remain stagnant. Public agencies can continue the status quo for many years without any outward indication that change is needed; in many cases, the public, elected officials, labor organizations, and fellow managers may not be aware that changes are needed to improve the quality or efficiency of service. However, the quality of service, equipment, training, strategy, tactics, and management practices fail if a fire or emergency service does not continually seek and embrace change. Good leaders must be willing to propose changes to correct any shortcomings in services. Leaders must recognize the ability of their organization to adopt change without forcing too much change too quickly, which drives people out of their comfort zones and promotes resistance.

Leaders must be patient and persistent. They must communicate the vision that justifies the change to the organization's members, the public, and government officials through words and actions. Permanent change does not occur until most of an organization's members believe in the new vision. To secure change, a leader must allow sufficient time to ensure that the next generation of officers embodies the new vision. When appropriate, adoption of the new vision can be a requirement for promotion (e.g., minimum education standards). If implemented slowly, these changes can be met by all members, thus increasing acceptance and perceived fairness throughout the organization. In this manner, change initiatives encounter less resistance and have an increased chance of survival. In addition, a leader must lead by example, recognizing how their actions influence others, both positively and negatively.

## Creating Change

Leaders at any level can help pave the way for change. A newly appointed leader can create an enlightened climate by setting the tone from the start, helping members to think of change as normal and expected. More experienced leaders can use a national or local incident to take the organization on a more progressive course. The first time a leader institutes a new change, it may help to start with an issue that is meaningful but will not solicit strong opposition; for example, an issue that the union or volunteer members would support, or not strongly oppose. Tradition in fire and EMS services is strong, and transformation may happen slowly. A leader cannot force change and expect that it will be accepted in all cases. The leader must follow a set of processes to create progressive organizations and allow them to adapt to changing circumstances and environments. Although changes can be implemented by issuing a new standard operating procedure (SOP) or order, to be completely successful at making the change last in the future, change needs to be applied using a plan.

After change is identified from the vision, leadership must take over to produce permanent change that can be successful in transforming an organization to better service. The following steps help a leader create change:

1. Identify the need for change and create a sense of urgency.
2. Create a guiding coalition.
3. Develop a vision.
4. Communicate the vision.
5. Overcome barriers and resistance.
6. Create short-term wins.
7. Institutionalize change.

## Step 1: Identify the Need for Change and Create a Sense of Urgency

To identify the need for change, a leader should perform or make arrangements for a group to perform a policy analysis as discussed in the chapter *Public Policy Analysis*. The results of the policy analysis identify the problem and the need for change, assess the urgency, and recommend potential solutions. From this process, the leader develops their vision.

With this vision, the leader must be able to convince 10–15% of the members that this is a good idea before formally starting the change process. This core of believers is necessary because some changes require adaptations of tradition and may contain sacrifices. For example, many fire departments have added EMS to their scope of services. This change requires most of the members of these fire and emergency services organizations to achieve some new level of training, such as emergency medical responder, emergency medical technician (EMT), advanced EMT, or paramedic certification. Although each situation is unique, fire and emergency services organizations have offered benefits, such as incentive pay and paid time off to attend classes, to offset the sacrifices that would have been necessary had the organization simply mandated the new levels of job performance.

In the case of many volunteer organizations, these sacrifices are not offset by any monetary incentives. Because volunteers are very dedicated to their organization and the public they serve, they try to fulfill the requests of their administrators despite the sacrifice. However, it seems that the new EMS demands on volunteer fire and emergency services organizations may be resulting in a reduction of members. Increased call volume training requirements are the main cause of this adverse impact, and is an example of an unintended consequence of change.

Fear of the unknown tends to magnify the doubts of members about potential sacrifice. To create a sense of urgency, the cooperation and willingness of a small group of core members is needed to support the change initiative. In a public service agency, the support of elected and appointed officials is also critical; the same 10–15% support rule also applies to them. The change leader needs supporters above and below them in the organization to support the change.

No matter how hard the change leader pushes, if others do not feel the same sense of urgency, the momentum for change will probably die far short of the finish line. In a paramilitary organization it is always possible to force change by issuing a new SOP, but if the SOP is not based on some credible justification supported by a guiding coalition, then the change may only be temporary until the next administrator takes charge.

However, mandating change is always possible when implemented with governmental regulatory power. In some organizations it may be possible to gather the support of a core group simply because of the members' loyalty to either the organization or a charismatic leader. Avoid falling into this trap of support by blind loyalty; make sure the justification is based on facts and the core members' support is real.

Unlike for-profit businesses that must respond to economic pressures and stockholder demands, fire and EMS services often receive fairly consistent funding from tax revenues. As such, the natural economic forces that create a sense of urgency in the business industry are not present in government agencies. Still, factors that may influence a fire service leader to instigate change include cuts in the budget that may cause salary or benefit reductions, layoffs, or the closing of fire stations or companies; lack of funding to replace old or unreliable fire or emergency medical apparatus; or some other threat to the membership. Major changes have occurred when career firefighters have accepted reductions in salary and benefits when convinced that the alternative would have been layoffs of fellow firefighters. The urgency level was high for a clearly visible, credible crisis.

Because fire and emergency services organizations receive funding from tax revenues, which generally are constant and reliable from year to year, it is common for the following subliminal message to be present: we are winners and must be doing something right, so relax and enjoy the status quo. Recent shocks to the US economy are pushing some fire and EMS organizations to reduce expenses, but they still lack the competitive influences of a market economy driving them to change. In the business world, where there is competition and free entry into the market, competitors are very quick to point out their better service or product.

Many of the situations that a business leader can use to create a sense of urgency are not available to the fire and emergency services organization. This makes the leader's job more difficult, but not impossible. The chief officer may have to look very carefully, and have the patience to wait for the opportunity to make a change. This opportunity may unfortunately come in the form of a crisis (e.g., a national catastrophe or death of a local rescuer) that ultimately catches the people's attention and increases the level of urgency.

In the public arena, it pays to be prepared for these tragedies. Have a plan and justification prepared. These are opportunities to pump up the urgency for change. For example, procuring needed equipment and personnel, funding training sessions, or creating mutual aid agreements may be easier after an earthquake or hurricane. In one case, the State of Florida's fire service identified a major problem in the ability of mutual aid fire companies to communicate by radio during several hurricanes and major wildfire incidents. As a result of these catastrophic tragedies, the state funded portable radio equipment that could provide compatible radio communications using portable handheld radios and a portable base station with a 100-foot crank-up tower **FIGURE 4-1**.

Another example of a tragic event that resulted in major change is the unprecedented terrorist attacks on the United States on September 11, 2001. Among the many resultant changes within the federal government, federal grants that were slated to be eliminated from the federal budget are now being funded again. It was agreed on by many fire service administrators that more resources were needed for first responders at every-day and catastrophic emergencies. These administrators then pulled together their visions and collectively applied pressure on their federal legislators, ultimately receiving the help they knew was necessary.

However, waiting for a tragic incident to initiate change is an unpredictable leadership strategy. Raising the public's interest during good times may not be easy,

**FIGURE 4-1** Radio communication tower.
Courtesy of Ryche Guererro

> **Words of Wisdom**
>
> "Perpetual optimism is a force multiplier"
> – Colin Powell
>
> Source: Powell, C. & Persico, J. (1995). My American Journey. New York. Ballantine Books, p. 335.

but it is certainly possible. Strong leaders can create awareness, helping people see the opportunities or needs without experiencing real losses. Leaders should help others understand the possible consequences of having inadequate staffing, insufficient training, or out-of-date equipment and the effect these shortcomings could have on the quality of service provided. In this way, the leader can create a sense of urgency through a hypothetical crisis rather than a real one. Members, elected officials, and the public need to be convinced that change is necessary to solve a problem or improve service.

## Step 2: Create a Guiding Coalition

Why can't the chief officer just order a change and have everyone comply? The answer to this question is found in the nature of organizations and the people affected.

A strong guiding coalition with the right composition, level of trust, and shared vision is one key to successful change. The coalition has the responsibility to recommend implementation of a particular change and monitor progress towards that goal. Coalition members should understand the need for change and agree with the sense of urgency **FIGURE 4-2**. For example, one contemporary issue involving change in many fire and emergency services organizations is the implementation of a physical fitness program. The facts that support this type of program are very compelling and based on solid research. In addition, from a common sense perspective, having members in good physical fitness is good for their health and provides an emergency services worker who can perform at greater levels of strength and endurance. To implement this change, the leader would want to identify those people who could be appointed to the physical fitness committee (guiding coalition). One technique that can be successful in identifying committee members is to implement a pilot program and ask for volunteers.

The committee members should have position, expertise, credibility, and diverse perspectives. Members should be optimistic, confident, competent, and result-oriented. They should have the courage to say what is on their minds even if it may be unique or could hurt somebody's feelings. They should be dedicated to providing quality service to the public.

**FIGURE 4-2** Guiding coalitions should embody expertise, credibility, and leadership.

**FIGURE 4-3** Team-building events and exercises may be useful for creating a unified sense of purpose.

All members of a guiding coalition must be committed to achieving the same goal for real teamwork to be possible. Team-building events and exercises such as off-site retreats may be useful for creating a unified sense of purpose and priorities among coalition members **FIGURE 4-3**.

Some opponents may claim that guiding coalitions are stacked with "yes members." However, the purpose of this committee is to help implement the change, and without committed supporters it would not function properly and may fail. As an overall strategy, the leader could make a potential change list; prioritize the items on it; and then choose one item that is major, clearly justified, and easier than others to achieve. Creating a new tradition that expects and looks forward to change helps facilitate future changes.

## Step 3: Develop a Vision

The role vision plays is one of the more important elements of a successful change. Vision plays a key role in change by helping to direct, align, and inspire the actions of large numbers of people. A vision is like a goal.

A good leader has a clear vision of where they want to take the organization in the future. This vision is a result of the leader's formal education, intuition, common sense, and professional training, coupled with years of experience, which allows him or her to envision progressive changes that they would like to implement. In most cases, the leader is the source of the change proposal. It may not be an original idea, but it is their vision of the idea that is first proposed and communicated to members and elected officials.

It is important to ensure that the vision is realistic and can be achieved using the organization's available or potential human and material (equipment) resources. When a change vision becomes clear, the leader may feel uncomfortable because it may cause some pain to other members of the organization. In the typical fire and emergency services organization, new members start at the bottom and are promoted slowly to the top, making many friends along the road. Although the leader may not want to jeopardize these friendships, they must be confident in and have the courage to stand by their vision, even if it goes against tradition or has the potential to disrupt personal or professional relationships.

## Step 4: Communicate the Vision

To achieve organizational goals, a leader must be able to communicate their vision to the members of the organization. The vision should be clear and concise to make this possible. Overly complex visions are not easily understood. Therefore, before attempting to communicate your vision to others, you may wish to write it down and have a person outside the field of fire or EMS review it. If someone with very limited knowledge or experience in the field can understand it, then members of your organization should be able to understand it as well. Remember that your vision will also be shared with government officials, the media, and the public, so keep it simple and straightforward. In most cases of major change, the approval of elected and appointed government officials is required and they must be convinced of the validity of the change.

Explaining your vision should take no more than 5–10 minutes. Analogies may be a useful way to help others understand. For example, to illustrate the need for adequate staffing, it may be useful to compare the organization to a sports team (e.g., "Imagine trying to win a basketball game with only four players—rather than the customary five—on the court."). As another example, to relate the need to purchase a new ambulance, one may wish to make an analogy to the familiar need to replace a car that has too many miles and years of use. By using analogies, you have a simple request and clear justification for your vision that others can understand. It is also important for a leader to communicate the vision by example, serving as a role model for the steps necessary to achieve a set goal.

Written and verbal communication can be helpful to communicate one's vision. Most people do not assimilate anything the first time they are given a new idea or message. To counter this issue, a common technique used in the advertising field is to use repetition and various media to communicate the desired message. This technique might also be beneficial for leaders communicating their vision.

Another effective technique of communicating the vision is leadership by example. However, if this is not done correctly, it can send a message that the leader is all talk and not serious about change. Once committed to a change vision, the leader must monitor every detail for signs or behaviors that are contrary to the new way of doing things. Watch for inconsistencies and identify them immediately. Again, the details can kill the proposed change. For a change program to work, it must have a high level of credibility and support from a substantial number of members, the public, and elected officials.

> **Words of Wisdom**
>
> "Great leaders are almost always great simplifiers, who can cut through argument, debate and doubt, to offer a solution everybody can understand."
>
> – Michael Korda, editor and author
>
> *Source*: Powell, C. & Persico, J. (1995). My American Journey. New York: Ballantine Books, p. 383.

Finally, the urgency rate needs to be high enough to find the time needed to effectively communicate the vision and complete the transformation. Fostering change can be a time-consuming process and leaders need to keep a close eye on the progress to ensure the effort is on schedule.

Creating SOPs is another effective method to communicate a vision. SOPs are often a response to real and perceived safety issues and the need for higher quality service including consistency. These are items that can cause liability exposure if not addressed by the department. For example, if there is a failure to provide for the safety of the members or to give adequate service to the public as measured by nationally accepted professional standards, there is an increased liability exposure.

Relevant government rules and standards should be referenced in written communication to relate the urgency for change. For example, if a fire department needed to purchase protective clothing, the leader should reference the requirement to comply with NFPA 1971, *Standard on Protective Ensembles for Structural Fire Fighting and Proximity Fire Fighting*. At times the chief officer may need to tackle more controversial changes, such as those relating to standards of training, incident command, staffing levels, and medical and physical fitness standards. These are admittedly aspects of change that can be costly or draw resistance. However, most of these changes can be phased in over time (as noted in the NFPA 1500 standard), which drastically reduces the potential impact and resistance.

## Step 5: Overcome Barriers

A leader must work to overcome barriers to change. Common barriers include tradition, lack of resources, insufficient staffing, and resistance from members, government officials, or even the public. Before instituting a change, a leader should develop a list of potential barriers and devise possible solutions for overcoming them.

The most important step of overcoming barriers is to identify them. At times, the barrier may be invisible to the organization and its administrators because it has been institutionalized into the organization's tradition. Therefore, the guiding coalition may prove to be an excellent resource for identifying problems and developing possible solutions.

After the barrier is identified, convincing members, elected officials, and the public that it is really an impediment to quality emergency service becomes the paramount task. If the barrier is clearly defined, generally its effect on emergency service becomes apparent. Identifying the problem is always the hardest part of any change process. Looking at the experiences of other fire or EMS services can be a useful approach. Great ideas can be found in small and large organizations, including those staffed by volunteers, career, and combinations of the two. Remember, what one department considers a barrier may already have been overcome by adopted practices or procedures in another department. Citing the experience of another department can be very helpful in selling a proposed solution.

One type of barrier that the fire and emergency services organization may encounter is that of structure. Local structural barriers vary from department to department. Generally, some of these barriers are legacies of how things are done in a major central city; the city's procedures become normal throughout the surrounding

> **Case Study**
>
> **Leading Change: A Safety Example**
>
> A fire rescue department issued driving regulations that contained a requirement that all emergency response vehicles must come to a complete stop at all red traffic signals and stop signs (as per National Fire Protection Association [NFPA] 1500 standard) during emergency responses. Each paramedic unit was equipped with an automatic speed recorder. A crash occurred one day when a paramedic unit failed to stop at a red traffic signal, striking a car that was proceeding through a green light and killing the driver of the car.
>
> The "stop at all traffic control signs and signals" regulations had been issued about 3 months prior, but had never really been enforced. It was common for many drivers to proceed through the stop signs and signals without coming to a complete stop. When these circumstances were relayed to the jurisdiction's lawyer, he recommended deleting the driving regulation, stating that if the department was not going to enforce the regulation, it would make the jurisdiction more liable.
>
> However, the chief was able to persuade the attorneys that the safety regulation was imperative and a plan was implemented to enforce the rule. By identifying a flaw in the department's management and deciding to enforce the safety rule, the chief communicated strong support for this safety change. This is genuine leadership.

## Case Study

### The Barrier of Tradition

Two fire departments share a common border. One department is completely paid and the other is a combination department. When dispatching fire companies, the paid department expects that each company will respond without delay, staffed by a minimum of four firefighters per company. This is the typical organization of many completely paid fire departments.

On the other side of the border, the combination department was created by consolidating several volunteer departments into one county fire department and hiring other paid members. Generally, the paid firefighters are assigned to stations to offset periods when adequate numbers of volunteers are not available. When alerted to a call, this department has no guarantee of the number of members available, the time needed to respond, or even if they will respond at all (although the latter would be rare in most cases). Although a growing number of volunteer organizations have recently been requiring on-duty staffing by their members, this department does not require their volunteer members to spend on-duty time at the station or to be on call in the neighborhood, because this was never undertaken by the original volunteer departments. Although a need for paid personnel has been acknowledged, the department continues to maintain as it always has (tradition).

suburbs even though times change. For example, when many central cities first created their paid fire departments, labor was relatively inexpensive. They were able to hire and staff their companies with large crews and many stations. Today, however, this is not normal, with the high cost of career firefighters and EMS personnel.

Another barrier that occurs in fire and emergency services organizations relates to supervisors who oppose change. Supervisors to a large degree are a product of their experiences. These supervisors may be very good at taking commands and managing emergency situations; however, they can present an enormous obstacle when major changes are pursued. Some do not like change and may see the change as a personal attack on their safe, predictable, and comfortable environment. Easy solutions to this sort of problem are difficult to find.

Sometimes special interest groups, such as labor unions and volunteer associations, also take on this characteristic of resisting change. If influential supervisors or groups are not confronted early in a change process, they can undermine the entire effort. For powerful organizations, placing representatives on the guiding coalition may take care of some concerns or at least temper resistance. By including representatives as part of the process, they are exposed to the justification for the change, making it more difficult for them to strongly oppose it. Consensus-building techniques can also be very helpful in these situations.

A leader frequently encounters some resistance to change from staff, government officials, and the public. Staff members may be comfortable with the existing policy and have difficulty veering away from tradition, or they may fear that the change is not good for them personally or the organization as a whole. However, for a change to be institutionalized, a leader must work to overcome these barriers of resistance. This process may not be easy; implementing significant change takes time and patience. Because of this reality, one must be careful when celebrating or rewarding short-term wins. Although the opposition may acquiesce to the new change, there is the possibility that they might reassert their resistance at any time. Until the changes have truly been institutionalized into the organization's culture, which may take many years, permanent change can be very fragile.

One should also look for interdependent systems that require additional changes. Whenever a new item is found that needs changing to support the original change, this insight helps solidify the original change. This process is called "facilitating," and the leader should be the main person to help make possible the change using their influence. When new obstacles are discovered, the leader must help find new solutions to overcome the obstacle.

## Member Resistance

Change can cause a fear of losing some existing benefit or privilege. It also threatens the investment in the status quo (tradition). The longer tenure a person has, the stronger may be the opposition. For example, change may conjure up fears of losing status, money, authority, friendships, personal convenience, or other economic benefits. Change expert Lane Wallace (2009, p. 52) explains why some people cling to an opinion even when faced with overwhelming opposing substantiated proof

> [S]ocial science researchers have found that people employ "motivated reasoning" to fend off evidence that their strongly held beliefs are wrong. Many people feel that they are their opinions, and hate to lose arguments. So, when confronted with new, troubling information, ideologues selectively interpret the facts or use "contorted logic" to make the conflicting evidence just go away.

In this sense, resistance can be associated with a personality trait and, therefore, may be near impossible to change. It should be expected that there will always be some members that oppose the change.

Even when the change is not in the members' short-term best interest, the leader should be able to explain the benefits and personal fulfillment from making the shift and the negative consequences that would emerge without the change. Leaders can help reduce members' resistance by:

- Starting a comprehensive education and communications program.
- Incorporating members into the decision-making process.
- Facilitating meetings where members can voice their opinions and concerns.
- Encouraging members to deal appropriately with apprehension and stress associated with the change.
- Rewarding members' efforts to overcome resistance and embrace change.
- Consistently and vigorously enforcing the rules and regulations that define the change.

When these efforts, combined with inspirational leadership and creative incentives, are not enough to persuade a member, discipline may be necessary. The final stage of the disciplinary process is separation from the organization. Faced with this consequence, most members comply given a reasonable amount of time to conform. Be sure that you do not rely on voluntary compliance or institute rules that are not enforced; this invites failure. Even in the largest organization, disciplinary action may only be necessary for a few members; word spreads quickly that the leader means business. Although the leader should primarily aim for voluntary compliance, disciplinary enforcement may be necessary to ensure compliance.

This also may hold true with managers. A leader should approach resistant managers and try to convince them to buy in to the new vision. If the supervisor does not provide voluntary compliance, disciplinary action may be necessary after a written warning against interference, insubordination, or noncompliance is issued. Although disciplinary action may be possible, separation from the organization may not be an option because of union agreements. Think through your plan and consult with the organization's attorney before implementing a disciplinary action.

## Step 6: Create Short-Term Goals

While working toward long-term goals, a leader should have intermediate checkpoints to periodically monitor progress. For example, a person who is working toward significant weight loss by dieting and exercising can weigh in and celebrate each new pound that is lost. These short-term goals are necessary to carry the momentum and urgency for a long-term project. This is especially true with any behavioral change. For example, if an EMS service has adopted a goal of being more customer-oriented, then positive behaviors exemplifying good customer service should be encouraged, recognized, and reinforced.

Short-term wins are also very effective at keeping the urgency at a high level. They can be incentives, such as recognition, money, or both. If these positive incentives do not work, then disciplinary actions, following the standard local procedures, should be taken at these short-term benchmarks for those who fail to comply. For example, the first notice could be verbal to the employee and a written note inserted into their personnel file; second notice would be written to the employee specifying the infraction, time for compliance, and consequences of noncompliance; and eventually a final step would be termination.

Short-term goals should have the following three characteristics: (1) visibility to the entire organization, (2) clear and unambiguous outcomes, and (3) outcomes that are a direct consequence of the change. It is important to have plans for accomplishing short- and long-term goals and to have patience and perseverance on the path toward achieving these goals.

### Words of Wisdom

"Most people won't go on the long march unless they see compelling evidence within six to eighteen months that the journey is producing expected results. Without the short-term wins, too many employees give up or actively join the resistance."

–John Kotter, author of *Leading Change*

*Source*: Kotter, J. P. (1996). *Leading Change*. Boston: Harvard Business School Press.

### Case Study

**Institutionalizing Change**

The leader of an EMS agency decided to transition to a 1:1 model that requires an ambulance staffing configuration of one paramedic and one EMT. This system made sense because paramedics were also staffed on fire engines. A 1:1 model would increase efficiency by providing two paramedics at the patient's side and reducing the need for a dual medic transport unit. If a second paramedic was needed for patient care en route to the hospital, the fire medic could ride along.

However, a few ambulance paramedics expressed discomfort with having an EMT for a driver and partner, believing that the EMT may not have the skills required for advanced life support partner collaboration. The EMTs were also hesitant, only having had experience with non-emergency transfer work, and expressed reluctance to accept additional responsibilities and increased expectations. Local politics and attitudes complicated matters, because the relationship between the public (fire) service and private (EMS) agencies was

*(continues)*

> **Case Study** *(Continued)*
>
> strained. The leader recognized that the change would be difficult for fire and EMS agencies, but realized that the economic situation required a more efficient and effective system of emergency response. The leader was confident that with proper education and training, all involved would be comfortable in their new roles.
>
> The leader followed seven steps for institutionalizing the change:
>
> 1. Identifying the need and expressing a sense of urgency based on the economic demands
> 2. Forming a guiding coalition to assist in the development of a vision
> 3. Creating the vision for a 1:1 model
> 4. Writing a new SOP outlining the details of how this model would work and the steps for transitioning to this new system
> 5. Identifying a plan to overcome the barriers and resistance felt by the paramedics and EMTs
> 6. Setting up a series of short-term goals that were widely visible, clear, and unambiguous, and direct results of the change
> 7. Formalizing the change to the new model of care when hiring, training and educating new employees

## Step 7: Institutionalize the Change

For a change to become permanent, the underlying culture of the organization must change. The culture of an organization is very difficult to address directly because it becomes an invisible part of the normal day-to-day operations and has a powerful influence over the behavior of members who are expected to conform. In most organizations, individuals are selected and trained to meet the expectations of the existing culture. Shaping of a new member's behavior, attitudes, and values is done routinely by the incumbents without a conscious effort; however, there are steps a leader can take to overcome this barrier.

First, the leader and guiding coalition should discuss the rationalization, justification, and evidence that support the change, recognizing how the culture has served the organization and acknowledging that its traditions are no longer efficient, safe, or useful. Next, the leader should aim to bring on new hires that represent the values of the new culture and to promote members who are open to the new changes. Then, the leader should carefully monitor progress across the organization to determine the implementation's success and the need for any corrections. By listening to members and practicing management by walking around (discussed in the chapter *Management*), a leader can identify problems and make appropriate corrections. Do not wait until forced by the opposition. They may use these errors to bring down the entire change project.

If the opposition identifies problems and forces modifications to the change program, the status quo forces will have won a battle, and they will be harder to defeat the next time. After a problem has been identified with the change initiative, correct it as soon as possible. Admit the mistake, unintentional consequence, or misunderstanding before others force their own modifications.

Fostering change can be a time-consuming process; leaders need to keep a close eye on progress to make sure the effort is on schedule. Finally, the leader should lead by example, exemplifying the new cultural norms. Cultural change is the final step in institutionalizing a change. The true test of a leader may occur only after he or she leaves the organization: if the new practices and vision remain and the organization does not regress to past practices, then the full transformation has been completed.

> **Case Study**
>
> **AIDS and the Change to EMS Culture**
> In the early years of EMS, there was little credence given to disease transmission and cross contamination. Field responders were often covered in the blood and bodily fluids of their patients, and they often wore little if any protective gear. When the first case of AIDS was reported in 1981, there was little information about how this virus was transferred, but there was a fear that contact with an infected person's blood would lead to certain death. The fact that the disease had a long incubation period left many responders wondering if they had already been exposed.
>
> AIDS changed the landscape and practices of health care. There was a lot of uneasiness as providers continued to care for patients during the massive research effort to understand this disease. In 1992, Occupational Safety and Health Administration standard 29 CFR 1910.1030 (Bloodborne Pathogens) became effective for all employers with employees "reasonably anticipated" to have occupational exposure to blood and other potentially infectious materials. This Occupational Safety and Health Administration standard required employers to determine exposure, create an exposure control plan, perform annual awareness training, and implement engineering and work practice controls. As a result, EMS developed a new standard of patient care and a dedicated effort by employers to meet the new mandates. Providers began using latex gloves, cleaning contaminated equipment, wearing HEPA respirators, and using safer sharps equipment.
>
> *(continues)*

## Case Study (Continued)

Initially, there was reluctance to change the culture of EMS, even in the face of potential death. Some providers thought the gloves were annoying and refused to use them. Some agencies were slow to adopt work practice controls until they were mandated. As new providers joined EMS, they were trained to a new, safer standard, but some of the seasoned EMS providers believed the new requirements were unnecessary.

Changing organizational culture and the behavior it creates takes time and effort. An administrator must be willing to use all the strategies at his or her disposal—including harsh discipline—or the safety of the members and service to the public suffer.

## Leading Change at All Levels

Even if not the administrator of the organization, you may be able to lead an effort for change. If presently in a role of middle or lower-level management and having a vision that requires change to accomplish, you may wish to keep these ideas to yourself or limited to those associates and friends who will not be threatened by the thought of change. In some organizations, being known for new ideas that threaten fellow members or superior officers may cause you to be overlooked for promotion.

Still, this does not mean compromising personal professional standards. A future leader must be careful to seek out those who are open to change for support.

You may wish to start by finding a senior supervisor who would be open to the change and meeting with them privately to explain the need for the change. Depending on the topic and size of the organization, the leader may form a guiding coalition of supporters to propose a pilot program and test the change. Recognize that approval may have to come from the senior officers, lawyers, elected officials, and member groups.

In many cases, the most effective place for leading change is at the top. Even the chief officer may be limited by elected officials, city managers, or, for volunteer organizations, those who elected the officer. And it is not unusual for employee organizations to have influence. By following the change step process, leaders will have their best chance of leading an organization through significant change.

Leadership is truly a lifelong learning project. John Kotter, the author of *Leading Change*, points to five actions the leader can take to successfully lead change: (1) take risks, (2) self-reflect, (3) solicit the opinions of others, (4) listen carefully to others, and (5) be open to new ideas (Kotter, 1996). The more knowledge you have, the easier it is to find suggestions and ideas to overcome opposition. Ask questions and reach out for the opinions of others.

Organizational change is never easy, and you need plenty of patience to follow the seven-step change process outlined in this chapter, but progressive change is the only avenue to creating a truly competent professional fire and emergency services organization. Keep in mind that the process begins only after the problem has been identified. Be very careful to take the time and effort needed to ensure that the analysis and its conclusions are supportable when challenged. This is the hardest step and the most critical. Again, patience and perseverance are the keys to success.

## CHAPTER ACTIVITY #1: Leading by Example

A metropolitan fire department adopted a physical fitness program that required mandatory participation and a yearly test. The program, which covered new employees only, required participants to sign a condition of employment document stating that they would not be able to stay employed as a firefighter if they did not pass the annual test. Because these firefighters had to pass this test to graduate from their recruit school, they were all capable of passing the test the first day in the fire station.

Once each year, all battalion chiefs (BCs) had to oversee a physical fitness test for the firefighters in their battalions. One particular BC was a strong believer in keeping fit for both personal and professional reasons and was also able to successfully pass the yearly test.

During each shift the BC would participate in the department's physical fitness program and on many occasions would join one or more companies during their training. It was well known by all firefighters how the BC felt about physical fitness. For the yearly 1.5-mile run test, the BC would advise the firefighters that they would have to beat him to the finish line. The BC would then run at a pace to finish in 12 minutes, which was the pass-fail mark. In the 3 years that the BC was assigned this field command, he never won a race. In this period of time, all the firefighters passed the test.

*(continues)*

## CHAPTER ACTIVITY #1, (continued)

### Discussion Questions

1. Is this BC helping to lead change? If yes, explain in detail.

2. Is it fair that the new employees have to pass this test and not the existing firefighters? Discuss fairness in terms of the new recruits, fellow employees, the department, and the public.

3. Find (using an Internet search) and analyze three separate ideas or arguments that a leader could use to support or oppose a physical fitness program for firefighters. Provide complete reference sources.

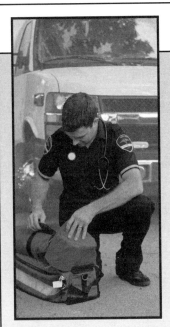

## CHAPTER ACTIVITY #2: Communicating the Vision

Your ambulance company is in contract negotiations with the county for your 911 transport contract. Even though your company has been the exclusive provider in this area for a long time and has a record of exemplary service, it may not be enough. The final decision will weigh heavily on who turns in the lowest bid. Competing companies will do anything they have to in order to win this large county contract.

### Discussion Questions

1. How can you make a maximum effort to win points in the areas in which you have more control?

2. Would the use of social media aid as a means of communicating with the public be helpful or problematic in this case?

3. What other methods can your department use to convince the system and the community that your company is the best choice?

## References

Kotter, J. P. (1996). *Leading change*. Boston: Harvard Business School Press.

Powell, C. & Persico, J. (1995). *My American Journey*. New York: Ballantine Books.

Wallace, L. (2009). *Why fact can't compete with belief*. Retrieved from http://www.theweek.com/article/index/101112/Why_fact_cant_compete_with_belief# Volume 9 (2009-10-09)

# CHAPTER 5

# Financial Management

## Fire and Emergency Services Higher Education (FESHE) Course Objectives

### Module II: Core Administrative Skills
The students will:
2. Describe the integrated management of financial, human, facilities, and equipment and information services. (pp 57–74)

### Module IV: Leading Change
The students will:
5. Describe how an organization can respond to current or emerging events or trends. (pp 73–74)

### Module V: CRM—A 21st Century FESA Responsibility
The students will:
5. Identify direct and indirect costs associated with fire. (pp 66–69)
6. Analyze economic incentives that encourage and discourage fire prevention. (pp 61–66)
7. Describe the role of fire and emergency services in the economic development and neighborhood preservation programs of the community. (pp 60–61)

## Knowledge Objectives

After studying this chapter, the student will be able to:
1. Understand the importance of a budget in fire and emergency medical services (EMS) administration. (pp 57–58)
2. Explain the phases of a budgetary cycle: planning, submission, review, and administration. (pp 58–59)
3. Identify the roles of the financial manager and the public in the budgetary process. (pp 59–61)
4. List common sources of agency revenue. (pp 61–64)
5. Outline different types of taxes and fees and how they can generate revenue for a fire or EMS agency. (pp 61–62)
6. State alternative revenue sources, such as government bonds, investments, and borrowing. (pp 62–64)
7. Discuss the financial considerations of providing fire and EMS as free public services. (p 64)
8. Explain techniques for increasing funding, such as providing evidence from national standards, performing a cost-benefit analysis, and defining potential sources of funding. (pp 64–66)
9. Know how to track expenditures. (pp 66–68)
10. Describe purchasing policies, total cost purchasing, cooperative purchasing, and legal purchasing considerations. (pp 68–69)
11. Understand the importance of financial accountability and the process of financial audits. (p 69)
12. Discuss the difficulties involved with making fair and ethical budget cuts. (pp 70–71)
13. List ways administrators can try to protecting their agency's budget. (p 71)
14. Explain relevant tax considerations that affect fire and EMS administration. (pp 71–73)
15. Know how the global economy can influence the budgets of local agencies. (pp 73–74)

# Introduction

Financial management is the process of translating financial resources into services while providing fiscal accountability. To make the best decisions for their organization, fire and EMS administrators must have a basic understanding of the budgetary process, knowledge of their agency's sources of revenue and expenditures, and an awareness of relevant tax considerations. Understanding these concepts is extremely important because one of the main criteria government officials and the public use to measure an organization's success is its ability to achieve its goals within its allotted budget. Although chief officers may not be experts on taxes, they must have a good understanding of all tax issues before proposing a tax solution to help fund a new program or enhance an existing one.

This chapter presents an overview of the process of fiscal management along with tools and techniques that fire and EMS administrators can use to identify the real effects of budgetary changes and ensure adequate funding.

# Budgets

An administrator is responsible for planning and tracking the agency's budget—its revenues and expenditures. The administrator needs to be very cognizant about every penny in his or her budget to achieve a moral, ethical, and legal duty to the public, elected officials, and members of the department.

## Types of Budgets

There are several different budget procedures. Each type of budget has its strengths and weaknesses. The most commonly used in the field of fire and EMS is the *line-item budget*. Line-item budgets are useful for monitoring expenditures and tracking financial accounts, although they are generally not as effective as other methods for measuring outcomes. *Performance budgets*, *zero-based budgets*, and *program budgets* are more useful for this type of goal. Administrators who wish to explore these techniques should review the chapter *Public Policy Analysis*, which discusses measuring outcomes in detail.

Line-item budgets are essentially a statement of used and available funds broken down into categories. The categories start with the major agencies, such as police, education, and fire and rescue. Next, the budget is broken down in each department to represent major program areas, such as prevention, operation, and EMS. In each area, costs are listed by major categories, such as salaries, retirement, and maintenance.

This budget type is used in the request for funding on a yearly basis. After the budget is approved, budget reports are issued on a monthly basis to monitor the expenditures and detect any deviation. These reports contain lines of funds for each category and read from left to right with such categories as Approved, Current Month, Year-to-Date, and Balance Available.

Each month, the administrator should examine the budget reports carefully. If any items are unbalanced (e.g., if more than half of allocated overtime funds for the year was spent in just 1 month), the administrator should examine the reason for this imbalance and make necessary changes to the budget or agency practice, such as by requesting supplemental appropriation, transferring funds from other areas of the budget, or cutting back on overtime for the rest of the fiscal year. The fiscal year, which is a full year period, starts on October 1st for the federal government.

When an agency's budget changes, the administrator can easily add or detract an incremental amount of money to a line-item budget to reflect changes in revenue or costs. For example, if the tax revenues increase or decrease for the next fiscal year, it is easy for the government accountants to simply cut all budgets by a specific percentage. This relieves them from cutting a specific program that may be viewed as a popular public service.

To illustrate the complexity of a reduction in funding, consider the following example. The accountants in the finance department of a fire and emergency services (FES) organization have proposed to cut the budget, and the chief is given the latitude to propose a reduction plan. The department has 100 employees and the new authorization allows only 90 employees. In implementing this budget cut, the chief may choose to cut all of the fire prevention and training staff positions, which total 10 employees. This proposal would keep all existing fire stations open and seems to be the least controversial plan.

However, as the months and years go by after the cut, the real impact of eliminating fire prevention and training will become evident as firefighters lose their proficiency and injury rates increase. There will also be a greater number of more severe fires as a result of neglecting fire protection features in buildings and lack of quality proficiency in firefighting. As exemplified, politically expedient cuts in budgets can be a lot more complicated than they initially seem.

---

### Case Study

**Budget Cut for Training Programs**

An example of a negative outcome caused by cutting training programs occurred on Memorial Day weekend, 2011 for the City of Alameda Fire Department in Northern California. A 52-year-old man, allegedly wishing to commit suicide, walked into the cold water of the San Francisco Bay. Although 911 was called and the Alameda City Fire Department responded, firefighters and police refused to go into the water to save the man. An hour later the man succumbed to hypothermia.

In previous years, the fire department for this island community had a comprehensive water rescue

*(continues)*

> ### Case Study (Continued)
>
> program that included shore-based and surface-based rescue techniques. After several years of struggling to balance the budget, however, the water rescue program was abandoned and decertified. That day in 2011, the fire department paramedics were following a department policy created in 2009 that prohibited personnel from performing water rescues because they were not currently certified to do so.
>
> In the wake of this disaster, the chief changed the policy to give incident commanders more discretion on using personnel and equipment based on specific circumstances. The department's training division reimplemented water rescue training, and two shallow water rescue boats were put into service. Unfortunately, however, damage was done that can never be completely mitigated. This case exemplifies the importance of considering the consequences of budget cut decisions.
>
> *Source:* "Drowning Death Pushes Alameda Officials To Address Water Rescue Training." KTVU.com. Web. 2011. Retrieved from http://www.ktvu.com/news/news/drowning-death-pushes-alameda-officials-to-address/nD63s/

A further example of externalities is when a building owner considers the installation of an automatic sprinkler system. In many cases the developer opts to forgo the extra expense of a fire sprinkler system. However, if a sprinkler system is installed in a building, a potentially large and catastrophic fire in the community may be prevented and lives and property may be saved.

Furthermore, a fire of large magnitude may overwhelm the local fire department's resources and present a hazardous situation for firefighters. If the fire destroys the owner's building, the municipality loses a substantial amount of real estate tax, not to mention the potential for a number of local people to be without jobs if the building is a business. One of the major justifications for automatic sprinkler ordinances is the elimination of target hazards that would have this type of adverse economic impact on the local community. These are benefits to the entire jurisdiction in addition to the property or business owner. Again, when market forces fail to bring a demand for the public good, this becomes a justification for government regulation. This subject is discussed further in the chapter, *Government Regulation, Laws, and the Courts.*

## Budget Process and Planning

### Budgetary Cycle

What is described here is the general budget process; some elements might start years in advance, such as capital improvements or increasing staffing. After determining the most appropriate type of budget, an administrator can begin the budgetary cycle, which consists of the following general phases:

- Planning
- Submission
- Review and approval
- Management

### Planning

The process of budget planning, which is discussed in greater detail in the chapter *Public Policy Analysis*, may begin with the selection of a task force made up of representatives from the agency, the public, other agency heads, and elected and appointed officials. The task force should consider the agency's sources of revenue, operating costs and expenditures, and any relevant guidelines or standards to determine the most appropriate budget for the organization. Administrators may receive guidance from the city manager, county manager, or budget director regarding expected increases or decreases in the anticipated revenue or expenditures for the coming year, particularly in relation to salaries and cost of living increments. This is a good venue for the administrator to test out his or her vision and get valuable feedback.

### Submission

After the proposed budget has been created by the advisory committee, the administrator needs to prepare and submit a formal budget to the budget director, legislature, or mayor or other elected official. This proposal is the administrator's responsibility and need not be identical to the advisory committee's recommendations. Although the recommendations are to be strongly taken into consideration, they are not mandatory and the administrator has the final say.

As part of the submission, the administrator needs to provide justifications for each major section of the budget. Any new items or enhancements call for detailed explanations and possibly their own separate budget forms. Consider reviewing successful budget increase requests from previous years; this can help the administrator submit a request that has the best chance of being approved.

### Review and Approval

In most cases, the proposed budget submission receives a preliminary confidential review by the designated government officials, although this review process varies depending on state and local laws. Some areas may require that this preliminary proposed budget and justifications be made public, but in either case the administrator can anticipate receiving private input from the city manager, public safety director, or budget director. The administrator should carefully consider this input and make every effort to resolve any differences before moving on to the next phase or making budget wishes known publicly.

After this preliminary review, the agency's budget is consolidated into one municipal budget, which is analyzed for arithmetic accuracy and compliance with municipal or agency policies and objectives. If the total requests are above the estimated revenues, it is typical for the personnel in the budget office to make recommendations on budget cuts for individual agencies. The chief officer should stay closely involved with this process and use every opportunity to express support for the fire department or EMS budget. It is ideal to have good relationships with officials in the budget agency to help the process of negotiation and compromise go more smoothly. These relationships are not built up overnight but are generally the result of long-term friendships.

After the budget has been reviewed and approved by the municipal executive officer and the budget agency, the budget goes to the legislative body for approval. Budget hearings are scheduled and made available to the public. At this time, the chief officer and department staff members should be prepared to answer and defend any inquiries from special interest groups, the media, citizens groups, and labor organizations.

Remember, at this point the budget is now a proposal of the senior elected or appointed official, and any attacks on this budget needs to be defended. The chief officer will be asked to appear before the legislative body to officially present the department's budget. The officer should be very familiar with the budget and its justifications. To prepare for this appearance, it is useful to role-play with senior staff and try to answer the most difficult questions that staff members can think to ask regarding the budget. The administrator will also want senior staff to attend the budget presentation to address specific questions in their areas of expertise.

**TABLE 5-1** summarizes stages of budget review.

## Management

After the budget passes legislative review, it is officially adopted. The next step, management, is the final element of the budget cycle. In this phase, the manager monitors revenues and expenses over time and revises the budget if necessary. He or she may also be asked to prepare periodic budget reports as required by the executive officer, elected official, or budget department head.

## Key Players in the Budgetary Process

In addition to the many people involved in this ongoing budgetary cycle, the administrator must be in contact with several other key decision-makers in the budgetary process including the agency's financial manager and other agency managers.

## Financial Manager

The financial manager's role in an FES organization is a busy one. After the budget has been adopted and funded, one budget cycle ends and the next cycle begins. While administrators are busy with the never-ending budgetary cycles, they must rely on a financial manager to alert them to any potential problems that may interfere with the agency's budget plan. In larger organizations, this may be a separate individual, whereas in smaller departments it may be the duty of the senior administrator.

Problems that may need to be addressed include unexpected overtime, increased costs in capital items, tax revenues below predictions, insufficient fee collection, or unfunded mandates. For example, government regulations and professional standards may be changed to require updating of or new safety equipment. This was the case in the early 1970s when self-contained breathing apparatus was required to be updated to "positive pressure." This regulation, which protected the wearer from any air leaks inside the face piece, was not combined with any funding from the state or federal government. Therefore, the local governments had to pay the bill.

If the budget was approved on the basis of a fee collection, the financial manager should keep a very close eye on this collection process to make sure it does not fall below expectations. A revenue shortfall in fee collection is likely to directly impact the agency. A revenue shortfall in

### TABLE 5-1 Stages of Budget Review

| Stage | Reviewing Party | Notes |
|---|---|---|
| Preliminary review | Designated government official (city manager, public safety director, budget director) | • Confidential, or may be made public depending on state or local laws |
| Municipal review | Municipal executive officer, budget agency | • Budget is analyzed for accuracy, as well as compliance with municipal policies and objectives<br>• Recommendations are given for individual agencies in the municipality at this stage |
| Legislative review | The legislature has public meetings to discuss the budget and receive input from all groups listed. | • Budget hearings are made available to the public<br>• Chief officer presents the budget and answers questions from special interest groups, media, citizens, and labor organizations |

a municipal tax can be remedied, however, by a reduction in any or all of the municipal agencies' budgets.

Financial managers should also follow any proposed legislation or regulations from state and federal governments that might affect the agency's funding. To stay up-to-date with legislation and regulations, the financial manager may rely on an automatic referral system or a network of other financial managers, FES organizations, and professional organizations (e.g., the International Association of Fire Chiefs or the National Highway Traffic and Safety Association) that publish this information.

The National Highway Traffic and Safety Association's "Emergency Medical Services Agenda for the Future" includes information on EMS legislation and regulation and system finance. The National Registry of EMTs (www.nremt.org) published a document called "The National EMS Scope of Practice," which highlights scope of practice concerns. The National Registry of EMTs also establishes standards for EMS certification and licensure. State websites have documents specific to state legislation and regulations. For example, California has an organization called California Fire Chiefs (www.calchiefs.org), which is a statewide organization that follows legislation specific to fire service and fire EMS operations. The financial manager should review these types of publications and report in writing the existence of any proposed legislation or regulations that may have a financial impact on the department.

Monitoring the community for demographic changes that may affect the services requested and tax revenues available (e.g., aging communities experiencing a decrease in population and average taxpayer income) can be critical to avoiding financial surprises. This information is readily available from the US Census Bureau. Year-to-year changes are monitored at the local, county, and state level by government agencies.

## Bureaucrats and Other Agency Managers

The administration of an FES organization is one part of the government bureaucracy. Bureaucrats from other agencies are also in competition for tax revenues. These bureaucrats are expected to implement the programs approved by elected officials. However, as noted by University of Washington's Professor of Economics Neil Bruce, "[I]mplementing a political program is not like executing a computer program. The bureaucracy is not a simple machine that blindly follows orders" (Bruce, 1998, p. 150). Therefore, bureaucrats end up with a lot of latitude to determine how to implement their programs.

In some cases, bureaucrats from several agencies can cooperate for a common goal such as increasing tax revenue. Raising taxes to fund ongoing and new programs does not occur routinely, but did occur in New York City when the real estate tax rate was raised by over 18% in 2002. The additional revenues reduced the proposed cuts to the Fire Department of New York for 2003, even though the department had to close six fire companies.

Most government agencies have very few standards for the services provided to the public. For example, if the police chief, a bureaucrat, proposes a budget increase to cover the cost of new police officers, the chief typically justifies this by reporting that the department would be able to reduce some type of crime with more personnel. The police chief may select a crime that has recently received some notoriety. Because there is no standard on the number of police officers needed, the chief's request will probably be judged by the elected officials on the justification provided, the influence of the police chief, and the potential for gaining votes in the next election. Public fire protection and EMS are fortunate to have national standards for their service levels, which are not held by police and education. In general, the following three areas are covered quite comprehensively in the national standards:

1. Training for firefighting and EMS personnel: National Fire Protection Association (NFPA) professional qualifications standards for firefighters and Department of Transportation standards for emergency medical technicians and paramedics
2. Safety of personnel: NFPA 1500, *Standard on Fire Department Occupational Safety and Health Program*, and Occupational Safety and Health Administration (OSHA) Fire Brigade and Respiratory Protection standard
3. Deployment and staffing of fire and EMS units: NFPA 1710, *Standard for the Organization and Deployment of Fire Suppression Operations, Emergency Medical Operations, and Special Operations to the Public by Career Fire Departments*

The more chief officers rely on national standards, a formal planning process, or cost-benefit analysis, the better chance there is to convince elected officials to support new and existing programs.

## The Public and Their Representatives

Recognizing that the government must choose between many conflicting interests and striving to present strong arguments to support the fire or EMS department's budget requests are important roles of fire and EMS administration. Without valid guidance, governments do not always do what is right, efficient, and in the best interests of citizens. If successful, fire and EMS administrators have convinced the elected and appointed government officials to spend taxpayers' money on the department's needs.

In addition to convincing government officials responsible for this decision-making, administrators must also aim to convince the public of the justification for the department's funding needs. Although most communities use elected representatives to make budgetary decisions, some towns use a process of direct voting in which citizens can directly approve or oppose budgets, ordinances, or bond issues. It is assumed that when all else is equal, voters vote according to their pocketbooks.

Most elected representatives behave in a like fashion. If they watch the financial situation of the voters very closely, they have a good chance of being reelected.

When requesting additional funding, administrators must convince voters or their representatives that increased spending is of value to the public. The justifications should contain a very clear explanation and evidence that the department will be able to increase the quality of FES with the new funding. In general, public polls show that most citizens believe that fire and EMS departments are doing a good job. Therefore, it may be difficult to convince the public that there is a need for additional resources without a convincing argument that the requests are justified.

Just as when speaking to government representatives, the administrator must clearly explain how the extra funds benefit the community by enabling the agency to increase the quality of its FES. For example, if the administrator is requesting funds for a new station and increased staffing, he or she has to demonstrate that the increased spending decreases response times. Several FES organizations may join forces to appeal to the public. A cooperative, regional approach may be best to gain public approval for services, such as hazardous materials, terrorism planning, response to local catastrophic events, and technical rescue teams.

These regional approaches help convince the public and elected officials that the department is looking for budget solutions that consider all costs. Most taxpayers practice economical managing of their own finances and expect the same from the government.

Administrators should seek out influential private citizens in the community and work to cultivate an ongoing friendly business relationship with as many of these important people as possible. In many cases, these private citizens have as much, if not greater, persuasive power over the decision-making process of local government than the elected officials. Remember, statistics may be the best way to demonstrate the benefits of existing service levels and the impact of reductions, so it is a good idea to identify and document all benefits of current services to the community and know this information well.

# Revenues

To protect an agency's budget and obtain funding for new programs, administrators should have a solid understanding of common sources of revenue. The main sources of revenue for most fire and EMS agencies are government taxes, such as property tax, income tax, and sales tax. Increases or decreases in these taxes may be caused by economic variations or just represent normal fluctuations of these sources. For government officials to approve an increase for a specific new or expanded program, the administrator must be able to convince them that there is an overwhelming need for the new program and present a solid justification that would be supported by the voters.

# Taxes

Fire and EMS agencies may receive money from a variety of different taxes, including property tax, income tax, sales tax, and other taxes. *Property tax* (or real estate tax) is the most common funding source for local government programs, particularly in states that do not have a state income tax. This tax is assessed as a percentage of the total value of property and improvements, or in some cases as a rate based on income for commercial property. One big advantage of the property tax is that it is relatively easy to collect. Nobody can hide a house or a store from the tax collector.

For a property that has structures on it, there is a direct connection with fire services because there may be a need to extinguish a fire in the building. Residential property taxes are assessed at a flat rate based on the value of the property, as opposed to a progressive tax like the federal income tax, which increases the higher the income of the individual. All property owners pay the same percentage based on the value of their property. If a homeowner lives in a million dollar house on a horse farm, that owner pays a lot more tax than a family living in a mobile home on a small parcel of land. The greater the value, the greater the tax.

Commercial property taxes may have different rates and can be based on the income of the business and are therefore subject to wide fluctuations. When property values fell in 2008 during the downturn in the housing market, so did tax revenues. This kind of reduction usually does not occur quickly because the tax changes only after the property is reassessed by the tax collector's office. Some states assess properties on a multiyear basis; therefore, the impact is not realized for several years. There is a lag between the cause and the effect until property values can be reassessed. This is also true on the other end of this trend, when the property values start increasing. This type of variation in tax revenues can be predicted ahead of time by monitoring the sales price of properties in the community.

*Income tax* is the federal government's leading source of funds, followed very closely by the payroll tax (Social Security and Medicare). Income tax is a progressive tax, which means it is taxed at a higher percentage for higher wage-earners. In 2012 the Internal Revenue Service reported that in 2010 the top half of US taxpayers paid 97.6% of all federal income tax. Included in the top 50% were taxpayers who had an adjusted gross income (after deductions) of $34,338 or greater, which includes many paid firefighters. (Internal Revenue Service, 2012)

Many states have income taxes. Several of these states link their state income tax to the federal calculations for taxable income. Some cities also have income taxes. Because income tax revenues fluctuate with the economy, this source of funding is affected very quickly by significant economic changes. Furthermore, this type of tax has one major growing problem: tax evasion. One study reported that 18–19% of income is not reported to the Internal Revenue Service (Cebula & Feige, 2008).

*Sales tax* is a tax applied to consumer purchases, such as electronics and vehicles. It is affected very quickly and is the most widely fluctuating revenue source, because personal spending levels vary greatly based on season and individual considerations. The sales tax is a common source of revenue at the state and local level. Most states have a sales tax, although there are five states that do not: Alaska, Delaware, Montana, New Hampshire, and Oregon. For the other states, there are differences in the percentage rate and the items covered from state to state. For example, some items that may be exempt are food, clothing, and medicine. In several states, either a county or city government can add a local sales tax on top of the state's sales tax.

One of the first things people do when faced with the loss of a job or the threat of being laid off is to stop spending on items that have a sales tax, such as cars and large appliances. Another issue is the growing number of consumers that shop on the Internet. Most Internet sites do not collect sales tax for states unless they have physical presences. This is a growing problem for states that rely on the sales tax.

Some states greatly benefit from the sales tax, especially those that have a lot of tourists. Tourists can pay a substantial amount of tax to help finance state and local governments. For example, state and local governments may aggressively pursue tourism to counteract the cost of public schooling, which is a major cost for the state and local governments. There are many examples of state or local governments reducing or eliminating real estate taxes on a new amusement park or tourist attraction in anticipation of the sales tax receipts and new jobs.

There are several other types of taxes, such as fire service taxes, use taxes, and government fees for services. A *fire service tax* can be a set value for all residences or can vary depending on the square footage of the dwelling or the *fire flow formula*, which calculates the approximate amount of water needed to extinguish a fire based on the combustibility and area of the structure.

A *use tax* applies only to the use of specific goods or services, such as taxes on motel rooms, rental cars, and tourist attractions; these funds are applied to offset the cost of providing public services to tourists. One of the most common use taxes is the road tax, which is added to the price of fuels used by motor vehicles. Consumers see this as a fair tax because it is directly based on the use of the roads. The more miles driven or the poorer the fuel economy of the vehicle, the more tax that is paid. However, this tax is not collected when vehicles use electricity to power their travel. This is a new problem being studied by policy makers.

## Fees

Government fees may be applied to offset the cost of specific services when those services benefit the user directly, such as road tolls, driver's license fees, and professional licenses. Specifically with regard to providing EMS services, an ambulance transport fee can be a substantial source of revenue. Some fire departments become sold on taking over EMS transport after calculating the amount of revenue that can be realized from transport fees. Insurance companies, such as Medicaid and Medicare, usually pay most of the transport fees. However, there are many examples of these fees not covering the full cost of EMS transport. Fire organizations must factor in all of these elements and then decide whether ambulance transportation fees are a significant revenue generator or just another financial and liability burden. In addition, if firefighters are used as EMS providers, there may be a conflict with required in-service EMS and basic firefighter skills refresher training, call volumes, physical fitness sessions, and prefire planning duties. In busy departments, there simply may not be enough time to attend to all fire and EMS duties.

In fire departments, a fee for fire prevention plan review and fire inspection may be an alternative revenue source **FIGURE 5-1**. When applied, the fire department may serve as the government's enforcement authority for required inspection of fire protection equipment. In these instances, inspection fees should not be used as a way to acquire additional funding for the agency's overall needs, but just to cover the actual costs of the inspections. It is a conflict of interest to use the government's authority to require inspections and mandate a fee at the same time.

## Government Bonds and Short-Term Borrowing

Although most of a fire or EMS agency's operating costs are covered by taxes and fees, additional funds, such as those necessary for major capital improvements, are generally provided by loans through the sale of *government bonds*. A capital item is typically an item that has a long life expectancy, such as a fire station. Fire stations have a 30- to 50-year life expectancy and are very expensive; they are a perfect example of a capital item purchase that must be funded by loans from the sale of bonds.

**FIGURE 5-1** Some fire prevention agencies charge fees for plan review and inspections.

State and local governments sell bonds for a low interest rate at a set interval for up to 30 years with tax-free interest. Many local and state governments have either an official or a recommended debt ceiling. Often this is expressed as a percentage of the total revenues (e.g., bond debt payments should not be more than 10% of tax revenues).

Several governments have suffered financial crises because of inappropriately using bonds to bridge the gap between necessary expenditures and incoming tax revenues. For example, if the city has only enough funding to cover fire services through June and tax revenues are distributed in July, the city could get a one-month government loan to cover its expenses for that month. This can end up being a troublesome practice, because the 1-month loan extends to 2 months, then 3, and so on. This was one of the questionable financial practices that brought New York City to the brink of bankruptcy in 1975. In cases such as this, the government is borrowing to cover operational costs such as salaries. This is a poor practice, and the options for the elected officials are to either cut costs or raise taxes.

The federal government is able to borrow to cover its excess expenditures over revenues, most of which is caused by operating expenses, not capital improvements. This federal debt is funded by Treasury bills, Treasury notes, Treasury bonds, and Treasury Inflation Protected Securities. Theoretically, federal debt never has to be repaid; only the interest payments must be paid each year. Therefore, the federal government and its elected officials can vote to spend money, but do not need to vote for taxes to pay for the spending or cut spending.

A few larger departments have used a bond sale to replace all (or a large portion) of their units at one time. In these situations, the department has all brand new equipment in the first year, but as time goes on, all of the apparatus ages simultaneously, increasing the probability of repairs, breakdowns, and out-of-service time. Because most apparatus divisions are staffed and funded with a constant number, there would be no work in the early years, because the apparatus is new and under warranty. With all the apparatus aging at the same time, the maintenance cost would be substantial in later years with many vehicles in the shop for repair.

A replacement plan that budgets the replacement of apparatus on a scheduled basis is easier to justify. After approval for this type of funding is gained, it is easier to achieve approval for the next fiscal year. For example, if last year's budget contained $250,000 for a new engine, then next year's budget request would just include an incremental increase to cover the increased cost for 1 year. It is far easier to gain a 3% increase in the budget than to go back and justify each new apparatus purchase.

However, smaller FES agencies may also use money from the sale of bonds to purchase fire rescue apparatus. With the cost of fire and rescue apparatus at levels of $250,000 to more than $1 million, it is very difficult for small departments to be able to save the amount of money needed to pay for them. Because these departments buy only one new fire rescue apparatus every 5–20 years, a bond may be a good method to raise the needed capital. Larger departments generally have purchase plans to buy a certain number of apparatus each year.

It is important for fire and EMS administrators to remember that bonds are a form of borrowing and that any time money is borrowed, there is an additional cost associated with the loan. It also is important for administrators to recognize that in many jurisdictions, voter approval by ballot is necessary to sell government bonds. This process of voter approval can take a long time and is not guaranteed, although fire bond referendums historically have high approval rates among voters.

Whenever revenue is generated by the sale of bonds for a specific *capital improvement project*, these funds cannot be used for any other purpose. In some cases, dependent on local or state laws, funds may be transferred from one project to another project as long as both are capital improvement projects. For example, if excess funds remain at the end of the construction of a new station, these monies may be used for a rehabilitation project at a different station. Funds for a *turnkey project*, which means the project is immediately ready for use, can be used toward station construction, furnishings, and, in some cases, new apparatus. However, funds from bond sales cannot be used to cover salaries because these costs are part of the operating budget.

## Investments

Investing funds is another source of revenue for governments. Governments have the ability to earn interest on their bank accounts. Many sources of revenue for governments, such as taxes, are due to be paid by a certain date. For example, the federal income tax is due by April 15$^{th}$ each year. Most of this revenue is needed to pay salaries and other fixed costs throughout the year. Because these funds are not needed immediately, the government can invest in short-term investment instruments. Many larger governments hire financial managers to invest these funds.

Some investments can be very risky, especially those that report high returns. One notorious example occurred on December 6, 1994, when Orange County, California became the largest municipality in US history to declare bankruptcy after its risky financial investments unexpectedly lost most of their value. Generally, however, governments tend to stay with low-risk investments. For example, the federal Social Security Trust Fund can invest only in federal government bonds (T bills). Even in times when some investment funds were reporting up to 30% returns, the Social Security Trust Fund was only making 5–6% on its surplus funds.

Caution is always necessary with government revenues. Many states are cutting their budgets and dipping into emergency funds to make ends meet. In some situations, this is compounded by the reduction of investment income in retirement accounts. When states should be contributing more into retirement accounts to offset the losses in stock investments, they have less tax revenues. This is problematic.

### Other Forms of Borrowing

If government bonds are not available because of the debt ceiling, an agency may lease apparatus, with the option to purchase it later. *Lease-purchase contracts* are an option when the need for new equipment or apparatus is critical. In addition, a lease-purchase contract might have a better chance of approval from the legislature because the costs are recorded as a line item in the operating budget instead of a capital purchase. This concept is similar to purchasing a new vehicle; an individual may purchase it by using a loan with a monthly payment, leasing it with a monthly payment, or paying cash for it up front.

## Techniques for Increasing Funding

After an administrator has a basic understanding of the agency's sources of revenues, it may be useful to consider various approaches to increase available funding. These techniques include providing supportive evidence from national standards, performing a thorough cost-benefit analysis, and defining potential sources of funding.

### Provide Evidence from National Standards

The FES administrator must recognize that his or her organization is part of the government bureaucracy and is in competition with other agencies for tax revenues. Most of these other agencies have very few standards for the services they provide to the public, because output from these agencies is very difficult to measure. However, fire and EMS agencies have had standards to benchmark services provided and to justify their budgets since the 1970s and 1980s. Therefore, fire and EMS administrators are at an advantage when they can support their requests for additional funding with evidence about the benefits it will bring.

Several relatively recent national standards such as NFPA 1500, *Standard on Fire Department Occupational Safety and Health Program*; NFPA 1710, *Standard for the Organization and Deployment of Fire Suppression Operations, Emergency Medical Operations, and Special Operations to the Public by Career Fire Departments*; National Highway and Traffic Safety Administration, *National EMS Education Standards*; and OSHA, *Fire Brigade, Hazardous Materials, and Respiratory Protection*, can be used to focus budget requests on resulting quality and safe service to the public. These standards address training, professional qualifications, personnel safety, and the deployment and staffing of fire and EMS units. When chief officers use

---

### Chief Officer Tip

**Fire and EMS as Free Public Services**

In most communities, fire and EMS services are provided as a free public service. Unfortunately, because there is no direct charge to the consumer, these public services are often requested more times than they are needed. This results in an excessive demand, causing delayed response times when units are not available. The higher the call volume, the greater the chance of the closest unit being busy on another call. As a result, FES organizations struggle with false alarms caused by faulty equipment and excessive EMS calls caused by request for service for nonemergency medical problems. To try to reduce excessive use of fire services for nonemergency situations, some jurisdictions have adopted an ordinance that fines businesses that have multiple false alarms to encourage them to fix malfunctioning alarm systems.

Many departments that provide EMS transport are struggling with heavy demands for service. This service is essentially free to the patient in most cases. Even where there is a fee, it is either paid by medical insurance or Medicaid/Medicare, or not paid at all. A free public service can create a market inefficiency when people request the public good (emergency ambulance service) when it is not really needed.

Most funding for fire and EMS services is provided by the property tax. However, many properties do not pay any property tax, because it is normal to exempt federal, state, and local government properties along with nonprofit organizations, such as churches. Because FES are a public service afforded to all property owners, visitors, and residents regardless of their tax contribution, property owners who do pay property taxes essentially subsidize free services for those properties that are tax exempt. In jurisdictions where governments or nonprofit organizations have large holdings of property and buildings, there is a concern that these organizations unfairly do not pay for the emergency services they receive.

There are some solutions to generate revenues while still providing a public service. In some cases, agreements have been achieved that require an annual fee in lieu of property taxes. This situation helps fund the entire Washington, DC, government, including the fire and EMS department. In other cases, a donation of property for the construction of a station has occurred, for example, at some large universities. In addition to these efforts to generate revenue, government regulations that encourage owners to purchase and maintain smoke alarms and sprinkler systems may be necessary to keep demand for fire services down.

these national consensus standards as part of a formal budget planning process, they increase government and public support for new and existing programs.

## Perform a Cost-Benefit Analysis

If a survey was performed about the value of FES, there probably would be a great variance in the valuations. For example, a family enjoying a nice quiet evening in front of their television, with no fire or smoke in their home, would probably express a low value for having an FES organization ready to respond and rescue them in a fire emergency. Take this same family and place them into a 13th floor apartment with a fire blocking their exit to the stairway, and there would be a new higher value for the FES organization.

In most cases, it is very difficult to measure benefits or direct outcomes from FES programs. What makes this cost-benefit analysis difficult is that it is hard to assign a value to an individual's life or quality of life. Moreover, the outcome—saving lives and property—is often the result of a complex set of circumstances. The following is a hypothetical set of circumstances that led to a death in a house fire:

1. Fire prevention regulations did not require smoke alarms to be installed in existing dwellings.
2. The local FES organization did not offer any home fire safety inspection services or the free installation of smoke alarms.
3. The occupants of the dwelling did not see any value in purchasing and installing smoke alarms.
4. The occupants of the dwelling did not recognize the hazard of careless smoking habits in the family room.
5. The first arriving fire unit was understaffed or personnel were not adequately trained.
6. When the victim was eventually located and brought to the front yard, the FES organization was not equipped to provide EMS.

It could be argued that correcting any one of these six items could have potentially saved this fire death victim. However, anything less than 100% effort in all these key areas can lead to loss of life and property.

Still, there are some benefit evaluation measurements that may be helpful. For example, in this specific case, the costs are associated with correcting all six of the items listed. The benefit is described as saving victims and property from a structural fire.

In the business world, the benefit is new or increased profit; however, the vital piece to this analysis is to have benefits that are measurable or can be described in common sense consequences. For example, an administrator is considering equipping the next fire engine purchased with a Compressed Air Foam System (CAFS). The costs are easily obtained from the manufacturer. The benefits need to be described in subjective terms because sound numbers are impossible to calculate. Some of the advantages of CAFS are lighter weight hose and increased firefighting efficiency of the foam over plain water. It is not possible to be able to quantify these advantages as a result of the great number of variables that are typical in any hostile fire situation. However, CAFS has many positive benefits to firefighting that can be described as potential benefits rather than hard numbers.

## Define Sources of Funding

To increase the likelihood of a spending request being approved, the fire or EMS administrator can point to specific areas of available funding. Sources of funding may include increases in tax revenues or fees. The United States Fire Administration guide, *Funding Alternatives for Emergency Medical and Fire Services*, provides information on locating and implementing traditional and nontraditional methods of funding.

In addition, several federal grant programs offer financial assistance to fire departments in the United States. All of these programs are under close scrutiny each fiscal year, especially now that the federal debt is a major issue with voters and elected representatives. There are four separate grant programs. The Assistance to Firefighters Grant Program is a 1-year grant provided directly to fire departments and non-affiliated EMS organizations to enhance their firefighting and emergency response needs. Fire Prevention and Safety Grants provide assistance for local, regional, state, or national organizations for fire prevention and safety, primarily directed at high-risk target groups including children, seniors, and firefighters. Staffing for Adequate Fire and Emergency Response Grants are awarded directly to volunteer, combination, and career fire departments to help the departments attain 24-hour staffing by hiring career firefighters or to help recruit and retain volunteer firefighters.

Finally, the Department of Homeland Security offers fire departments Assistance to Firefighters Fire Station Construction Grants to build new or modify existing fire stations. Initially designed as a stand-alone program, they are now folded into the US Department of Homeland Security's consolidated grant program. The United States Fire Administration has a webpage (http://www.fema.gov/welcome-assistance-firefighters-grant-program) that helps departments with their grant requests. Evaluation of grant applications is done by fire service peers.

EMS grants may be obtained for equipment, apparatus, and education through the U.S. Health Resources and Services Administration and other state or local organizations. Insight into how to be a successful grant applicant can be obtained by contacting other departments that have been successful and asking them for advice.

Another possible source of funding discussed previously in this chapter is cost recovery programs that bill insurance companies for services provided. Ambulance companies may bill for transport fees, or hazardous materials teams can file for reimbursement through the transport vehicle's insurance coverage or OSHA's

*Emergency Response Operations for Releases of, or Substantial Threats of Releases of, Hazardous Substances Without Regard to the Location of the Hazard.* Some departments charge a fee for fire and EMS services at motor vehicle crashes to the at-fault driver's insurance company.

Administrators should be prepared to recommend a funding source from a new tax or an increase in an existing tax or fee if asking for substantial additional funds. Proposing a tax increase should be saved for those occasions when attempting to make substantial improvements in the organization and its services or when implementing a master plan. For example, in a department that needs three new stations for adequate coverage, an implementation plan may detail the opening of one new fire/rescue station a year for the next 3 years. This also requires the hiring of new personnel each year. The proposal for funding may include step increases in a tax to generate an additional number of dollars each year.

Along with a proposed funding source that is a tax or fee, be prepared to explain all the ramifications and who is affected. In general, if the taxpayers are convinced that there is added value or service from the increase in the tax or fee, they will probably support the increase.

# Expenditures

## Salaries and Benefits

In challenging economic times, a steady stream of stories in the media point out some of the most outrageous examples of salaries and benefits of public sector employees. This news can be very troubling for the public, especially those without jobs or generous benefits. However, this was not always the case.

Many years ago it was common for most government workers to make very poor salaries and, in the case of firefighters, work very long hours. As a reaction to pressure from employees and the ever-increasing use of public sector unions, it was popular for elected officials to grant generous increases in benefits, especially those that did not have to be paid in the near future such as pensions, medical insurance, and cash-outs for vacation and sick leave. At the same time, public sector employees were gaining increases in wages and, over many years, their salaries ended up above those in private industry.

Elected officials and the public are now examining some private sector practices to reduce cost of employees' salaries and benefits. An example of reducing the costs of a benefit, while actually providing a healthier employee, is a policy that Safeway Inc., one of North America's largest supermarket chains, instituted in 2005. This plan rewards employees for meeting certain health criteria. These criteria are based on the notion that 74% of all healthcare costs are confined to four chronic, mostly preventable, conditions: (1) cardiovascular disease, (2) cancer, (3) diabetes, and (4) obesity (Burd, 2009). To determine whether an employee receives the incentive, Safeway tests employees' tobacco usage, healthy weight, blood pressure, and cholesterol levels. A plan like Safeway's could "ultimately reduce our nation's health-care bill by 40%" (Burd, 2009). In fact, during the first 4 years after the plan was implemented Safeway was able to keep its per capita healthcare costs flat, whereas other American companies' costs increased 38% (Burd, 2009). Similar programs could be very valuable in providing increased health and safety for firefighters and EMS personnel.

## Tracking Expenditures

FES administrators should be able to spot mistakes or unplanned expenses as soon as possible so that spending can be adjusted. A system should be set up that keeps track of expenditures on a monthly basis. At the beginning of the fiscal year, the administrator should calculate monthly anticipated expenditures. Some expenses, such as salaries, are predictable, whereas others may be one-time or variable expenditures, such as overtime.

**TABLE 5-2** shows a sample line-item budget for 4 months of the fiscal year. Because this is only one-third of the annual budget, the benchmark for its expenditures is one-third of the total (or 33%). On the bottom line, the report points out that 33.01% of the department's budget has been spent or encumbered, which indicates a healthy budget position. However, several line items are over the 33% mark (e.g., printing, electricity, and travel). If the bulk of these areas of expenditure had been anticipated in the first 4 months, this would not be a problem. However, because a substantial amount of travel is projected for the upcoming months, the administrator should closely monitor the situation and may need to cancel travel plans or transfer funds from other line items. Other expenditures that are below 33% (e.g., heating, which is not yet in season) may take place later in the year, so this surplus is only temporary.

At the end of each month, the administrator should review the actual expenditures for the month and compare them to the expected amounts. A running total should be calculated along with any variance to be sure the agency is not spending more than budget allocations allow. To ensure that adequate funds are available for the purchase of goods or services, the administrator can use an *encumbrance*, which sets aside designated funds until the invoice is received and payment is issued. In this way, funds are earmarked for a particular expenditure and are not confused with other available funds or confused as a surplus for transfer to another department. An encumbrance is an accounting procedure that indicates that the money has already been committed to be spent even though the actual payment may be made sometime later in the fiscal year. In some cases, the encumbrance can be paid in the following fiscal year, allowing the use of those funds from the current year. For example, if an administrator encumbered funds for CAFS but there was a delay

## TABLE 5-2  FES Department Budget April Expenditure Report

| Object | Description | Budget Plan | Current Month | Year-to-Date | Encumbrances | Balance Available | % |
|---|---|---|---|---|---|---|---|
| 102 | Salaries, permanent | $695,271.00 | $57,939.25 | $230,975.90 | $0.00 | $464,295.10 | 33.22% |
| 103 | Salaries, overtime | $48,668.97 | $4,055.75 | $13,294.27 | $0.00 | $35,374.70 | 27.32% |
| 104 | Health/welfare insurance | $75,228.32 | $6,269.03 | $24,999.00 | $0.00 | $50,229.32 | 33.23% |
| 105 | Dental insurance | $11,402.44 | $950.20 | $3,907.65 | $0.00 | $7,494.79 | 34.27% |
| 106 | Retirement | $108,462.28 | $9,038.52 | $36,795.00 | $0.00 | $71,667.28 | 33.92% |
| 107 | Vision care | $2,781.08 | $231.76 | $939.27 | $0.00 | $1,841.81 | 33.77% |
| 120 | Medicare | $2,363.92 | $196.99 | $786.46 | $0.00 | $1,577.46 | 33.27% |
| Total Category 1 | | $944,178.01 | $78,681.50 | $311,697.55 | $0.00 | $632,480.46 | 33.01% |
| 201 | General expense | $3,500.00 | $605.00 | $421.00 | $0.00 | $3,079.00 | 12.03% |
| 241 | Printing | $1,300.00 | $137.50 | $550.00 | $0.00 | $750.00 | 42.31% |
| 251 | Communication | $8,000.00 | $608.00 | $2,505.00 | $0.00 | $5,495.00 | 31.31% |
| 261 | Postage | $2,400.00 | $177.00 | $792.00 | $0.00 | $1,608.00 | 33.00% |
| 291 | Travel | $500.00 | $0.00 | $0.00 | $225.00 | $275.00 | 45.00% |
| 331 | Training | $800.00 | $0.00 | $0.00 | $275.00 | $525.00 | 34.38% |
| 341 | Facilities operations | $4,500.00 | $0.00 | $996.00 | $0.00 | $3,504.00 | 22.13% |
| 361 | Electricity | $16,000.00 | $2,310.00 | $9,240.00 | $0.00 | $6,760.00 | 57.75% |
| 371 | Heating oil | $18,000.00 | $591.00 | $2,110.00 | $0.00 | $15,890.00 | 11.72% |
| 410 | Professional and special services | $45,000.00 | $3,712.53 | $4,418.27 | $12,000.00 | $28,681.73 | 36.49% |
| 431 | Data processing | $1,800.00 | $148.50 | $594.00 | $0.00 | $1,206.00 | 33.00% |
| 503 | Uniforms | $500.00 | $0.00 | $0.00 | $500.00 | $0.00 | 100.00% |
| 524 | Vehicle operations | $49,000.00 | $3,791.16 | $15,348.66 | $0.00 | $33,651.34 | 31.32% |
| 569 | Equipment rental | $12,000.00 | $0.00 | $3,960.00 | $0.00 | $8,040.00 | 33.00% |
| 571 | Unallocated | $0.00 | $0.00 | $0.00 | $0.00 | $0.00 | 0.00% |
| Total Category 2 | | $163,300.00 | $12,080.69 | $40,934.93 | $13,000.00 | $109,465.07 | 33.03% |
| Total Org/Prog | | $1,107,478.01 | $90,762.19 | $352,632.48 | $13,000.00 | $741,945.53 | 33.01% |

in the manufacture and delivery, the contract may be extended at the same price, allowing delivery and invoicing to occur in the next fiscal year.

Keeping real-time spending records for each month facilitates the planning of any future and actual seasonal peaks (e.g., winter heating oil) or one-time costs occurring in a later month. If, for example, the purchase of protective clothing for wildland firefighting was planned for the eighth month of the fiscal year (2 months before the next wildfire season), these funds could not be encumbered until the order is actually placed. The administrator would note this future purchase in a personal spreadsheet.

It is also a good idea to keep personal records of expenditures to double check computer reports from the budget department. Data entry mistakes are always a possibility. When dealing with number codes to designate the department and line-item category, hitting the wrong number key can put expenditures into the wrong agency's budget. A procedure of double keying, where two different operators enter the same data, helps to eliminate most of these errors. The administrator should also verify computer reports for timeliness and accuracy to ensure that the budget department is not late in documenting expenditures or revenues so there is no confusion about the actual amount of available funds. For example, if the department has just received an order of fire hose costing several thousand dollars, but it has not shown up on the latest budget summary report, the administrator needs to ensure his or her records are accurate.

Finally, it is a good idea to keep a list of the previous year's total expenditures for each month to identify any significant changes that may have caused variances in the present budget cycle. Any unexpected costs that were the result of new mandates, labor contract changes, severe weather, or large-scale emergencies may require the administrator to recalculate the budget or request additional funds. Look for any surpluses to cover these costs.

At the end of the fiscal year, many government agencies attempt to spend any unspent funds. In many cases, these purchases are not needed but are used to assure that there is not a surplus to be given back to the general fund. To avoid this practice, budget department personnel at the direction of the city or county administrator may temporarily suspend capital purchases around this time of the year.

## Purchasing Policies

Because purchasing policies vary widely among different governments and their agencies, the administrator must become familiar with local rules and procedures; failure to follow these rules can lead to disciplinary procedures. Policies may determine the following:

- Limits on spending for various services and commodities
- Limits on purchases made without a formal bidding process
- Standard procurement of expendable items
- Contract items on a master purchase order list
- Policies for emergency procurement during a disaster incident
- Policies regarding petty cash
- Procedures for inventory of property or equipment
- Required approval for capital items or travel
- Standard service contracts
- Lists of approved vendors

If there is no formal policy for a particular purchase, it is prudent to find out what the past practice has been and to document the details of the purchase with the budget analyst assigned to the department. It is also wise to consider how the transaction would look to outside observers (e.g., local media) before making a purchase.

Purchase policies may be based on the useful life span prediction for an item. For example, if the department has four fire engines, and the chief can convince the governing body that the useful life of a fire engine is 12 years, then a good policy dictates purchasing a new fire engine every 3 years to ensure that no engine is over its useful life.

## Total Cost Purchasing

Another calculation that the administrator should consider is *total cost purchasing*. The total cost of an item considers both the cost of purchase and any other associated costs, such as maintenance and downtime (including the cost to rent or reserve associated equipment), and any surplus value the item may have at the end of its useful life. The total cost of an item is especially useful to compare bids for major purchases.

When comparing bids from different vendors, it is not always best to go with the lowest bidder. If the low bid vendor's equipment has high maintenance costs or worse than average down time, the actual total cost of the purchased item may be more than that of another bidder with a slightly higher purchase price. Remember, the business of emergency services is about always being ready to respond. An unreliable piece of equipment should not be acceptable. It may be useful to contact peers to obtain a list of other agencies that have purchased the same equipment and ask about their experiences with a particular manufacturer.

## Cooperative Purchasing

One way to reduce the total cost of an item is to find opportunities for cooperative (or joint) purchasing of equipment or supplies. In some areas, such items as fire hoses, personal protective equipment and gear, and even apparatus may be on a regional or state purchasing bid list. In some cases, purchases may also be made from the US Government Services Administration's Cooperative Purchasing Program. These joint arrangements consolidate purchasing power across several departments, resulting in lower total costs. *Cooperative purchasing* can save money because companies typically give a discount on bulk purchases. However, cooperative purchasing may also result in a few disadvantages, such as the need for compromise regarding local preferences of design or manufacture, a potentially longer delivery time, and a lack of control over the contract process. Most of these disadvantages can be overcome by detailed review of the specifications, noting where they differ from the department's previous purchases. If the variance is truly significant, a request for a modification to the supplier or manufacturer may be obtainable. The administrator should be involved in the process for the development of the consolidated bid specifications to get better results for their agency.

## Legal Considerations

There are many legal considerations to the purchasing process, and the administrator needs to understand the rules, policies, and laws for the local jurisdiction. For example, in some jurisdictions, an affirmative action plan requires a certain percentage of the purchasing dollars be awarded to minority- and female-owned businesses. Unfortunately, some of these practices may cause additional dollars to be spent from the budget and inferior equipment to be purchased.

When discussing these issues with government officials, argue from the perspective that the department needs the best equipment or supplies that can be purchased within reason. Stress that emergency services to the public require the most reliable equipment and supplies possible. In some cases, the US Supreme Court has found these set-asides unconstitutional; however, some administration and elected officials strongly support these practices. Therefore, it might be wise to avoid discussing the legal aspects or ethical considerations and instead review these practices on a case-by-case basis.

In other instances, relationships with previous contractors, suppliers, or manufacturers may affect the bidding and awarding of contracts. For example, in the past a department has only purchased Smith fire trucks, but now Jones fire equipment has presented a more appealing bid in response to the latest request for proposals. The department may prefer Smith fire trucks because the department has standardized on this equipment, allowing the department to store only one manufacturer's parts, and the department's mechanics have become very familiar with maintaining and repairing this brand, which simplifies the maintenance program. Take care not to allow passion or loyalty for a particular brand, model, or contractor to interfere with the purchasing of equipment. Total cost purchasing calculations may be useful to help administrators make an unbiased, cost-effective decision in these instances. Whenever there is a question, the administrator should consult with municipal attorneys for legal counsel before awarding a contract.

## Accountability and Auditing

Financial accountability is confirmed through the auditing process. *Financial audits* are commonly performed on an annual basis to ensure that financial records are accurate and complete. As part of this process, an auditor examines assets to confirm that they are in their assigned place and checks the approved budget for congruence with actual expenditures.

Periodically, the department will be audited for its use of public funds. Auditors confirm that public funds have not been misappropriated, embezzled, accidentally doubled, or stolen and that there was no failure to follow procurement policies. The consequences of any improprieties or negligence can include prosecution, loss of funding, a negative reputation, and additional regulations.

The fire or EMS administrator should be alert for any random or scheduled audits and check with predecessors or government financial officials to obtain a briefing on the last audit. This is one of those areas where it pays to be very detail oriented. If there is an occasion when the purchasing policy was not followed and there was a good justification, make sure to thoroughly document the situation and all the details. This documentation can be extremely helpful when auditors question emergency spending or other unapproved purchases.

In addition to financial audits, a fire or EMS agency may also be asked to participate in performance audits, such as compliance or procedural audits. A *compliance audit* checks for adherence to a legal or policy mandate. This type of audit can include elements of fiscal management: for example, if a department charges a fee for fire prevention plan review, the audit may review fee calculation, collection, and storage. The compliance audit may also include a review of the agency's petty cash account for small purchases to ensure compliance with relevant spending guidelines. *Procedural audits*, designed to evaluate management policies and procedures, also can include financial policies and procedures.

Government officials may hire outside consultants to complete these audits. Administrators should always comply with the auditing process; their cooperation ensures a fair review. Remember that no department is perfect, and the audit may uncover areas for improvement. Auditors' comments and suggestions can provide a valuable opportunity for additional growth, new projects, or new funding possibilities. Look for ways to turn these suggestions into new requests for additional funding, if necessary. The administrator would be wise to create a management team to work with auditors to answer their questions and explain the rationale for certain decisions, policies, or procedures.

There are occasions when audit reports are used to justify increases in the budget. If an administrator encounters strong resistance to budget requests, he or she may ask that an outside consultant provide an independent professional opinion. It is a good idea to review the request carefully before making this proposal; if the consultant recommendations do not support the budget requests, the request may backfire and the administrator will lose credibility with superiors. Again, this strategy should be used only when one is absolutely sure that the request is unquestionably justified.

### Chief Officer Tip

**Disaster Purchasing Plans**

Agency plans and budgets are designed to cover routine call loads, and may include extra funds for unanticipated larger incidents; but in general, fire and EMS budgets cannot afford to set aside funds for major disasters. When Oklahoma was hit with a catastrophic tornado in 1999, EMS agencies formed ad hoc emergency rooms until their supplies ran out **FIGURE 5-2A**. When massive wildfires hit San Diego in 2007, responders tried to battle the flames and care for 500,000 evacuees in an already overloaded EMS system **FIGURE 5-2B**. In these types of situations, agencies do whatever they can to acquire supplies, including going to noncontract vendors or pulling from other nonaffected areas. At some point, a state or federal disaster declaration can be made to use government supply caches. Disaster purchasing plans are essential to provide guidelines in these emergency situations.

**FIGURE 5-2** Purchases for major disasters may need to occur during the disaster. A. Oklahoma tornado, 1999. B. San Diego wildfire, 2007.
A. Courtesy of FEMA/Andrea Booher B. © Ted Soqui/Corbis

## Making Budget Cuts

If forced to cut the department's budget, it is a good idea for the chief to have a plan that protects the highest priority items. It may be very tempting to choose items that are the least controversial to cut, but in the long run these types of cuts may have a distinctly adverse effect on quality of service. For example, when faced with mandatory cuts in their budget, many chiefs of FES organizations cut training or prevention items to keep all existing fire stations open. This can be short sighted; the long-term effect of cutting in these areas can be substantial. It may take a year or more, but cuts in training impact the quality of services to the public, as firefighters lose their level of proficiency and injury rates increase.

An article in *Firehouse* magazine explains the typical budget cut situation (2010):

> We have never encountered, and we don't know anyone who ever has encountered, a budget meeting where funds were reduced while raising performance expectations. We typically impose that upon ourselves. We want to do better every year...In the face of budget cuts, most fire departments do everything possible to maintain the same or similar levels of service. What they ought to be doing is reducing services every time the budget is reduced.

In the FES, "Boy Scout syndrome" is sometimes used to refer to personnel who try to help others at all times. Others call this the "can do" attitude. It is the optimistic view that goals can be accomplished even when resources and members are not adequate or safe for the job. However, if cutting services is unavoidable, how are budget-cutting decisions ultimately made?

## Adaptive Policies

Budget reductions may be disguised by dangerous and ill-advised adaptive policies, such as brownouts or reduced staffing. *Brownouts* are a strategy in which fire companies go unstaffed on a rotating basis. In most cases the object is to meet a budget reduction without closing any fire stations. This type of cut is almost invisible to the public unless the closest company to a tragic fire happens to be the one out of service. This practice is dangerous because it is impossible to anticipate if a fire will occur in the first due area of a temporarily closed station. An alternative measure may be to reduce staffing. Fewer firefighters per company may be able to keep costs down. However, mini-pumpers or fast attack units can never adequately replace fully staffed companies. For safety and efficiency, national standards and rigorous scientific research have substantiated that a minimum of one officer and three firefighters are needed on duty at all times; this minimum is also in compliance with NFPA standards and OSHA regulations. Therefore, neither brownouts nor reduced staffing are recommended practices.

One safe alternative for nonmetropolitan departments is shared specialized services with other surrounding departments. For example, if three departments are bordering each other, one could staff a ladder truck, one a heavy rescue unit, and the third a hazardous materials company. Some states have organized and funded hazardous materials companies to cover large areas protected by smaller departments.

When faced with major reductions to tax revenues, many jurisdictions choose to just cut everyone's budget the same percentage. This is a modified zero-based budget approach. *Zero-based budgeting* is a financial decision-making technique that reverses traditional budgeting. In traditional incremental budgeting, only variances from past years are justified, based on the assumption that the "baseline" is automatically validated. In zero-based budgeting, every line item must be approved as if it was the first time it was requested.

Zero-based budgeting is based on the assumption that the present level of funding may not be justified. Funding levels evolve historically by each agency fighting for its fair share and may not match present-day needs. In addition, existing funding may not reflect the relative desires of the public and its elected officials. It may be argued that if last year's budget was a fair distribution of tax revenues, the same percentage split should also be fair for the following year. Zero-based budgeting challenges this assumption and requires agencies to justify budgets from the ground up.

Another technique that could be used inside the agency is an "across the board" percentage cut. Depending on the size of the department, distributions vary. The following is an example of a department that has to make a 10% budget cut:

- Fire prevention division has 10 inspectors: eliminate one inspector.
- Training and safety division has 20 members: eliminate two members.
- Operations division has 82 members: eliminate 8 members. This elimination would probably require the closing of one company along with its station to keep staffing at safe levels for the other companies.

The screams can be heard throughout the fire department when a fire station is proposed to be closed. For the safety of the public, a sign should be placed on the station that says CLOSED. That would keep any citizen from stopping by with a heart attack or other emergency expecting to get help. It would also be a daily reminder to anyone passing the station that it was closed for lack of funding. If the station was truly justified by minimum response time criteria, funding for staffing would most likely be found in the near future. However, if the department deeply cuts nonemergency services or staffing per company, it may never get these resources back.

## Common Areas for Budget Cuts

The most common areas cut in FES budgets are staffing, stations, salaries, and retirement benefits. For the FES administrator, it may seem unfathomable to make cuts; the thought of reducing services may seem totally unacceptable. Yet, at some point, all administrators are faced with the difficult task of finding ways to reduce the agency's budget. Although closing stations or reducing services may seem like the last option from a personnel or community perspective, these options may be considered along with other cuts. The chapter, *Ethics*, discusses these dilemmas further.

Cutting retirement benefits is a very controversial endeavor to reduce the budget. Often these retirement plans are already underfunded because of unrealistic union contract agreements or pensions that were negatively affected by stock market downturns. Even though experts project that individuals should have 80% of their working income to maintain a comparable lifestyle after retirement, this seems like an impossible dream to many firefighters and EMS responders. Lower retirement ages, longer life spans post-retirement, and a large baby boomer population reaching retirement age add to the current financial pressures.

An Associated Press article warned, "States—and the counties and cities that comprise them—may be forced to reduce benefits, raise taxes, or slash government services to address a $1 trillion funding shortfall in public sector retirement benefits" (Associated Press, 2009). Indeed, some governments have now resorted to cutting retirement benefits, increasing the minimum retirement age, making employees pay more into the retirement system, and providing more robust oversight and investment rules.

## Protecting the Budget

Although budget cuts may be necessary, the administrator should make a concerted effort to protect the budget as much as possible. This effort involves building political alliances and promoting the department's programs to key political officials and influential citizens. In some respects, the fire and EMS administrator is a salesperson who must constantly sell the department's products and services. Part of this sales pitch involves educating the public. The administrator should make regular public presentations using speakers to convey the important message of the department's services to public interest groups and organizations. When the government calls for budget cuts, the administration will need as much support as possible.

The chief officer should have an internal audit of the budget done to identify any items that could be deemed unnecessary or unjustified. These items may not undergo scrutiny during good financial times, but when budget cuts are necessary, the presence of any unjustified items rightfully decreases the cost-conscious reputation of the department, leading to additional oversight and cuts. If administrators can eliminate these items ahead of any budget crisis, their reputation of being a professional public servant may help the department hold onto more of its budget during hard times.

## Tax Considerations

"Taxes are what we pay for civilized society," said Supreme Court Justice Oliver Wendell Holmes, Jr., in 1887. But, what is a tax? What types of taxes are there? When is a tax fair? What impact does the tax have on the local economy?

Because taxes are the major source of revenue for most fire and EMS departments, it is worthwhile for administrators to have a basic understanding of important tax considerations that can affect their agencies. The major concern is that the tax base or service fees on which fire and EMS agencies depend may not be adequate to fund increases in costs or other growing demands. Administrators may see increases in taxes as the best solution, but state and local governments are limited legally and practically in their ability to increase these taxes. Even when increases are possible, citizens and government officials are often not supportive of tax increases unless they are convinced about the possible consequences of not taking action, such as reductions in vital emergency service and response times. Administrators must recognize that businesses, special interest groups, and the taxpaying public are all affected by changes in taxes and also recognize that tax changes have a significant effect on the market economy.

It should also be noted that increases in taxes do not always lead to the forecasted increase in revenue. Every product and service has a relationship with supply and demand **FIGURE 5-3**, which is affected by its overall cost including taxes or fees. When the price rises, the demand falls. Therefore, if the government increased taxes significantly, the demand for the taxed good or service would decrease and tax revenues might even fall. For example, when the government increases a tax on cigarettes, less people may purchase cigarettes, thus leading to decreased tax revenue and a decrease in smoking. Similarly, if the government significantly raises taxes on new housing, by impact fees or new regulations, such as earthquake and hurricane protection, rates of construction and home purchases may fall, leading to a decrease in tax revenue. Still, the overwhelming reasons for purchasing a particular home are the characteristics of the neighborhood and the schools; therefore, unless the new total cost is substantially increased, the increase in taxes or costs does not have much of an impact on home purchases.

It is important to consider who is being taxed (e.g., the consumer or business). In 1996, Congress reinstated the 10% federal excise tax on airline tickets. Even though the price of airline tickets should have gone up 10% to cover the new tax, the airline tickets initially rose only 4%. Businesses typically have to absorb some of the tax levied on their products and services. In this case, the airline companies lost 6% in revenue as a result of the airline excise tax. However, this result normally does not hold true over a longer period of time. Eventually consumers see prices increase.

In this case, and for all taxing issues, businesses do not actually pay the tax. The loss in revenue is passed on either to the investors in reduced dividends or to employees through reduced salaries. If you follow money to its source, you find out who actually pays the tax. For example, when buying a car from a large auto manufacturer, federal taxes, which are estimated to be about 23%, are actually included in the cost of the car. Therefore, in this case, the buyer pays the tax.

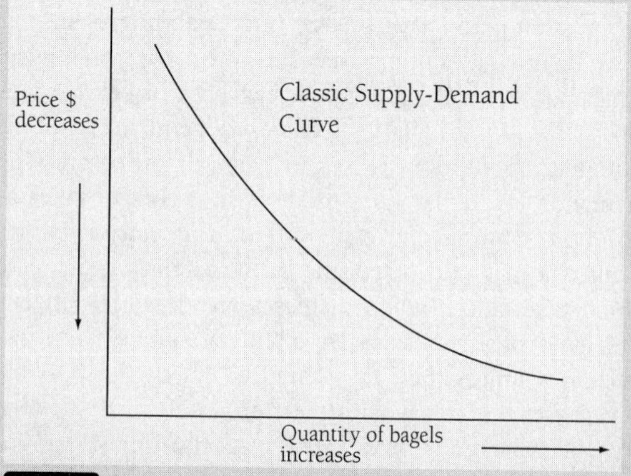

Similarly to taxes on goods or services, the government cannot simply increase property taxes significantly without backlash from businesses and home owners. In most states, property taxes are assessed on an annual basis, but in some states, such as California, rates are adjusted only when the property is sold. Rising values and property tax rates have resulted in tax revolts across the country.

Property tax rates also affect developers' decisions as to where to construct a new building. This can be of great concern to the appointed and elected officials who prefer to attract new development and tax revenue to the municipality. Some jurisdictions reduce, suspend, or eliminate property taxes to attract a major industry. Sometimes other economic factors, such as the availability of public water and sewer, are of greater concern.

There are situations in which higher property taxes may seem worthwhile. For example, parents with school-age children may be willing to pay more taxes to be in a city with an outstanding public school system or lower crime rates. Similarly, retirees and senior citizens may be willing to pay more to be in areas with convenient and quality medical facilities. The quality of fire and EMS services is normally not a consideration in selecting an area to buy a home.

In some areas, a dedicated property tax is used exclusively to fund FES, eliminating the cross-agency competition for available funds. In some emergency service districts, an elected board of commissioners is responsible for managing the tax revenues. Often, these commissioners may be firefighters or EMS providers, which can create complications of bias with regard to financial oversight and decision-making.

Administrators should be aware that property taxes are the major source of revenue for state and local governments, and for their FES departments. Therefore, if additional funds are needed to support existing programs or start a new program, the property tax and user fees probably are the primary sources of funding. Taxes that require minimum enforcement effort, such as the property tax, are those that work best at the local and state level. Increases in taxes or fees are more likely to be approved when the program they will fund is shown to increase the quality or quantity of fire or EMS service and the tax or fee is viewed as fair, equitable, and easy to enforce.

## Tax Incidence Analysis

There are *flat* (or *regressive*) *taxes* (e.g., sales tax, applied to all citizens in a certain area equally) and *progressive taxes* (e.g., income tax, applied at a varying scale with higher wage-earners paying a higher percentage). When recommending that a progressive tax be adopted, keep in mind that the people in the highest 10% of the income bracket may have significant influence in the local political process and pay a major portion of federal income

**FIGURE 5-3** Classic supply and demand curve.

tax already. However, flat (or regressive) taxes may face opposition from those with fixed incomes who may not be able to afford the additional expense. Some jurisdictions have revised taxing ordinances to reduce or limit the burden on the truly needy, such as exempting homes under a certain assessed value from property taxes or application of a fixed fee tax per dwelling unit for FES. Remember that a tax proposal is still evaluated for its fairness. As part of this analysis, the resulting benefits or public good from the increase are compared to its additional cost. If the benefits derived create an increase in service or an improvement in the quality of the service, then there may be stronger support from taxpayers.

There are some taxpayers in this country who have a distrust of government and believe there is never a good justification for an increase in taxes. However, remember that the fire service is trusted more than any other public institution and may be able to overcome the public's apprehensions. If there is a chance of overcoming these negative government feelings, it is with a strong justification of improving service. There must be a clear cause-and-effect relationship between the new program spending and increases in service or the quality of the service. Arguments that are ambiguous or subjective often fail.

## Interstate Considerations

States cannot legally tax anything involved in interstate commerce. They also cannot discriminate against non-residents or out-of-state businesses in their application of taxes. If taxes in one area become excessive, citizens or business owners may elect to move to another state. Therefore, there is some competition to keep taxes low to attract and keep businesses and residents. Areas with large tourist industries often have low taxes for their residents, because taxes on hotels, entertainment, and rental cars are paid by nonresidents.

## Tax Avoidance and Reduction

A desirable feature for a tax system is low cost of administration and compliance. For every tax there are taxpayers who try to minimize or circumvent tax payments through avoidance and evasion. For example, sales tax may be avoided through Internet or catalog sales or by driving to other states to make purchases. In general, property taxes are the hardest to avoid as long as the assessment on the property was done correctly and legally.

## Local Impact of a Global Economy

Most FES organizations are funded at the city or county level. The tax revenues at the local level can be impacted not just by the local markets, but also by state, national, and global markets and economic trends. Anything that hurts the local economy impacts the collection of these taxes. The exact timing of any economic impact on the local budget can vary depending on the local tax type and collection process.

More and more of the US economy is affected by global influences and financial trends. Problems with the stock markets in Indonesia, Japan, Russia, Brazil, parts of Europe, and other countries have had substantial impact on the US economy. Recently, terrorist attacks by organizations and individuals from other countries have had a substantial negative economic impact. Trade agreements open up US markets to competition with foreign products.

The United States is the world's largest debtor nation. Although this situation does not seem to have any real impact on the economy at the present time, keep this potential adverse economic situation in mind. A recession occurred during the period between December 2007 and June 2009 mainly because of the practice of approving mortgages to buyers who were not financially qualified to make the payments. When the price of homes started falling, they could not sell their homes, resulting in foreclosure. To compound the problem, these mortgages were bundled into investment packages that were sold to banks and other investment companies. The financial companies caught holding mortgage-based investments at the time of the subprime mortgage crisis lost so much value that some needed a bailout from the federal government; others went into bankruptcy. Numerous risky investment strategies and some outright dishonest practices added to the magnitude of the financial crisis. This crisis continues to have a major adverse impact on the real-estate tax at the local level.

> ### Facts and Figures
>
> **The US Financial Problem**
> In the article entitled "The US Financial Problem Explained," private investor Jim Slater cites the following US financial statistics from 2011:
> - US Tax revenue: $2,170,000,000,000
> - Federal budget: $3,820,000,000,000
> - New debt: $1,650,000,000,000
> - National debt: $14,271,000,000,000
> - Recent budget cut: $38,500,000,000
>
> Slater then goes on to compare these statistics to a household budget, by removing eight zeroes from the above numbers:
> - Annual family income: $21,700
> - Money the family spent: $38,200
> - New [annual] debt on the credit card: $16,500
> - Outstanding balance on the credit card: $142,710
> - Total budget cuts: $385
>
> *Source:* Slater, J. "The US Problem Explained!" Web 2011. Retrieved from http://www.jimslater.org.uk/the-us-problem-explained/

If the local community relies on industry or manufacturing for a good portion of its tax base, there may be trouble for these companies in the future and trouble for taxes. The decline in manufacturing jobs is a natural evolution of the labor market in this country. As the costs for labor have risen in the United States, the costs of labor for manufacturing in other countries have stayed at a lower level.

A similar trend can be seen today as US companies downsize and re-engineer themselves. Many of the jobs lost are in the middle management ranks. Simultaneously, the American educational system is receiving low marks for its ability to produce graduates that excel in high technology and the skilled, educated workers needed to compete in the global market.

The US economy is highly regionalized. An older city that contains a large number of heavy manufacturing firms may experience a greater reduction in local taxes than the suburban area that surrounds the city. Older cities in the United States have already seen reductions in city services as a result of declining tax revenues. Social Security and Medicare are projected to run out of money in the future. Some of the options to maintain them are to either raise taxes or reduce benefits. Because most federal elected representatives are not anxious to take either option, these problems may not be resolved until a crisis occurs, and then change will be unavoidable.

Be aware that major changes are always possible in the economy at the local, state, and national level and may happen without notice. As a fire or EMS administrator, you must be prepared to defend your agency's budget to the municipal administration and other department heads if a downturn in the economy occurs. Use of sound financial practices and strategies puts emergency services agencies in the best position to provide their valuable services, regardless of local, economic, or political factors.

## CHAPTER ACTIVITY #1: Budget Increases

In the article "A Master at His Art," a unique technique for gaining budget increases is exemplified in a story involving former Los Angeles County fire chief, Keith E. Klinger.

According to the article, a rapid vegetation fire in Bel Air, California, in 1961 destroyed nearly 600 homes within 72 hours. Throughout that night, Chief Klinger actively led the response efforts, never taking a break or slowing down. The very next morning, he drove straight over to the Los Angeles County Board of Supervisors' weekly meeting. Although he wasn't on the meeting's agenda, Chief Klinger walked right up to the podium and, still covered with soot from the night before, gave the board a descriptive report on the fire that was taking over the mountains.

Without hesitation, Chief Klinger then proceeded to describe a 10-year plan for fire protection improvement. Although the cost of the plan was expensive, the board could not refuse it due to the current predicament. The chief had developed the plan prior to the Bel Air fire, but his effectiveness came from waiting for the best time to present it to the powerful elected officials.

*Source*: Page, J. O. (1998, May). A master at his art. *FireRescue*.

### Discussion Questions

1. Identify an existing need in an FES department.

2. Quantify the resources, both staff and equipment, needed to implement the program.

3. Provide a list of all costs.

4. Provide an overview of potential support or opposition to the program.

5. Describe a hypothetical or real situation that could support an emergency request to fund the program.

## CHAPTER ACTIVITY #2: Budget Adjustments

Central City is an older city that has grown slowly over the years by annexing several pieces of adjoining land FIGURE 5-4. The city covers approximately 60 square miles and is a typical, medium-sized city with diversified light industry, a busy commercial area outside the central core, a community college, one hospital, and several medium-sized assembly buildings in the downtown area. More recently, the city annexed an area primarily consisting of expensive homes in the northwest side in order to bring in more tax revenues to cover existing shortfalls.

The Central City Fire Department operates six engines and two ladders (Stations 1 and 5) under the supervision of two district chiefs. The department responds to life-threatening medical calls, but it does not transport patients. Firefighters work a three-platoon system. The following page includes the Central City Fire Rescue Department organizational chart FIGURE 5-5 and a table containing the characteristics of the fire stations TABLE 5-3.

The firefighter's union contract calls for a 53-hour workweek with an additional 3 hours of overtime pay. Therefore, each firefighter on shift works 56 hours per week. These costs are already built into the present budget. In addition, the budget contains enough funds to pay 5% of each overtime shift to fill vacant positions due to sick and annual leave.

A master plan was completed recently and the following major recommendations were made to the department: (1) increase the staffing of all companies to 4 on-duty persons to meet national safety standards; (2) construct a new fire rescue station in the northwest section of the city to reduce response times to the city average of about 4 minutes; and (3) start a new fire safety education program by increasing the staffing of the Fire Prevention Division by 2 personnel.

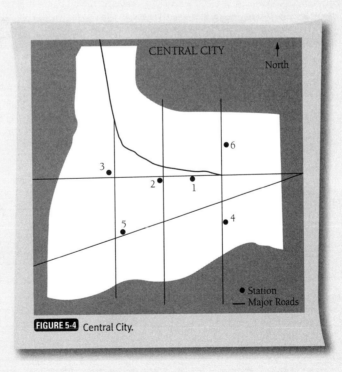

FIGURE 5-4 Central City.

*(continues)*

# CHAPTER ACTIVITY #2: (continued)

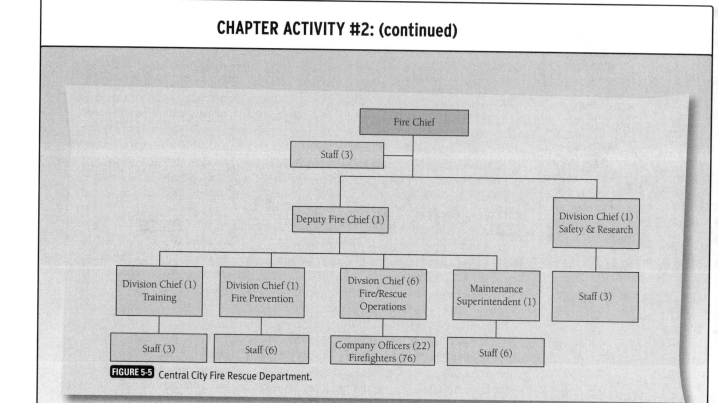

FIGURE 5-5 Central City Fire Rescue Department.

### TABLE 5-3 Central City Fire Stations

| Station | Data Constructed | Average Response Time | Population Served | Square Miles in First Due |
|---|---|---|---|---|
| 1 | 1937 | 2:45 | 13,000 | 7 |
| 2 | 1943 | 2:33 | 9,000 | 4 |
| 3 | 1957 | 5:50 | 36,000 | 16 |
| 4 | 1963 | 4:10 | 22,000 | 11 |
| 5 | 1971 | 4:35 | 22,000 | 13 |
| 6 | 1982 | 4:05 | 20,000 | 9 |
| Total | | 3:54 | 122,000 | 60 |

## CHAPTER ACTIVITY #2:

### Discussion Questions

Consider the following two scenarios and answer the discussion questions that follow each scenario. It is helpful if you can approach this activity without any biases. Many FES members prefer not to cut areas where they are presently working. Where appropriate, use reference material to justify any cuts or additions.

*Scenario #1*: You have been instructed to cut your budget by 10% and specifically advised that you must cut 13 positions.

1. Where would you cut positions and how would you justify your proposal?

2. How would you select and measure "outcomes" or consequences of your plan?

3. What type of reaction would you expect from: (1) the union president, (2) the public, (3) the press, (4) the city manager, and (5) the city council?

*Scenario #2*: Starting with the beginning level of staffing, you have been instructed that your budget will be increased by 10% and specifically advised that you must hire 13 new employees.

1. Where would you add positions and how would you justify your proposal?

2. How would you select and measure "outcomes" or consequences of your plan?

3. What type of reaction would you expect from: (1) the union president, (2) the public, (3) the press, (4) the city manager, and (5) the city council?

## CHAPTER ACTIVITY #3: Disaster Planning

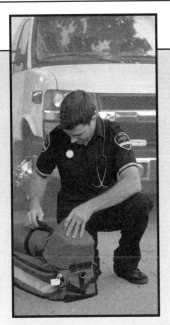

Some of the most expensive types of fire incidents are those that require both fire suppression and patient care. In the San Diego wildland fires of 2007, 500,000 people were required to evacuate, many of whom needed acute medical and trauma care. Even though local departments often know what types of natural disasters they should expect, they can't plan for any specific event. In the case of San Diego, the EMS system was quickly overloaded. Convalescent home evacuations brought patients with special needs into ad hoc evacuation sites like schools, churches, and civic centers.

### Discussion Questions

1. What are some of the things that make disaster planning difficult?

2. Does your fire district have a cache of disaster supplies or a disaster purchasing plan?

3. What other resources are available in a larger incident?

4. Describe any existing mutual aid agreements your department has with adjoining departments. Would this be sufficient for a major disaster? If there are none, describe a proposal for a regional agreement.

# References

Associated Press. (2009). Study: States must fill $1 trillion pension gap.

Bruce, N. (1998). *Public finance and the American economy.* Reading, MA: Addison-Wesley.

Burd, S. A. (2009, June 12). How Safeway is cutting healthcare costs. *Wall Street Journal.* Retrieved from: http://online.wsj.com/article/SB124476804026308603.html

Cebula, R., & Feige, E. (2008). *America's underground economy: Measuring the size, growth and determinants of income tax evasion in the U.S.* Retrieved from http://ideas.repec.org/p/pra/mprapa/29672.html

"Drowning Death Pushes Alameda Officials To Address Water Rescue Training" (2001). KTVU.com. Web. Retrieved from http://www.ktvu.com/news/news/drowning-death-pushes-alameda-officials-to-address/nD63s/

Internal Revenue Service. (2012). *Individual income tax returns.* Washington, DC: Statistics of Income Division, Unpublished Statistics.

# CHAPTER 6

# Human Resources Management

## Fire and Emergency Services Higher Education (FESHE) Course Objectives

**Module I: Leading and Managing Purposefully with a Community Approach**

The students will:

4. Identify local, state, and national organizations that will be beneficial to your department. (pp 95–96)
5. Describe how to take a proactive role in local, state, and national organizations. (pp 95–97)
6. Identify effective skills for developing a cooperative relationship with fire and emergency services personnel as well as public officials and the general public. (pp 90–92)

**Module II: Core Administrative Skills**

The students will:

2. Describe the integrated management of financial, human, facilities, and equipment and information resources. (pp 79–99)
6. Recognize the formal and informal dynamics of public organizations and describe strategies to ensure success. (pp 95–97)

## Knowledge Objectives

After studying this chapter, the student will be able to:

1. Understand the importance of well-trained and well-equipped professional firefighters. (pp 79–80)
2. Understand the function and operation of human resources personnel. (p 80)
3. Recognize ways of providing diversity in the department. (pp 80–83)
4. Examine the legal issues relevant in hiring firefighters and recruiting volunteers. (pp 83–89)
5. Understand and review the disciplinary process, along with legal issues that may arise. (pp 90–92)
6. Examine the process of job analysis and validation. (pp 92–94)
7. Recognize the influence and understand the operation of public sector unions. (pp 95–97)
8. Understand the concept of motivation and review ways to retain members. (pp 97–99)

## ■ The Most Valuable Resource

The members of a fire and emergency services (FES) organization are the most valuable resource in terms of costs and results **FIGURE 6-1**. Without well-trained, well-equipped, and well-prepared members, the essential tasks at an emergency scene cannot be accomplished effectively, safely, or competently. Luckily, a 2007 study revealed that firefighters ranked number two in job satisfaction. "Most American adults are employed and their job is not only their main source of income, but also an important life domain in other ways" (Smith, 2007, p. 2). In addition to a network of supporting personnel, resources, including stations, emergency apparatus, safety equipment, and specialized tools, are another component critical to success on the emergency scene.

FIGURE 6-1 Alachua County (Florida) Fire Rescue Engine Company.

The following depicts the cost of one engine company amortized for 1 year:

- Staff: Five firefighters (minimum crew of four plus one to cover leave impact) at $42,000 salary each plus fringe benefits (~40%) for three shifts (53 hours/week) = $882,000
- Facilities: One fire station ($3 million using a 30-year bond [6% interest] = $217,947 annually) and one engine ($300,000 using a 20-year bond [6% interest] = $26,115 annually) = $244,062 per year

The personnel costs for this engine company are more than three times the costs of the capital items needed to respond to emergency incidents. These costs do not include the overhead expenses of staff and administrative personnel. Behind each piece of emergency equipment are people who need to be hired or recruited, trained, evaluated, and motivated to do the job effectively and professionally. Fire and emergency service personnel management prepares individual members to do their jobs well.

## The Function and Operation of Human Resources

The human resources (HR) department is the agency within the local government responsible for recruiting, hiring, training, evaluating, and paying employees. The employees that make up the HR department have specialized training and education, but generally have an initial limit of knowledge regarding FES personnel issues. Instead, these employees are support staff for line agencies, such as the police, fire organizations, and schools. HR uses merit rules and regulations along with union contracts to provide direction for individuals carrying out the job duties. Merit rules are legally adopted by the jurisdiction and contain provisions for employee pay, benefits, discipline, and job descriptions. HR can also play a major role in acquiring additional funding for personnel and selecting the best people for various jobs.

It is very important to have a close relationship with the director of the human resources office of the municipality. Make an appointment to meet the director and get to know them personally. The first step in forming this relationship is to familiarize HR personnel with the unique nature of FES operations. The FES organization needs new members who have the ability to:

- Learn complex skills
- Acquire knowledge
- Have courage
- Be physically fit
- Not mind getting extremely soiled and sweaty
- Endure temperature extremes (hot and cold)
- Tolerate working with or around critically injured or deceased victims
- Operate in a family-like situation for up to 24 hours
- Be team players

If HR employees are not fully aware of the need for these unique abilities, they may unintentionally hire individuals who cannot perform the essential job functions when faced with other policy goals dictated by the city administration.

Educated HR employees have a better understanding of FES personnel issues. One way of accomplishing this education is to take them to an FES station or the training academy so they can experience the working environment (minus live fire or medical emergencies) and job tasks. HR employees should don the protective clothing and self-contained breathing apparatus, carry the trauma bag, and perform various job functions. Some of the same multimedia materials, lectures, and demonstrations that are used in public relations or risk reduction education programs can be very helpful in educating these employees.

The chief officer should have as much influence over personnel management functions as possible. For smaller organizations, an FES officer should be assigned part time to coordinate with the HR office; a larger department should have permanent staff assigned full time. These officers should have some formal education in personnel management. A civilian employee with the appropriate guidance and knowledge from the chief can carry out this function in some departments.

## Diversity in the Department

Complete diversity reflects a blend of gender, race, age, education, country of origin, and other characteristics, such as personal values. In the past, the typical FES organization was composed of people with similar characteristics—those who joined were expected to conform. In today's world, each member wants and deserves to be valued as a distinct individual. People from diverse backgrounds bring diverse values and experiences to their roles.

Starting in the 1970s, coalition groups began to support the rights of particular categories of people. These groups are very vocal in their identification of perceived or real discrimination and demands for corrections. Today, most FES organizations have actively recruited minorities and women to compensate for memberships that had been predominately white and male. In some cases, these organizations were under court orders or consent decrees to provide affirmative action programs to increase the numbers of minorities and women.

There is disagreement about the use of affirmative action, preferences, set-asides, and quotas—strategies used to hire minorities and women in many FES organizations. New legislation and contemporary court rulings have changed the laws of the past. For example, judges used to rule discrimination cases based on a lack of a specific percentage of the underrepresented group compared to the general population served. Now, discrimination in the FES organization cannot be proved solely based on these percentages.

Many FES departments use the diversity of the municipality as their benchmark for diversity goals. This method may skew the diversity statistic because the local city may have large numbers of one group. It may be better to use a larger area to represent the region where potential new members are recruited. The minimum area should be a county; larger departments might use several counties.

## Affirmative Action Cases

### The Boston Fire Department Affirmative Action Case
Some argue that the affirmative action points system may lead to widespread lying on applications to schools and jobs. This was the case in 1990 for the Boston Fire Department. Six firefighters from the department were found to have lied on their applications, claiming false minority statuses.

Two of these firefighters, twin white brothers Phillip and Paul Malone, had initially failed their qualifying examinations for the department. However, on subsequent applications, the brothers claimed that their great-grandmother was black, thereby qualifying for minority outreach standards. It was not until 10 years later that their lie was discovered (Vlahos, 2003).

### The University of Michigan's Affirmative Action Case
A 2003 US Supreme Court case upholding affirmative action policy at the University of Michigan law school was a surprise to those who believe affirmative action should not be used routinely to provide diversity. What makes this issue even more confusing is that the Court also ruled that the undergraduate policy at the University of Michigan was not legally correct and must be changed. "In that admissions process, which the university now has to revamp, the school literally gave extra points for applicants who were black, Hispanic or Native American" (Vlahos, 2003). In fact, the diversity points had the potential to outweigh points given for academic achievement. However, a more recent affirmative action case may affect the decision made in 2003. In *Fisher v. University of Texas*, Abigail Fisher claims she was denied admission to the school because she was white. Although the results of this case are not yet known, the US Supreme Court's decision has the potential to modify and overturn the University of Michigan case.

### The New Haven, Connecticut, Affirmative Action Case
On April 22, 2009, the US Supreme Court sided with firefighters claiming they should have been promoted based on examination results for lieutenant and captain positions. The results of the promotional examination were discarded when the city noticed that no African-Americans scored high enough to be promoted.

The firefighters that should have been promoted filed suit, alleging that the city violated Title VII of the Civil Rights Act of 1964, the Equal Protection Clause of the Fourteenth Amendment, and the First Amendment of the US Constitution. The firefighters argued that the examinations should have been race neutral and solely based on job performance and duties. They claimed to have been punished for their race. In June 2009, the US Supreme Court sided with the firefighters (Doyle, 2009).

## Diversity Selection in Practice

What does all this mean for the typical department? The following hiring scenarios seem to comply with the latest US Supreme Court decision and others:

- A policy can be used to rate each applicant based on scores or ratings from written and other examinations. The person with the highest score is chosen first, followed by others, and strictly based on their ranking. The US Supreme Court did not say that this traditional way of selecting new firefighters could not be used legally, but it must use job-related criteria and must not discriminate.
- Court orders must be followed if it has been determined the institution has discriminated. Affirmative action programs must be provided, resulting in a mandatory plan and quota.
- A compromise policy called banding takes into consideration both points of view. In a banding program, each applicant is put into a band and this list of applicants must be exhausted before going to the next band. One common cutoff for the top band is 85%. After the applicants take the examinations, they are put on a traditional list and ranked from highest to lowest. Then new members are selected from the top band first.
- In actual practice, the department may offer jobs to women or minorities in the band and then offer jobs to white males in a ratio to achieve diversity. This helps prevent what some would characterize

as reverse discrimination. Legally, there is no reverse discrimination, only discrimination. Some affirmative action programs have this undesirable feature.

- Banding with a cutoff of 85–100% uses a rationale identical to the typical grading schemes in school. For example, an "A" grade is actually a numerical grade of 90–100%. This approach takes into account the fact that human beings are consistently inconsistent in test taking within a small band. On any particular day, the numerical grade received and the position on a list can change for each individual within a predicted range. For example, one day's performance may see a person at the top of the list, but another day they may be 13th, although both scores may be in the top band.

To make banding work well for diversification, not everyone in the band can be offered jobs because the mix in the band is not likely to be consistent with policy goals. For example, if there are 100 people in the band, selecting 20–30 people provides a diverse group. To ensure this outcome, the department has to recruit and encourage minorities and women to apply for the positions. An affirmative action officer normally handles this job.

If the department has vacant positions and the ideal diverse mix cannot be achieved, the positions have to be filled. There are numerous and varied situations in which this may occur, such as with smaller departments that typically have fewer applicants. The administration must try its best to meet the goal, and if it cannot, which is not unusual, the positions must be filled with qualified applicants from the band.

It is typical in many departments for the chief to have the option of choosing from the top three on a list for a promotion. This option allows the chief to completely skip over someone who did well on the promotion process but who is not found to be acceptable to the chief. It is a compromise with unions who often want the first on the list to be the first promoted. This chief's option is generally not used to support diversity but to acknowledge the chief's professional judgment on the qualifications for promotion. Regardless, all individuals in the band must be promoted before promoting candidates on the next lower band.

It is never a good idea to lower the job requirements to select applicants outside the band, or to select applicants who cannot meet minimum requirements, such as physical fitness standards. There are three reasons to turn down applicants who do not meet minimum firefighting job requirements:

1. Firefighting is a very hazardous job; skills and physical strength can be a matter of life or death.
2. If firefighters are not able to adequately perform essential job functions, the public receives unsatisfactory service.
3. It is frustrating and demoralizing to the firefighters who meet the requirements. They may need to make up for the incompetence of others.

Emergency medical services (EMS) jobs may have similar, but slightly different minimum standards. The following are three reasons to turn down applicants who do not meet minimum EMS job requirements:

1. Candidates need to be in good physical condition and able to lift at least 125 pounds, or a weight required by the company. Even with advanced equipment like hydraulic gurneys, the emergency medical technician (EMT) or paramedic may have to move a patient out of a cramped bedroom or down a flight of stairs without injuring the patient or himself or herself.
2. Candidates must show competence with a basic written didactic knowledge test and a skills return practical scenario. The company must confirm that candidates have an adequate job-specific knowledge base and can adequately perform patient assessment skills.
3. Candidates must display an attitude indicating they would be well accepted as a partner and patient care technician. They must respond to questions with honesty and without ego, suggestion of bias, or display of negative attitude. Compassionate and respectful interaction with partners, public safety personnel, emergency room employees, and the patient and family is critical.

## Diversity Sensitivity Training

Many agencies have started offering sensitivity training to educate employees to be aware of personal differences. This training works to achieve social harmony among a group that may have little in common. For example, one FES department started an affirmative action program to recruit minorities for firefighter positions. The affirmative action officer decided to go to some local churches where the membership was mostly minorities. The officer found many applicants who successfully passed the written and physical fitness examinations.

The new firefighters were assigned to permanent fire stations after completing recruit training. Typically, the crew becomes a second family for the firefighter. Although differences in personal values may exist, many share these values or tolerate differences in values of crew members. Because the new firefighters were recruited from churches, they had strong religious values and personal beliefs that existing members did not share, in many cases. The result was a distinct lack of socializing for most of the shift.

Sensitivity training can go a long way in melding these diverse personalities to work together in harmony. For some minorities and women, this training can make the difference between feeling welcome and feeling alienated. Individuals who feel alienated could become less motivated on the job, and may show evidence of carelessness resulting in accidents. In addition, they may resign soon, because they feel out of place, which becomes a big loss to the organization that spent many hours training the employee.

Generally, diversity should be accepted and encouraged. Society shows that different cultures can peacefully coexist, and the workplace should attempt to reflect that as much as possible. The public can more easily relate to a workforce that is representative of their society. Furthermore, diversity provides a broader perspective in organizational problem solving.

While attempting to achieve this diversity, the department should not lower standards of performance. Acceptance of the new diversity of employees must be supported by minimum standards for job competencies. For these people to become part of the family, they must be able to perform the minimum job requirements. In the emergency services profession, members must rely on each other to work in teams and, more importantly, if one member needs help, any other member must be able to come to their aid.

It is unethical to hire or accept members into FES organizations who are not capable of performing critical job competencies. The public trusts that FES personnel can perform emergency functions safely and efficiently. If the personnel cannot perform to professional standards, they will fail the public and fellow members. There is strong support for this viewpoint as expressed in the International Association of Firefighters/International Association of Fire Chiefs Candidate Physical Ability Test: "Diversity should never come by lowering validated entry standards. Rather, it should come from actively recruiting qualified men and women candidates from all racial and ethnic backgrounds for careers in the fire service" (International Association of Firefighters, 2011).

Although the private ambulance industry does not have a standard physical agility test like the Candidate Physical Ability Test, lighter physical agility tests may be required by an employer. In addition, some states require a background check to rule out misdemeanor or felony convictions that would preclude the employee from working in the patient-care setting.

Some FES organizations, in their efforts to do the right thing regarding diversity, have hired or accepted members who cannot meet minimum job performance criteria. These situations are very challenging to resolve, requiring an administrator with a good understanding of leading change (see the chapter *Leading Change*).

## Recruitment and Selection of Firefighters and Volunteers

It is a good idea to ensure that the company is committed to providing diversity in the organization as new employees are hired or new volunteer members are accepted. It is rare that a formal quota system can be legally used. Quotas are allowed only when a court has ordered a remedy to a proved past discrimination practice.

Therefore, the department has to set up a recruitment and selection process to attract a diverse mix of applicants. For example, because minorities and women tend to be underrepresented in most FES organizations, the chief officer should assign a member the responsibility of recruiting minorities and women to apply for jobs or membership. There are many techniques for accomplishing this goal, but the department should consider only ethical and easily implemented techniques. The chief officer should consult with the HR office staff for more options.

Additionally, a professional presentation should be created or acquired to help educate and attract applicants. Locations or specific communication channels should be identified to reach the selected audience. For example, visits to women's athletic events at local universities can locate good prospects who are physically qualified to be members.

**Fire and Emergency Services Workforce Issues** A common workforce issue concerns the understanding that the typical FES organization is different from other occupations. Most of those differences are because FES work is extremely time-sensitive, risk-oriented, and team-based, with members in close contact with one another for long periods. The duration of the typical FES shift and the closeness of the club-like surroundings of the career and volunteer FES organization make it imperative to manage diversity effectively.

> **Chief Officer Tip**
>
> **Hiring and Recruiting**
> Hiring and recruiting are important issues to the human resources of both career and volunteer departments. Not only do the policies and procedures regarding hiring and recruiting need to comply with all applicable legal and ethical issues, they also must ensure that potential members of the department are well qualified and fit the job description of the position to be filled. NFPA standards such as NFPA 1001, *Standard for Fire Fighter Professional Qualifications*, and NFPA 1582, *Standard on Comprehensive Occupational Medical Programs for Fire Departments*, can be helpful references in the hiring and recruiting process.

## Legal Issues

This section contains up-to-date information from recent Supreme Court and legislative actions **FIGURE 6-2**. The law is constantly changing. Many of the recent Supreme Court cases have been decided by a slim majority (5–4), so replacements for retiring justices may swing the Supreme Court in a different direction.

Two aspects of legal issues affect FES: the recruitment and retention of members, and emergency operations. This chapter discusses only personnel management issues. Legal representation should be consulted for any changes in local, state, or national laws or interpretations that may make the legal opinions in this section inaccurate.

**FIGURE 6-2** The U.S Supreme Court building.
© Jason Maehl/ShutterStock, Inc.

## Hiring Issues

One of the most important labor functions is the hiring of new personnel or the acceptance of a new volunteer member. A mistake in this process can leave the department with an undesirable member who must either be kept or separated. Separation is never an easy or pleasant process.

The screening process checks skills, knowledge, and physical abilities that are needed to master essential job performance functions. For firefighters, the necessary skills for most job functions are not mastered until recruit training is completed. It is still essential to test for the basic skills that are needed to master the training program. Tests that indicate a minimum ability in reading, writing, math, and cognitive skills should be used.

This issue is illustrated by the experience of St. Louis in 2007 where "more than 70 percent of about 1,350 applicants failed the test [reading and math]" (Wagman, 2007). Three years earlier, in 2004, the city was sued by a group of African-American firefighters who challenged a promotional test for captain and battalion chief, claiming discrimination against African-Americans. A judge ruled that the city did not discriminate; however, the experience left the city wary of future legal battles.

Therefore, after receiving the examination results in 2007, the city decided to scrap the test to avoid more controversy. However, Chris Molitor, president of the International Association of Firefighters Local 73, argued, "I don't think it's asking too much for someone to read, write, and do basic math. This test was designed to eliminate those people who could not perform those basic functions. It sounds to me as if the test was doing its job" (Wagman, 2007).

Psychological testing is another area that deserves consideration in light of incidents of destructive behavior and arson by FES members. In 2009, Cincinnati started requiring firefighters to undergo psychological evaluations before being hired. This decision was in response to an increased number of firefighters with criminal and disciplinary problems. To keep the public respect, firefighters and EMS personnel need to be upstanding role models. Some basic testing categories to use are:

- Psychological
- Aptitude and achievement
- Personality
- Integrity and honesty
- Physical fitness

To reduce the likelihood of liability, FES organizations should use a cutoff score on screening examinations that indicates the minimum qualifications necessary for successful job performance. To compensate for normal variances in all testing results and to guarantee that the applicants selected can perform at minimum levels of competency physically and mentally over a 20- to 30-year career, the minimum cutoff should be above the theoretical lowest scores. It is not a good idea to hire or accept marginally qualified applicants; it is not fair to the applicants, the fellow members, or the public they serve.

## Reference Checks

Many companies have adopted a strict "name, rank, and serial number" approach to requests for information about past and present employees. The reluctance to give negative information about employees' performance and disciplinary employment records has backfired in some cases, causing liability exposure for these organizations. Furthermore, there is now increased risk of negligent hiring and retention of employees, especially in organizations with public trust.

For example, a new employee was hired as a paramedic after a complete background check. A past employer failed to provide information that the individual had stolen jewelry at the scene of a medical emergency incident. The previous employer was held liable for withholding information that was relevant to the employee's behavior on the job.

In response to this type of scenario, the state of Florida recently enacted a statute providing legal protection to employers who furnish information about present or former employees. This protection provides immunity from civil liability when sharing information about employees. The state attorney general can provide guidance about similar statutes in other states. The employer is covered unless it can be shown that the information was knowingly false or deliberately misleading, which would be a violation of the employee's civil rights, and the employee could seek appropriate legal action. The following are guidelines for providing reference information:

- Designate one individual in the personnel department as the only contact for requests.
- Require that a written request be made on company letterhead. Telephone the requesting agency to confirm all requests.
- Disclose only documented job performance information, not subjective evaluations.

- Do not disclose any information regarding discrimination complaints or medical/disability information that may reflect on equal employment opportunity status or a protective category.

## First Amendment: Freedom of Speech

In 2006, in the case *Garcetti v. Ceballos*, the Supreme Court placed an entire category of speech outside the protection of the First Amendment: statements made by government employees in the course of their official duties. Justice Kennedy, when writing for the majority, said, "[W]hen public employees make statements pursuant to their official duties, the employees are not speaking as citizens for First Amendment purposes, and the Constitution does not insulate their communications from employer discipline" (Garcetti v. Ceballos, 2006). Fire officials need to be especially careful of this restriction during poor economic times when they may be tempted to blame a fire death on inadequate funding. It should be noted that this restriction does not cover whistle-blowers.

## Civil Rights

Several acts of Congress have been based on the 13th and 14th Amendments to the US Constitution guaranteeing equal treatment or protection against discrimination based on race, religion, gender, or national origin. These acts are generally referred to as civil rights acts. In the FES profession, violations of these federal laws have resulted in monetary awards and court-ordered remedies typically labeled as affirmative action plans.

This area of law has been slowly evolving and is different today than it was 10 or 20 years ago. This subject is also complex, and legal advice should be obtained when considering a change in policy. A brief summary of the present situation follows.

A statistical imbalance between the percentage of a protected class in the workplace and the percentage in the general population is an indicator of a possible problem. The organization must study this imbalance carefully to see if any discrimination has taken place. In the past, this type of imbalance was prima facie evidence that discrimination had occurred. Now it is only a possible symptom, not conclusive evidence.

If a department enters into an affirmative action plan voluntarily (in writing or not) without a court order, and that plan contains specific percentage goals to hire minorities and women, the department may be open to civil court action based on reverse discrimination. Affirmative action plans can be used if they are designed to encourage and attract underrepresented groups to apply for employment or membership or pursue promotional opportunities.

## The Americans with Disabilities Act

The Americans with Disabilities Act (ADA) provides civil rights protection to people with disabilities and guarantees equal opportunity in employment, public accommodations, transportation, government services, and telecommunications. The act, which is enforced by the Equal Employment Opportunity Commission (EEOC), is designed to protect individuals from discrimination based on a disability that substantially limits a major life activity. Employers must make a reasonable accommodation for a mental or physical disability of an individual, unless it would pose undue hardship on the employer's operation.

In most cases to date, reasonable accommodation for disabilities has been shown to create undue hardship if applied to public safety jobs. Therefore, it is very important to document job performance requirements along with the medical, physical, and mental abilities needed to perform the job performance requirements.

Because of the intimidating nature of a federal regulation and the different analyses by many lawyers, some FES organizations have hired or accepted new members who physically or medically were not capable of performing the job of firefighter or EMT. Organizations should be very careful when getting advice about the ADA and research the many articles that have been written specifically about FES workers. Very simply, if standards for job performance are based on verifiable requirements, they are acceptable and legally defensible. Even though job standards may eliminate some people with disabilities, the standards are valid and meet the exceptions stated in the ADA.

Furthermore, organizations should be careful not to fall into the trap of lowering the minimum requirements to accommodate a perceived exposure to legal action based on the ADA, civil rights, or another federal law designed to protect against discrimination. The goal of all FES organizations is to give the best service possible within financial and practical limits of human performance, both mental and physical.

In early 2011, the EEOC began working on new guidelines to define disability. Some experts have estimated that these new guidelines will label millions of more workers as disabled. Attorney Condon McGlothlen explains, "It's going to be very difficult for employers to argue in just about any case that an employee is exaggerating their disability or that the person isn't genuinely disabled" (Bream, 2011).

**The ADA and Hiring Practices** During the hiring process, applicants cannot be given a medical examination or asked any specific questions concerning any medical or physical disabilities. However, applicants can be screened using mental and physical ability tests as long as they are validated as needed for essential job functions. For example, for EMS jobs requiring an ambulance driver's certificate, a physical examination may be required to determine adequate hearing, visual acuity, and basic mobility. In addition, drug and alcohol testing is allowed before a job offer.

After an employee has accepted a job offer he or she may be required to take a medical examination that could be used to rescind the job offer. In addition, it is very important, especially in FES organizations, to have an extended probationary period during which the employee can be separated without cause. It may take many months to evaluate a firefighter, EMT, or paramedic in actual emergency situations or those situations that occur infrequently, such as a large structural fire.

The probationary period should run at least 1 year after the successful graduation from a recruit firefighter school. The department then has a whole year to evaluate the new employees and be able to separate any who do not meet minimum expectations. The private sector is less standardized; successful orientation and field evaluation generally follows a probationary period lasting about 6–12 months.

Any standards for separation should be validated for essential job functions and each employee should be made very familiar with the rules. Legally, probationary employees have very few rights. It is better to make sure everyone, especially the probationary members, knows what the rules are, the justification for them, and how the members are evaluated.

It is also typical and preferable to have a 1-year apprentice phase where the new member learns additional knowledge and skills. For example, in the private ambulance arena, the candidate, after successful completion of preemployment tests and orientation, may be assigned to a Field Training Officer. Successful completion of this phase is based on competency rather than time, and it is generally shorter than in the fire service. The candidate must successfully meet required training goals, such as competent patient care and equipment handling, understanding of local policies and protocols, and driving skills.

## Injuries on the Job

Policies should be created to ensure documentation and verification of injuries that are reported on the job. Standardized forms should be submitted to the appropriate municipal official for workers' compensation.

High-ranking FES officials and the employee's immediate supervisors should visit the employee in the hospital or make personal visits to the home, especially in cases of serious injuries on the job (IOJ) events. Employees heal faster if they believe they are valued and missed. Supervisors should send a get-well card, make appropriate visits during recuperation, and provide the employee with the paperwork necessary for the workers' compensation claims and the payment of medical bills.

Supervisors should stay in close contact with the IOJ members until they are released to either light duty or full duty, keeping a constant dialogue to inform the employee of what is going on at work. If the department's official cannot visit in person, notes should be sent often. Finally, a small welcome-back gathering should be planned with the employee's close peers.

**Light Duty** To lower workers' compensation costs, FES departments often bring individuals back to a light-duty position if they are still recovering and are not ready for full duty. The temporary duty assignment should not consist of clerical work, such as answering telephones or filing. Instead, for those firefighters who are exemplary members, temporary assignments may be provided in such areas as fire inspection, fire safety education, or 911 dispatching.

This technique seems especially effective for employees who are on shift work. If the administrator brings the injured worker back on light-duty day work, 9-to-5, there seems to be a strong correlation to the injured employee wanting to return to full duty faster than expected. In fact, this policy is effective even when the department can bring the employee back to duty only for short periods at the beginning. The important point is to get the employee back in uniform and back to work as soon as possible. Otherwise, employees may get used to a sedentary lifestyle and lose the motivation to get back to work.

While the member is on light duty, provide him or her with written guidelines covering work hours, tardiness and absenteeism, rest periods, and personal appearance. Specify the procedure for requesting leave, including IOJ leave for medical appointments and physical therapy.

Some employees in the FES organization have attempted to use ADA and its reasonable accommodation feature to claim that an employer has the responsibility to create a light-duty job. However, this is not a requirement of ADA; light-duty positions do not have to be offered to employees who cannot perform the essential functions of their job.

For example, suppose an employee is injured on the job and is covered by workers' compensation. After a recovery period, medical evaluation states that this person will never be able to fully recover to perform the essential functions of a firefighter. The employee may want to work in another position that does not require strenuous physical work, such as in the dispatch office. However, this would require the department to pull one full-time firefighter position from the staffing of fire and EMS units.

Instead, the employer may offer a job in another agency that the employee could manage with his or her permanent limitations. In this case, the same rules as a new hire apply. The employee must be able to perform the essential job functions.

## Pregnancy Issues

If a pregnant employee cannot perform the essential functions of her job, should the department provide an alternative light-duty job? In most cases the answer is yes, this is a good idea. However, there is no absolute requirement to provide a light-duty assignment. If the general practice is to provide other employees (especially males) who have temporary disabilities with light-duty positions, then it is advisable to do the same for a pregnant employee.

A pregnancy creates extra weight, changes in equilibrium, and loss of agility. Any of these situations can cause injuries and failure to be able to complete essential job functions. Generally, the department can order a pregnant employee to take a job performance test or be evaluated by the department's physician at the beginning of the third trimester and earlier if there are any medical complications or observations of substandard job performance. The department should adopt a written policy with input from the physician, the union, and the department's legal advisor.

Previously, employers have prevented pregnant workers from doing their jobs simply because of a concern for the well-being of the fetus. There is some evidence that elevated temperatures and the high concentration of carbon monoxide typically found in firefighting causes medical harm to the fetus. In EMS, the biggest risk to a pregnant woman comes from lifting patients (especially bariatrics), dealing with combative or assaultive patients, or working at potentially unsafe scenes, such as motor vehicle accidents. There is also concern for communicable disease exposure, especially with such a disease as rubella, which can harm a developing fetus. However, the Supreme Court has held that the well-being of the fetus is the concern only of the woman.

## Family and Medical Leave Act

The Family and Medical Leave Act (FMLA) of 1993 has affected many FES agencies. FES companies with minimum staffing levels commonly must hire additional personnel or provide overtime for leave impact. Studies have shown that up to 20% additional personnel must be hired to compensate for leave in traditional sick, annual, and IOJ leave provisions. To maintain four people on duty without overtime, the department needs to hire five firefighters.

The FMLA mandates that employers with more than 50 employees must offer employees up to 12 weeks of unpaid leave per year for family responsibilities, such as the birth of a child; care of a seriously ill child, spouse, or parent; or a serious illness of the employee. The employer must give workers their previous position back when they return to work. Therefore, during the absence, the FES agency may have to cover the vacant position with overtime. Unless the employee is on some type of paid leave, the cost is the difference between the salary of the absent employee and the overtime costs, which generally run about 50% more. Larger organizations may hire additional personnel to cover the vacancies.

The additional cost to cover the shift with overtime may actually total a lot less than the additional 50% discussed previously. A firefighter's normal pay includes a base salary and benefits. Benefits may run up to 40% or more of the salary. Therefore, it costs the difference between regular pay and overtime, or about 10–15% more for each shift. However, if the firefighter is on paid sick or annual leave, the costs are 150% more because of the overtime needed to fill the empty position. Clearly, this can be a substantial unexpected cost.

If an employee needs IOJ leave to recover from a worker's compensation injury or is granted sick or annual leave for family obligations, the employer should notify the employee that the leave of absence is counting against FMLA leave. If this notice is not given, the employee may be able to apply for an additional 12 unpaid weeks off after he or she is found fit for duty from their worker's compensation injury or when sick or annual leave runs out.

In addition, the employer should have a written policy that combines FMLA leave with other paid leave when appropriate. For example, if an employee is off on sick leave for a serious medical condition, the FMLA leave can run concurrently. It is important to have this type of policy to reduce the chance of employees abusing time-off potential.

## Drug and Alcohol Testing

The FES organization should have a written policy prohibiting members from using, consuming, or being impaired by (legal and illegal) drugs or alcohol. When the policy is first implemented, all existing members and any new applicants should sign a consent form for testing that may be implemented in the following circumstances: preemployment or premembership; at random; after an accident or injury (as soon as possible); and suspicious activity or behavior.

If a member refuses to take a drug or alcohol test, a copy of the signed consent form should be produced and the member reminded that rejecting the test will result in dismissal. The alcohol test could use the same pass-fail criteria used for determining an impaired driver under the state's motor vehicle statutes. Refusal to submit to a drug or alcohol test or a positive result should be considered legal grounds for immediate dismissal.

## Sexual Harassment

The administrator should be aware of two types of sexual harassment: quid pro quo and hostile work environment. Both of these expose an employer to substantial liability that can result in large monetary awards. Therefore, it is important for an employer to set up policies, training, and procedures to prevent and eliminate this conduct in the rare instances when it occurs.

These cases can be very complex, and courts have found that a consensual sexual relationship is not necessarily a welcome one. One key to whether sexual harassment has occurred revolves around the issue of whether or not the sexual advances were welcomed. For example, at work, a member becomes friendly with another employee, and eventually asks that person out on a date. At that time, there is no knowledge of whether the advances are welcome. If turned down for a date and the other person makes it clear that there is no romantic interest, any future request for a date may constitute sexual harassment. This is especially true if the person proposing the date is the supervisor of the other employee.

For example, quid pro quo sexual harassment occurs when a person in authority controls an employee's future in the organization. This type of sexual harassment may occur when a supervisor offers the employee a promotion, raise, or other benefit in exchange for a sexual favor; it may also occur when a supervisor threatens demotion, transfer, or termination. Whether the consequences for the employee being harassed are positive or negative is irrelevant.

Hostile environment refers to employees' conduct of a sexual nature that may be offensive or intimidating. The simplest example of a hostile environment is allowing male employees to post pictures of women either naked or in sexually implied poses. The telling of sexually explicit jokes is another good example of a hostile environment. The EEOC has outlined the following factors for determining if a hostile environment is present:

- Conduct was verbal, physical, or both
- Conduct was frequently repeated
- Conduct was hostile or patently offensive
- Alleged harasser was a coworker or supervisor
- Others joined in perpetrating the harassment
- Harassment was directed at more than one individual

The following guidelines are helpful in preventing and defending the organization in sexual harassment situations:

- Implement a written policy of antiharassment with examples of prohibited actions.
- Provide training for all members including a description of conduct that is illegal or unacceptable.
- Have all members sign a receipt that they have taken the training and understand the policy.
- Clearly post the policy in work areas.
- Identify the persons or office to be contacted.
- Encourage members to report any potential situations that may lead to a violation.
- Vigorously investigate and discipline any violations.

**Dating Policies** It can be very difficult to recognize whether such actions as sexual propositions, sex-based comments, or leering and staring are welcomed. Some organizations have created formal policies covering these potential relationships at work. An outright ban has been found to be neither possible nor preferable.

In an FES organization where employees are assigned to a shift crew, it may be preferable to have a policy that requires the transfer of one of the partners in a consensual relationship. This also solves the problem of a possible quid pro quo situation where a supervisor is one of the partners. In the typical FES station, employees spend long shifts together and sleep at the employer's facility. A different policy may need to be used for staff employees working in an office.

Even in a volunteer department, dating can be disruptive to the relationships of the other members if there is any hint of favoritism. A strong chief officer and an oversight board can be very effective at preventing any adverse consequences.

## Fair Labor Standards Act

The core requirements of the Fair Labor Standards Act (FLSA) are fairly simple. Once covered, employees working more than a designated number of hours in a pay period must be paid overtime. However, public employees have some special rules. First, the employee can accept compensatory time at a rate of 1.5 times instead of overtime. Also, firefighters' work period is defined as a 28-day cycle, and the average hours per week before overtime are a maximum of 53 (212 in 28 days). These regulations are meant to accommodate the common 24-hour on and 48-hour off shift that many firefighters work.

This act has caused controversy in two areas. First, the definition of firefighter contained the provision that a majority of on-duty time be spent in fire protection duties. Some fire departments providing EMS assign firefighters to staff these units, causing them to work more than 40 hours per week. This qualifies as overtime for all other FLSA-covered workers (other than police, who can work up to 171 hours in a 28-day cycle) (U.S. Department of Labor, 2012).

In the past, there have been a number of firefighters assigned to full-time EMS duties who have filed lawsuits and won their cases. These were very costly to the jurisdictions, which had to fund large back-pay awards. With more and more fire departments offering EMS, this issue became a major problem. In response to this conflict, Congress passed a new law defining a firefighter's job requirements as including EMS, fire prevention, and other duties where life, property, or the environment is at risk. This definition now allows FES departments to use the 53-hour workweek as long as these workers are dual-trained in fire and EMS.

Other issues surfaced when it was noted that overtime could be awarded if employees volunteer in the municipality where they work. For example, in a large county fire department, an employee may work at Station 1 and be a volunteer member of Station 15. Because of the overtime interpretation, paid employees have been instructed not to volunteer at any FES station in their jurisdiction. This restriction has caused some hardships in combination FES departments.

In an appeals court decision, *Benshoff v. City of Virginia Beach* (1999), the court dealt with the issue of Virginia Beach career firefighters volunteering at independent volunteer rescue squads in their jurisdiction. These stations are completely staffed by volunteers and govern themselves. They also do not firefight. The court decided that the FLSA did not grant overtime to career firefighters volunteering at these independent volunteer rescue squads. This decision holds only in the US Fourth

Circuit Court of Appeals: Maryland, Virginia, West Virginia, North Carolina, and South Carolina.

## The Financial Impact of Lawsuits

A study released in 1999 by Jury Verdict Research points out the growing legal risks employers face.

- From 1994 to 1997, punitive damages were awarded in more employment disputes and employer negligence cases than in any other type of lawsuit.
- Median jury awards between 1996 and 1997 increased roughly 286% from $64,750 to $250,000.
- Punitive damages were awarded, in addition to compensatory damages, in 34% of discrimination cases and 38% of sexual harassment cases.
- The average compensatory damages awarded to victims of discrimination or sexual harassment increased from $188,347 to $212,598.
- Plaintiffs are more likely to prevail in discrimination and harassment suits than are defendants. In 1996, 51% of plaintiffs were victorious; in 1997, plaintiffs prevailed in 58% of these types of cases, indicating an upward trend.
- Recoveries by former employees are also significant. Awards to plaintiffs in constructive discharges and wrongful terminations averaged $461,745.

## Recent Supreme Court Cases

**Sutton et al. v. United Air Lines, Inc. (1999)** This case concerned two pilots who were refused employment by United Airlines. Even though both pilots met the Federal Aviation Administration vision requirements for the position, they failed to meet United's more stringent requirements. The two pilots had uncorrected vision of 20/200 but had corrected vision identical to United's more stringent requirements. They sued United for discrimination under the ADA. Not only did the Court support United's stringent requirements, but the Court also did not find the pilots substantially limited in any major life activity. Therefore, they were not regarded as disabled under the ADA.

This ruling has application for those departments holding medical examinations to specific standards, such as the National Fire Protection Association (NFPA) Standard 1582, *Standard on Comprehensive Occupational Medical Program for Fire Departments*. This standard contains several items that a potential member may argue permit an accommodation under ADA. Provided the department's medical and physical requirements are based on job performance standards and are essential to the job, they can be used even if they discriminate.

**Kimel et al. v. Florida Board of Regents (2000)** In this case the Supreme Court found the Age Discrimination in Employment Act not enforceable in state and local governments. Therefore, unless a state law prohibits age discrimination (and some have an exemption for police and fire), an FES organization is allowed to have a mandatory minimum or maximum hiring and retirement age. For example, the federal government requires its firefighters to retire at 57 years old. Because of this requirement, and because their retirement plan requires 20 years of service, the maximum hiring age is 37. In another example, a metropolitan department had a maximum age of 29 for hiring and a mandatory retirement age of 55 years old. These types of age requirements were viewed as discriminatory in the past, but are now allowed in state and local governments.

In explaining its decision to repeal the Age Discrimination in Employment Act, the Court stated, "Old age does not define a discrete and insular minority because all persons, if they live out their normal life spans, will experience it" (Kimel et al. v. Florida Board of Requests, 2000, IV C).

If the department is using a mandatory retirement age, the cutoff should be justified using hard data and studies. For example, the 2009 Firefighter Fatalities report from the NFPA states, "The rate for firefighters aged 60 and over was three-and-a-half times the average. Firefighters aged 50 and over accounted for two-fifths of all firefighter deaths over the five-year period, although they represent only one-fifth of all career and volunteer firefighters in the U.S" (NFPA, 2010, p. 6). This evidence seems to support a mandatory retirement in the 55–60 age range. However, these statistics have not been normalized for criteria, such as medical and physical fitness levels. It is likely that a firefighter who is medically and physically fit would have a different death rate than the general firefighter population. Therefore, a yearly extension based on a comprehensive medical examination and physical fitness test could be allowed.

A comprehensive research paper authored by the Ontario Association of Fire Chiefs in 2009 looked at the ramifications of ending the mandatory retirement age of 65 in Canada. This issue has not troubled the US fire service because many retirement systems have generous pensions for firefighters at relatively young ages. However, in the future this is likely to change. The Canadian study concluded that physical fitness declines with age, but that exercising regularly can slow the process. Therefore, the Ontario Association of Fire Chiefs suggests that "prior to the age of 40 firefighters take a physical fitness test once every five years, between the ages of 40 and 50 once every two years and over 50 once a year" (Ontario Association of Fire Chiefs, 2009, p. 44). The argument for physical fitness tests can be furthered by an NFPA report that states that in 2009, 19 out of 35 firefighters who were victims of sudden cardiac events had prior existing conditions, such as hypertension or other heart problems (NFPA, 2010). These medical problems should have resulted in treatment or forced retirement before the death occurred.

## Insubordination

To improve morale and workplace harmony, employers may attempt to silence employees who regularly spread unsubstantiated rumors, vociferously challenge management, or publicly attack an organization's practices and policies. In this endeavor, the administration should be careful not to violate a person's First Amendment rights (freedom of speech).

The government has a legitimate interest in promoting the trust of the public in the services it provides. Therefore, unsubstantiated public statements can be handled as insubordination. Also, any high-ranking officer can be ordered not to communicate with the public.

The preferable way to handle this situation is to assign a public information officer who normally gives all official statements to the media. Then, by written order, all members should be directed to refer any inquiries to the public information officer. This prevents confusion by the public when an official statement is released.

## Silencing Complaints without Violating the Law

First and foremost, psychologists recommend that the best method of reducing worker complaints is to create a safe, respectful, and productive workplace. The administration should listen to and take action, when justified, regarding the concerns and views of members. A formal process for employees to submit, anonymously if preferred, comments and constructive criticism should be provided. A post office box and an e-mail account should be used to receive this feedback. In addition, some departments use a telephone number connected to a voice message recorder.

It is also important that these concerns be acknowledged and attempts be made to correct unsafe situations. Some type of feedback should be given to employees, such as through an employee newsletter indicating the department's position and any attempts to solve the situation. If the members see attempts to solve the problem, they are more than likely to support the department's efforts.

For those complaints that are completely unfounded, the chief officer may want to follow this five-step procedure:

1. Remain objective, even in the realization that some members are chronic complainers and there is always a group that is opposed to any changes.
2. Focus on conduct when the complaints are disruptive in nature.
3. Determine if the member needs or could use some professional mental health services for truly disruptive behavior. Privately suggest the member contact the employee assistance program, because it can be helpful for his or her dissatisfaction.
4. Evaluate observations and supporting evidence to see if the behavior calls for disciplinary action.
5. Select an appropriate level of discipline.

## Public Sector Discipline

There is general perception that after a public sector employee is hired and finishes probation, it is impossible to terminate the person. This is not true. Public sector employees have "property interests" in their jobs requiring the employer to use due process in disciplinary procedures to separate an employee. Property interest is a legal term indicating that rights to a job are similar to and as strong as the rights the owner of a piece of real estate has. Before an adverse action can be applied, due process should be followed. The administration should follow these general guidelines:

1. Provide a written notice of charges, preferably a three-step process starting with a verbal reprimand unless the charge is of great consequence to the organization.
2. Include a complete explanation of the evidence and reasoning on which the charge is based.
3. Provide a meaningful opportunity for the employee to be heard by an impartial decision maker, preferably with a representative of the employee present, such as a union shop steward.

Additionally, consider using the following checklist to help guide the administration through the separation or serious disciplinary action:

- Is there sufficient basis for discipline?
- Is there appropriate documentation?
- Is the inappropriate behavior related to conduct, such as insubordination or tardiness, or performance, such as failure to perform a job requirement?
- Was the employee given all of his or her rights during the investigation?
- What is the employee's previous disciplinary record?
- Was the violation serious, and were there any mitigating circumstances?
- Have other employees been disciplined for the same actions; if so, what were their consequences?
- Were all the appropriate steps in the disciplinary process followed?
- Did all the notices clearly detail the observed actions and the violation of the law or policy?
- Did the employee have adequate time, representation, and opportunity to respond to the charges?
- Was the agency's attorney consulted before proceeding with the notice of disciplinary action?

## Fair, Reasonable, and Evenly Enforced Discipline

The first step to ensure that discipline is fair is to make sure written policies and standard operating procedures (SOPs) include a listing of potential inappropriate behaviors and performance standards. In some cases, a general statement and explanation may be sufficient, such as in the case of insubordination. In others, as in the situation of illegal drug and alcohol use while on duty, the standards should be specific. Generally, it is a good idea for the administration to keep these policies and SOPs to as few as possible. In the emergency services arena, it is always better to keep rules and regulations simple.

An ongoing training program to inform all members of inappropriate behaviors and minimum performance standards is mandatory. Members should be given adequate notice and time for compliance, or members may win appeals of the adverse action in many cases, allowing another chance and additional time to comply.

The department's disciplinary actions must also show a consistent and nondiscriminatory record of enforcement. For example, a department has an SOP requiring drivers responding to emergency incidents to come to a complete stop at all stop signs and red traffic lights. During the 3 years after the SOP was issued, the department made no attempt at enforcement. Then, a driver was charged with failure to follow the department's SOP (this is the first written enforcement record), and a penalty of 1 month's pay was proposed. Between the severity of the adverse action and the lack of any prior history of enforcement, the department will have difficulty sustaining this action if the employee appeals to an arbitrator or the courts. It is very important to have an active enforcement policy.

Also, a department should have a policy of due process that outlines a disciplinary procedure starting with a written warning and, with continued violations, resulting in substantial consequences. Appropriate notices of hearings and the ability to acquire legal or union representation need to be fair and realistic.

## Probationary Period

A newly hired FES employee should be put through a vigorous probationary program. Many potential employees present extensive qualifications, education, and experience to gain a job offer. This probationary period is a chance to gauge the effectiveness of previous education and experience.

An organization should have a structured evaluation in place containing all of the essential job performance criteria and standards. The assessments should be standardized at prescribed periods, including a formal written evaluation completed by the supervisor. All supervisors should be trained in the use of the evaluation systems and encourage specific comments and observations in the report.

The probationary period should emphasize feedback to the employee aiding the individual in becoming a better employee, but also designed to weed out those who are not qualified. A formal testing procedure for the basic essential skills and knowledge that the employee should have mastered in recruit training should be used.

Supervisors should be reminded of the grave consequences of providing charitable probation evaluations. Retaining an FES member who is not qualified has double-edged consequences. The public does not get the best service and the member is not able to operate safely, creating a hazard to himself or herself and the team.

A probationary member need not be told the reason for termination; the administration can simply state that the new member has not satisfactorily completed the probationary period. In some cases, it is better not to tell the probationary member the reason for termination, especially in writing. Although the employee has no winnable legal appeal, a written reason may unintentionally encourage a lawsuit. Anyone can instigate a lawsuit, even if there is no chance of winning. Probationary firefighters or EMS personnel should be aware that they have not satisfactorily completed the basic fundamental skills or behavior expectations.

## Terminations

This section discusses appropriate methods for conducting terminations, lowering exposure to legal challenges. Even with adequate evidence and support for a performance or disciplinary termination, if handled inappropriately, it can lead to legal disputes that can be costly and disruptive to the FES agency.

Members may be motivated to sue simply because they feel they were not treated fairly or with respect. The administration should approach the separation as a situation where there was not a good fit between the member and the job, rather than labeling the member as incompetent, lazy, or another derogatory characterization. This approach ensures terminated members feel better about themselves. Not everyone has the aptitude or mental and physical fitness to perform the demanding FES duties. They are not bad people; they are just in the wrong occupation.

Everyone (except the probationary member) has a right to respond to their accusers and have competent representation in the administrative process leading to termination. Even when it may seem like a clear case, the agency should follow all the policies that provide due process for the member.

Reductions in force (RIF) is another situation that calls for employee termination. It is absolutely imperative to have an established RIF policy and to make all members aware of the selections process. RIF decisions that may be discriminatory, such as selecting all members older than 50 years old, should be avoided. From an accountant's perception, terminating one chief officer may create a cost savings equivalent in salary and benefits to terminating two or three firefighters or EMTs. A task force of members should provide recommendations and adopt a written policy.

In the situation of a voluntary termination (the employee quits), the supervisor should gather as much information about the member's motivation as possible through an exit survey. The objective is to uncover any issues that could later support a legal claim of discrimination by the member. The supervisor should attempt to find out if any underlying reasons for the resignation had not been previously stated by the member, and whether these complaints should have been addressed by the agency, such as would be appropriate for discrimination. The following questions should be asked:

- Why is the member leaving?
- Would he or she consider returning to the organization?
- What plans does the member have in the near future?
- Does he or she believe that the FES should consider changes? If so, what changes and why?

## Constructive Discharge

Constructive discharge is a legal term indicating that the member was somehow convinced to resign because of intolerable working conditions. Some common example scenarios are:

- Hostile work environment (e.g., sexual harassment)
- Unsafe working conditions (check Occupational Safety and Health Administration and NFPA safety standards for compliance)
- Insufficient information about alternatives in disciplinary or performance actions
- Not given the option of continuing employment in lieu of an early retirement (sometimes used with RIF to reduce the payroll)

## State and Local Hiring Laws

Many states have statutes referencing the hiring and, in some cases, operations of FES. For example, in the state of Florida, an applicant must be a state-certified firefighter to become an employee of a municipality. The state specifies the curriculum using NFPA Firefighter I and II and a minimum of 360 hours of training. In actual practice, because the State Board of Education must approve the curriculum, these training programs are a minimum of 480 hours. These programs are delivered throughout the state by community colleges, vocational technology centers, and a few larger FES organizations. The state certifies these training programs and their facilities. In addition, each graduate of these firefighter training programs must take a state-administered skills test and a written examination for final certification. This process provides a pool of certified firefighters who become applicants for firefighter positions.

A minimum hiring requirement for EMTs and paramedics is possession of a basic license or certification and a valid cardiopulmonary resuscitation card. Some states require a current National Registry of EMTs certification, but other states use the National Registry of EMTs only as a first-time certification assessment. Hiring requirements for paramedics may also include locally required certifications, such as advanced cardiac life support, pediatrics, and trauma training certificates. Candidates must also be tested for competency on optional scope knowledge and skills practices required in their local area. In some counties or regions, the local EMS agency requires an orientation on county or jurisdictional practices.

Many merit or personnel laws regulate employment at the local level. In addition, union contracts are enforceable and, in most cases, have stipulations affecting salary and benefits. These are too numerous to discuss here, but are all legally enforceable.

Some states are known as right-to-work states because they do not provide for the organization of public unions. In these states, it is common to not have a union contract or formal discussion with unions or other employee organizations.

It is a good idea to meet with employee representatives on a regular basis and discuss such topics as salary, benefits, working conditions, and safety issues. This practice generally raises the morale of members.

A job analysis is a common method of defining job requirements. A good example of the outcome of a job analysis can be found in FIGURE 6-3. In fire and EMS, very detailed training and testing criteria are readily available. There are three distinct parts to a job analysis process: (1) hiring criteria, (2) recruit training and testing, and (3) incumbent evaluation. All three should be consistent with the job analysis.

There have been very few legal challenges to new employee requirements put in place by most FES organizations. For example, a very detailed and comprehensive basic skills test is generally required of a new recruit. However, incumbents are not tested periodically to ensure that they can still perform the basic skills that a recruit must complete. Because a new recruit is on probation, he or she does not have the same rights of due process that an incumbent possesses. There may come a time when a future court decision requires equal treatment. Although it seems like a fair and ethical policy to treat all firefighters and EMS personnel equally, it is common to have different requirements for incumbents.

Practically speaking, it is a good idea to upgrade new employee and recruit requirements before upgrading incumbents, and then phase in requirements for incumbents. They need time to relearn some skills.

## Validation

Many HR professionals recommend using incumbents, supervisors, and personnel analysts as sources to develop a job analysis. A good strategy includes using each of these sources in the process; however, one thing to keep in mind is that incumbents may not be able to perform all job functions, thereby skewing the results.

The effort to create a job analysis should start with review of a standard, such as NFPA 1001, *Standard for Firefighter Professional Qualifications*. The administration should also do some research and contact several FES agencies to get copies of what peers are using, asking if they have had any problems with their job descriptions and whether they are happy with the abilities of their new recruits.

> **FIRE AND EMS DEPARTMENT**
>
> **I.** Position Title: Volunteer or Career Firefighter
>
> **II.** Summary Statement of Overall Purpose/Goal of Position:
> Under the direction of the Fire Chief and general supervision of the station officer; respond to emergencies involving fire, medical, and environmental concerns. Responsible for the care, operation, and condition of fire and rescue apparatus and the station.
>
> **III.** Duties:
> - Respond promptly and efficiently to fire, rescue, hazardous materials, and medical alarms.
> - Operate at emergency fire and rescue incidents efficiently and safely.
> - Maintain proficiency in fire and emergency medical knowledge and skills.
> - Conduct tours, lectures, and video presentations. Display fire apparatus. Participate in public demonstrations at local school programs and various civic and city functions.
> - Inspect business, public, and private properties for hazards and code violations.
> - Drive and operate department apparatus in emergency and nonemergency situations.
> - Maintain equipment on apparatus to include daily, monthly, and annual testing. Test and rotate hose.
> - Complete daily, monthly, and annual reports on the testing of fire and medical equipment.
> - Complete 8 hours of department training each month.
>
> **IV.** Qualifications:
> At time of application screening:
> - Must be at least 18 years old
> - Must live within twenty (20) miles of the municipality
> - Must not have been convicted of a crime for which the applicant could have been punished by imprisonment
> - Must not have been convicted of an offense involving dishonesty; unlawful sexual conduct; physical violence; or the unlawful use, sale, or possession of a controlled substance
>
> Before being accepted:
> - Complete entry-level written test
> - Pass physical agility test
> - Complete and pass drug test
> - Successful driver's license check
> - Check for convictions and outstanding warrants
> - Enter initial training program
>
> Education: High school diploma or GED.
>
> Medical and Physical Fitness: Must meet department's medical and physical agility requirements.
>
> **V.** Working Conditions
> This job entails regular exposure to dangerous situations under disagreeable conditions, including smoke, heights, fire, fumes, heat, cold, emergency driving; frequent exposure to dangerous situations with medical emergencies; must meet the Fire Department's physical agility standards; must be able to wear and work in fire department breathing apparatus.

**FIGURE 6-3** Example of firefighter job requirements.

After this basic research is complete, the information should be consolidated into one comprehensive job description. This draft should be circulated to the incumbents and supervisors in the department. In addition, input should be actively solicited from labor organizations and the personnel office.

Because of the possibility that incumbents and their supervisors may not encounter critical essential job functions on a regular basis, they may not identify some of the job functions as being essential. Advise those questioned that a critical job function may have to be performed only once in a career. It becomes essential not because of daily use or need, but because it is how the public and the national standards expect a firefighter or EMS provider to be able to perform.

To use an example from the ADA regulation, the expectation that a lifeguard who works at a pool must be able to bring a heavy person back to safety from deep-water depths is described in detail in the job analysis. This job function is one of the most essential functions because of the life-saving potential, which is the ultimate purpose of a lifeguard. Lifeguards may never face this situation, but they must still meet this essential job requirement.

Incumbents can also be valuable when planning an implementation strategy. Testing incumbents provides a realization of where the incumbent's skill levels are to successfully complete the tests. An implementation plan can be proposed that allows a phased-in compliance for existing members.

## Job Classification

Most members in any FES organization are at the firefighter or EMT level. The normal perception is that the officers and chiefs are more important than firefighters, but this perception is not correct. The firefighters and EMTs are the people who get the job done at the emergency scene, and the organization should show its respect and acknowledgment of these special qualities, especially as each member gains additional training and experience. The department should continually emphasize that the FES profession is a noble pursuit requiring high levels of skill, courage, and physical ability.

Some FES organizations have an annual awards ceremony that singles out individuals for heroic achievements. Most members are capable of heroic acts, but only a few firefighters or EMTs end up at the right place at the right time. Therefore, rewarding all members for their potential to perform at these heroic levels of service may be warranted.

To reward competent members, the FES organization could use a three-tier promotion system for firefighters: (1) recruit firefighter, (2) firefighter, and (3) senior firefighter. Each new level would have its own set of training requirements, testing, and experience on the job. For example, to gain the firefighter level, the member would have to successfully complete recruit training (NFPA Firefighter I and II) and a 1-year probation program. The lead firefighter would require an additional 3 years of experience, retaking of all of the basic skills and knowledge tests, and completion of an apprentice program. Each new level would be a promotion with an appropriate increase in salary.

Similarly, the National EMS Education and Practice Blueprint has established four levels of EMS providers: (1) emergency responder, (2) EMT-basic, (3) EMT-intermediate, and (4) EMT-paramedic.

### Emergency Medical Responder

Emergency medical responders are often police officers who are trained to carry and use automated external defibrillators. They perform lifesaving interventions while awaiting additional EMS response, then transfer responsibility, ensuring a progressive increase in the level of assessment and care. They can assist EMTs and paramedics in the field and during transport but are subject to medical oversight. This job requires 48–60 hours of training.

### Emergency Medical Technician

EMTs provide basic emergency medical care interventions for patients who access the emergency medical system. In rural jurisdictions, EMTs may provide the highest level of prehospital care. The EMT is the minimum licensure level for personnel transporting patients in ambulances. This job requires 150–190 hours of training.

### Advanced EMT

Advanced EMTs provide limited advanced emergency medical care and transportation for patients who access the emergency medical system. The advanced EMT scope of practice includes emergency medical responders and EMT skills and competencies and limited advanced skills, including some pharmacologic interventions. Advanced EMTs may function as part of a tiered response system but are not generally used in systems that have paramedic level care. This job requires 150–190 hours of training beyond EMT.

### Paramedic

The paramedic's primary focus is to provide advanced emergency medical care for patients who access the emergency medical system. The paramedic's scope of practice includes basic and advanced skills with equipment typically found on an ambulance, and includes invasive and pharmacologic interventions. In many communities, paramedics provide a large portion of prehospital care and often represent the highest level of prehospital care. This job requires 1,000–1,200 hours of training.

## Recruitment

In most cases, when a jurisdiction advertises a vacancy in the FES organization, many applicants respond. However, these organizations often have a difficult time finding diversity among applicants, and women and minorities are often underrepresented (Edwards, 2010). There is now an effort in many recruitment processes to reach out to women and minorities to encourage them to apply for jobs in FES organizations.

In the past, recruitment was simply a matter of choosing the best of the best. Now the question is: Can highly qualified applicants be chosen to reflect the proper diversity mix? In any case, this should be one of the objectives of the department's recruitment efforts.

Selection of new members is one of the most important decisions for the quality and effectiveness of the agency. Hiring decisions can and often do have long-range impacts regarding the competency and effectiveness of the department. People who do not meet the minimum job requirements should not be hired or accepted as members.

## The Selection Process

Some of the legal ramifications of selecting new members were discussed previously in this chapter. Consideration should be given to cutoff scores and pass-fail requirements for cognitive abilities (speaking, reading, math, listening, and writing skills); medical tests; drug and alcohol tests; criminal background checks; physical fitness; and cigarette smoking. For example, in some cases, firefighter job applicants must sign an affidavit that they have not used tobacco products for 1 year before being hired. Many departments follow this up with requiring firefighters to sign a preemployment agreement to not use tobacco at any time during their employment. For EMS, this is usually a recommendation rather than a requirement, although company policies may vary on this issue.

Setting acceptable minimum scores high enough so that the department selects those who will be competent members in their ability to learn and perform the job is a good idea. Selecting marginally functioning members causes unacceptable performance in the future. Remember, these are selections for new FES members for a minimum 20-year career.

Another issue is the option of hiring either trained or untrained certified firefighters. Each option has it merits. Selecting trained firefighters reduces training costs to a brief orientation. This is a substantial cost savings because a new firefighter recruit might be in training for 12–16 weeks. Also, many smaller departments cannot hire enough new firefighters at one time to justify a rookie school. In addition, it is common for larger departments to find that their applicants are trained firefighters from smaller departments looking for better salaries and benefits, or certified volunteers trying to obtain paid positions (Edwards, 2010).

Some prospective firefighters may apply to a certified firefighter training center to become a student. The student pays his or her own tuition and other costs for the typically 12- to 16-week training programs. With successful completion of the Firefighter I and II course, applicants are eligible to take the state firefighter examination. After they successfully pass the practical and written examinations, they are state certified and can apply for a paid position in the state.

Some larger jurisdictions hire personnel and pay them to attend training programs. These departments prefer to train their own recruits so they can impart their organizational culture and the specifics of their department to the recruit firefighters.

In addition, some departments also prefer paramedic certification. Many potential career firefighters/medics must have both certifications to be selected for an opening. This trend is becoming more common throughout the nation.

Private EMS is simpler to get into. Although private schools may offer shorter EMT classes, many personnel attend semester-long courses offered at community colleges. With the naturally high rate of attrition experienced by private ambulance companies, they will hire right-out-of-class EMTs. Many companies structure their training program with the understanding that this may be the employee's first job.

Additionally, some ambulance providers have their own paramedic schools. It is an advantage to an EMT to seek out these companies, because they may provide employee discounts, or in some cases a scholarship to the company's program. There is also a likelihood that he or she will be placed in the company as a paramedic after successfully passing the course. This arrangement is mutually beneficial. The EMT gets a discounted rate on the training and the hard-to-find internship, and then a job. The company retains a well-trained (and grateful) employee who already knows the policies, procedures, employees, and the response area. In some cases, as part of the paramedic school agreement, the new paramedic may be bound to the company for a period of time.

The quality of training programs varies from one company to another, but they generally include an orientation, field training time, emergency vehicle operator training, and a probationary period. After the employee passes initial training, the company may offer further training opportunities or continuing education classes.

# Unions

## Public Sector Unions

Unions are groups of workers that have formed an association to discuss issues with management. Many volunteer departments act like unions in that one of their priorities is to take care of the volunteer members. Unions protect jobs and attempt to improve the salary and benefits received from local governments. They also promote companionship and respect for fellow members. This creates a strong loyalty to the union or volunteer organization that sometimes can be in conflict with the main mission or goal of the FES organization—namely, quality service to the public.

The relationship between unions (and volunteer organizations) and the representatives of the public (e.g., chief officers, mayors, county executive) is a system of checks and balances—the administration is watching out for the best service to the public at a reasonable cost, and the unions are looking out for the rights of their members. However, this is an oversimplification; there are many situations in which unions and volunteer groups have supported changes that were primarily for the benefit of the public.

Unions come in all sizes and types. Some are extremely strong and influential; others are mere social clubs. In right-to-work states, unions tend to be weak and more advisory in their actions. As an administrator, never underestimate the power of the local union or volunteer organization. You do not have to give in to all of their demands simply because they have influence; however, the chief officer must respect their opinions and give them a chance to voice their concerns and beliefs.

## Private Sector Unions

The National Labor Relations Act was enacted by Congress in 1935 to protect the rights of employees and employers, to encourage collective bargaining, and to curtail private sector labor and management practices that could harm the welfare of workers. When employers deny the rights of employees to organize and refuse to accept the procedure of collective bargaining, this can lead to strikes. Strikes are a critical problem for a company when it may risk response time compliance.

Some private sector ambulance companies have unions. There are many varied opinions concerning the pros and cons of union affiliation. Whether a union is effective depends on the union itself and the agreements made with the company.

## Strikes and Job Actions

In most labor unions, the strike is the ultimate tool that can be used to gain agreement for an improved salary or benefit. Without this negotiating tool, public safety unions are at a great disadvantage in discussions with the dominant governments.

Some states allow binding arbitration for public safety employees. This permits a fair and equal status for labor and management. However, in right-to-work states, it is rare for there to be binding arbitration. In these situations, it is common for there not to be a written contract or for negotiations to continue endlessly. This is very bad for the morale of members, which may have a negative impact on service to the public.

## Bargaining Units

It is very common for there to be two bargaining units—one for firefighters and one for officers. In many cases, both units are represented by the same union president or are in some way combined into the same organization. Representing two classes that may be in opposition can cause conflict of interest for the union.

For example, if both the worker and supervisor are members of the same labor organization, they may become friends in the social atmosphere that surrounds these organizations. In some departments, only the fire chief is not a member of the union. When differences surface, it can become a one-person administration versus everyone else. This can isolate administration and make it very difficult to negotiate solutions to problems.

It can be very difficult for supervisory or management personnel to separate their loyalty to the firefighters from the duties they must perform as a supervisor. In the day-to-day running of the department, and in the negotiation process, supervisory personnel may be unclear on where their loyalty should stand. This situation is exasperated by the nature of FES work. Personnel perform their duties in small teams that become very close professionally and personally. Firefighters often become very close friends.

Often, one of these friends may be promoted and become the others' supervisor. The problem with this may not revolve around being a part of the same bargaining unit but rather from being a part of the same social organization (for paid firefighters and volunteers). Labor and supervisors have several needs that are the same, including pay and working conditions. It could be argued by a union that this is clearly a case of equal goals and, therefore, supervisors should be part of the union.

One way to answer this predicament is to set up a policy that awards the same working conditions to both groups but insists on separate bargaining units. For example, when bargaining for pay raises, work hours, and other monetary benefits, the city could bargain with both groups with the understanding that the final outcome would be equal for both groups. For nonmonetary or management prerogative items, separate discussions with each union president would suffice.

Finally, personnel who have middle management positions (above the company officer) should not be in any bargaining unit. These individuals should be a part of administration and be taken care of accordingly.

## Local Government Representatives

When formal negotiations start, the management is represented by several people from the government along with high-ranking officials from the FES agency. These government officials may have a narrow point of view or goal, such as a target budget figure. They do not have the knowledge or experience to foresee any possible adverse effects on the emergency services that could result from a new union contract that saves the city money. This same lack of knowledge is evident if using arbitration or fact-finding services.

An indoctrination program should be provided to inform these officials about the actual functioning aspects of FES. For example, they may not understand why there is a need for four firefighters per company when 61% of the department's calls are EMS calls that require only two people (a cardiac incident may require five or six). Emergency services are very technical and require a high level of skills and knowledge. Only when members have been providing emergency services for many years do they get an intuitive feel for what constitutes quality service, including the safety aspects of the profession.

## Grievances

This is an important aspect of any labor agreement. This process allows for the fair resolution of misunderstandings of the contract or an appeal process for disciplinary actions. The union is obligated to represent the member to guarantee that fairness was used in the determination of any adverse action proposed or taken.

This process also keeps many issues from being litigated in the courts. Most courts will not overturn a grievance procedure, especially if the process ends at binding arbitration. Therefore, when complying with the provisions in the labor contract for settling grievances, the decision by an arbitrator is most often final.

When processing a grievance, make sure the administration has all the facts and is well prepared for the hearing. If these facts indicate that the administration made an oversight or error, this should be corrected immediately. If not, include a full review and representation by the city or county attorney. The results of an arbitrated grievance can set a precedent that is very difficult to change in the future. Do not take these cases lightly. They can have long-term effects on the department's operations and administration.

## Progressive Labor Relations

It is beginning to be more common to see cordial relationships between labor and management. In the past, each looked at the other as the enemy and would do anything to win in negotiations and grievance proceedings. Today, although each has to represent its own constituents, where there is a mutual interest, each is able to help out the other.

A lot of the previous mistrust was the result of confrontational behavior that was common in the private sector during the Industrial Revolution. Even violence by management and labor was not uncommon. As a result of federal laws and regulations, and the revelation that competing in the new global economy requires well-motivated workers, labor and management are finding ways to reasonably settle their differences without hurting the competitiveness of the industry or business.

This type of new cooperation has also resulted in improved communications. The chief officer should schedule meetings with union officials on a regular basis. In addition, visits to work locations to talk and listen to all members are very important.

There have been cases where the union or volunteer leaders have lost touch with their own people. In these cases, internal communications can break down. Labor and volunteer leaders may be using old paradigms or their own agendas as a basis for representing the membership. Although this is not a good situation, it is common.

## Getting to Yes

The book *Getting to Yes: Negotiating Agreement Without Giving In* by Roger Fisher and William Ury outlines a successful type of negotiation process that attempts to separate facts and issues from people and their emotions. "Principled negotiation methods focus on basic interests, mutually satisfying options, and fair standards that typically result in a wise and acceptable agreement" (Fisher & Ury, 1997, p. 128). The techniques described in the book can also be used to problem solve differences between spouses, parents and their children, and many other interpersonal relationships.

One great method to use in negotiations is to brainstorm for possible solutions. The ground rules should allow any suggestion to be recorded without judgment. The group then needs to go back to each idea and analyze its cost, complexity, and reasonableness. After a set of realistic possible solutions has been agreed on, the separate parties can then concentrate on them.

## Motivation

In every organization there is the potential for members who can pull the entire organization toward mediocrity if the administration fails to deal with the situation. Motivation is an important concept that can be used to create a satisfied member who will be a top performer for the organization.

To simplify this concept, the following two areas, when accomplished, seem to produce a motivated member: benefits (hygiene factors) and pride (motivators). Examples of benefits include fair salary and work hours, health insurance, workers' compensation, and length-of-service awards (volunteers) or pensions. Examples of pride factors include modern and well-maintained fire and rescue equipment; professional qualifications for membership; a rigorous training program; and, in general, anything that leads a member to feel good about his or her contribution to the FES organization. These factors can be reinforced by recognition from the organization.

Although relative, pride factors tend to be more important to the individual than benefits. For example, volunteers will serve without any pay if properly motivated. Some would argue that the additional time required for a volunteer to complete a firefighter comprehensive training program would be a disincentive to retaining members. However, in an article titled "Why Do Volunteers Volunteer?" Executive Director of Bainbridge Island, Washington, Fire Department, Neil Good, states, "Volunteers are attracted to the theoretical and science of firefighting and medical. If the department has an active and rigorous training program, they will probably have a bunch of motivated volunteers" (International Association of Fire Chiefs, 1995, p. 2).

A rigorous training program not only creates loyal and motivated members, but it may also lead to fewer disciplinary problems. Members who have a strong allegiance to the career or volunteer organization feel good about their ability to provide professional emergency service. Never underestimate a person's pride—it is a very strong human emotion.

## Retention Within Volunteer Departments

In past years, men were free to commit many hours of their time to the local volunteer fire company **FIGURE 6-4**.

**FIGURE 6-4** Volunteer firefighters.
Courtesy of Michael Rieger/FEMA

However, more recently, economic pressures have reduced the availability of members. This is especially true for older members who find themselves with new demands on their time as they start families. Furthermore, changing social values have influenced many men to take a more active part in the raising of children. Between the increases in demand and newer values, men are no longer staying as active in FES organizations after taking on the responsibility of a family. Some of this downward trend has been offset, however, by an increasing number of women volunteering.

For additional insight, the US Fire Administration report "Recruitment and Retention of Volunteer Fire Fighters" discusses the core problems of recruitment and retention in the volunteer fire service, and provides suggestions for solutions to these problems from volunteer fire chiefs and firefighters from across the United States. This free publication can be ordered from the US Fire Administration Web site (http://www. usfa.fema.gov).

People who volunteer have the basic need for recognition, belonging, and the satisfaction that comes from helping people and serving the community. Retaining volunteers is dependent on acknowledging these needs. Some helpful suggestions follow:

- Cut unnecessary bureaucracy in terms of application and retention.
- Give volunteers clear expectations and guidelines.
- Provide flexible schedules to accommodate complex work and family schedules.
- Use non-FES people for administrative tasks.
- Create training programs to defray the expenses of certification and relicensure expenses.
- Consider grant funding for obtaining and replacing equipment.
- Investigate community groups that may donate or fund equipment.
- Encourage all volunteers to be recruiters and take part in training teammates.
- Continually show all members they are valued.

One example of an exceptionally well-run volunteer organization is the Bethesda–Chevy Chase Rescue Squad (BCCRS) in Bethesda, Maryland. The BCCRS works well as a volunteer organization for a number of reasons. First, the organization requires that each member complete a county physical examination, with promotions dependent on members meeting the Certification Standards for Training, Experience and Credentialing Regulation (G. Giebel, 2012). In addition, each member is assigned duty 1 night a week. On this duty night, the volunteer must show up in uniform and perform the following tasks: check out the apparatus, clean the station, and attend a crew training session. By having assigned duty nights, the volunteers recognize that the organization needs them.

Furthermore, because of a high call volume of over 10,000 calls per year, several ambulances are staffed each night. During the daytime, the BCCRS staffs a medic unit, a basic life support unit, and a heavy rescue squad. The service provided by these volunteers is not only respected by other FES organizations, including the five closest fire stations that have only career staff, but the customer also receives an immediate response to a call for help.

Although these qualifications and requirements may seem like a lot to ask of a volunteer, the BCCRS awards "Life Membership" to volunteers who complete 10 years of service. All of these motivating factors contribute to the retention of the volunteers of this organization.

## Retention Within Combination Departments

According to the NFPA, combination departments make up 24% of all departments and protect 33% of the US population. These departments come in all sizes and, therefore, generalizations fail to describe many of them. NFPA attempts to separate combination departments into categories of "Mostly Volunteer" or "Mostly Career."

Many of these departments report conflicts between volunteer and career firefighters, the origin of which often stems from issues of group loyalty and job security. For example, volunteer members may be concerned that if more career members are hired, fewer volunteers will be needed. It is because of this reasoning that many volunteer departments wait long periods of time without adequate volunteer staffing before venturing into the combination solution.

The first career member of a combination department is generally the driver, works weekdays, and is expected to maintain the station and apparatus. This career member is not able to advance in his or her profession to an officer rank (reserved for volunteers only) but is relegated to remain a firefighter. There are, however, many firefighters who are satisfied with these duties, and the volunteer department should be very careful to hire such an individual.

Still, there may be times when the combination department needs a full career crew, such as for a weekday shift. For example, in the past, the New York City Fire Department has staffed some ladder companies for the high incident hours of 3:00 PM to 1:00 AM. This arrangement is also commonly seen for EMS units. At this point, all members should be required to meet the same training, medical, and physical fitness standards. If the career officer must be certified as a NFPA Fire Officer I, then this same requirement should pertain to the volunteer officer.

One organizational technique is to keep the two factions separated. For example, the BCCRS volunteer organization works flawlessly with several all-career fire stations in its area. Another example can be found in Baltimore County's Fire Department. This department is made up of 25 career stations and 33 volunteer stations. At fires and other emergency scenes, volunteers work side by side with career personnel under a single chain of command and set of operating procedures. This department protects over 800,000 citizens in 610 square miles. Interviewed fire officials report smooth operations without major conflicts. In addition, the volunteer departments are very active and healthy organizations.

Another important issue related to combination departments revolves around pride. Companies can be represented through the uniforms they wear and the markings seen on their fire apparatus. Even when away from the station and off duty, members often wear T-shirts proclaiming their company. This holds true for the career firefighter and is even commonly seen at the station level. Many career stations are allowed to create a logo for their company, which increases the pride and morale of the members.

Bearing in mind these issues of pride and loyalty, the following guidelines can be very helpful in reducing personnel conflicts in combination departments:

1. Maintain equal professional standards of training, education, and physical fitness among members.
2. Maintain stations and apparatus for each group. If that is not realistic, organize members in like groups with their own identification. When either group is staffing the apparatus, it should be clearly marked as so. This can be done by permanently marking assigned apparatus or using temporary removable signs.
3. Remember, Pride = Loyalty = Good Morale = Retention.

## CHAPTER ACTIVITY #1: New Officer

A newly promoted officer is assigned to a very busy engine company and on the first day, is greeted by an energetic crew of three firefighters who seem pleased with their new officer. In reviewing personnel files and contacting the previous officer, the new officer has acquired some background information on the crew and is assured that they are aggressive firefighters who know how to put a fire out.

Before the morning coffee is finished, the company is dispatched to a reported fire in a single-family dwelling. The officer did not get a chance to explain any expectations or preferred procedures for emergency incidents and assumed that the crew would follow departmental SOPs. When the company arrives, it is clear that this is a working fire. Because it is a daytime incident, the officer is relatively confident that the dwelling is unoccupied.

Before the officer can make a brief size-up, the crew goes into action. Two firefighters pull the 150-ft, 1.75-in hose line and extend it to the side door (the fire can be seen coming out the windows near this door). Before the officer can join the two firefighters extending the hose line, they have forced the door and entered the house.

When the officer catches up, one of the firefighters has gone off on his own to search for victims. The fire was confined to a room and its contents and is extinguished using a few gallons of water. After the fire, the two firefighters discuss with pride their ability to get in fast and put the fire out before anyone else can arrive on the scene.

### Discussion Questions

1. Describe any potential disciplinary problems in reference to safety or SOPs that would be apparent in your department.

2. Discuss your own personal thoughts about the situation.

3. In a similar situation, how would you instruct your crew in their actions during emergency operations? Be specific.

## CHAPTER ACTIVITY #2: Compliance with SOPs

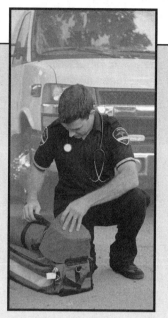

Few things in EMS are as important as paperwork. The patient care report (PCR) is not only a part of the patient's medical record, but provides information required to bill the patient or his or her insurance. A record of every patient encounter is required by law, and certain data points are mandatory whether the PCR is paper based or electronic.

In the past 6 months, an EMS agency that was using a paper-based PCR made the switch to an electronic format. The supervisor of the agency has faced a number of challenges ensuring that the SOP for completing PCRs is met. The SOP requires that PCRs be created on every call, that they are accurate and complete, and that they are submitted within 24 hours of the call.

Completion of PCRs is not an EMT or paramedic's favorite part of the job, and compliance can be a challenge. Specifically, the supervisor has a paramedic who is a good clinician, but hates doing PCRs. He is noncompliant with policy in terms of submitting them on time, and sometimes does not do them at all. This issue has become worse since the agency switched to the electronic format.

### Discussion Questions

1. How big of an issue is this?
2. What tactics would you use to help this employee become compliant?
3. Is this a clinical or an operational issue?

## References

Benshoff v. City of Virginia Beach, 180 F.3d 752 (1999).

Bream, S. (2011). *New rules would label millions of American workers as disabled.* Retrieved from http://www.foxnews.com/politics/2011/03/28/new-rules-label-millions-american-workers-disabled/

Doyle, M. (2009, April 22). A divided Supreme Court hears racial discrimination case. Retrieved from http://chat.anncoulter.com/phpBB3/viewtopic.php?t=83188&p=1796026.

Edwards, S. T. (2010). *Fire service personnel management* (3rd ed.). Upper Saddle River, NJ: Brady/Prentice Hall Health.

Fisher, R., & Ury, W. (1997). *Getting to yes: Negotiating agreement without giving in.* London: Arrow Business Books.

Garcetti v. Ceballos. (2006). US Supreme Court. Retrieved from http://www.law.cornell.edu/supct/html/04-473.ZS.html

International Association of Fire Chiefs. (1995, July 13). Why do volunteers volunteer? *IAFC On Scene*, 9, 1–3.

International Association of Firefighters. (2011). Fire Service Joint Labor Management. Wellness-Fitness Task Force. Candidate Physical Ability Test Program Summary. Retrieved from http://www.iaff.org/hs/CPAT/cpat_index.html

Jury Verdict Research. (1999). [Employment discrimination: Verdicts, settlements and statistical analysis]. Unpublished raw data.

Kimel et al. v. Florida Board of Regents, 528 U.S. 62, 120 S. Ct. 631, 145 L. Ed. 2d 522, 2000 U.S.

National Fire Protection Association. (2010). *Firefighter fatalities in the U.S.—2010.* Quincy, MA: National Fire Protection Association, Fire Analysis and Research.

Ontario Association of Fire Chiefs. (2009). Discussion Paper: Managing the end of Mandatory Retirement [age] in the Fire Service. Retrieved from http://www.oafc.on.ca/lib/db2file.asp?fileid=1389

Sutton et al. v. United Air Lines, Inc., 527 U.S. 471 (1999), 130 F.3d 893, affirmed.

U.S. Department of Labor. (2012). *Police and firefighters under the Fair Labor Standards Act (FLSA).* Retrieved from http://www.dol.gov/esa/regs/compliance/whd/whdfs8.htm

Vlahos, K. B. (2003). Race-based policies raise fraud concerns. Retrieved from http://www.foxnews.com/story/0,2933,95647,00.html

Wagman, J. (2007). City drops firefighters' entry test, *St. Louis-Post Dispatch.* Retrieved from http://mirroronamerica.blogspot.com/2007/09/st-louis-fire-department-entrance-exam.html

# CHAPTER 7

# Customer Service

## Fire and Emergency Services Higher Education (FESHE) Course Objectives

### Module I: Leading and Managing Purposefully with a Community Approach

The students will:

1. Describe the role of the fire/emergency medical services department as a part of the community government and comprehensive plan. (pp 102–110)
2. Explain the importance of a good working relationship with public officials and the community as a whole. (pp 102–105, 107–110)
3. Assess ways to develop a good working relationship with public officials and the community. (pp 102–110)
6. Identify effective skills for developing a cooperative relationship with fire and emergency services personnel as well as public officials and the general public. (pp 102–110)

### Module II: Core Administrative Skills

The students will:

3. Explain the importance of public access to government operations. (pp 102–103, 107)

### Module V: CRM—A 21st Century FESA Responsibility

The students will:

1. Assess the importance of integrating fire and emergency services into a community's comprehensive plan. (pp 102–110)
5. Identify direct and indirect costs associated with fire. (pp 102, 103–104, 107–108)
6. Analyze economic incentives that encourage and discourage fire prevention. (pp 107–108)
7. Describe the role of fire and emergency services in the economic development and neighborhood preservation programs of the community. (pp 102–103, 107–110)

## Knowledge Objectives

After studying this chapter, the student will be able to:

1. Understand the connection between market failures and public safety monopolies. (p 102)
2. Examine customer surveys and demographic changes. (pp 102–103)
3. List and describe the primary services that an FES organization should focus on when planning customer service efforts. (pp 103–105, 107–110)
4. Identify issues relating to organizations that provide EMS. (pp 105–106)
5. Understand the importance of the customer's distress and grief after a fire or other emergency that has resulted in life or property loss. (p 110)

## Overview

In the business community, customer service equates directly to survival and increased profits. Those companies that can identify what the customer wants, and then focus on these desires, are able to increase sales and profit. Companies have gone out of business by failing to keep an eye on the customers and their needs.

Going out of business is rare for the typical fire and emergency services (FES) organization. There are some exceptions however, especially in the emergency medical services (EMS) business where there are a significant number of private ambulance companies. Generally speaking, organizations that practice quality service are those that provide the best service to the public and are able to secure adequate funding from elected officials. FES organizations need to provide the upmost in customer service and should be cognizant of this fact.

## What Is Private? What Is Public?

Emergency services come in different forms. They may be provided by private or public organizations. The differences between these two forms raise a series of questions. Who should provide customers (the public) with emergency services? Should FES be privatized, or is this the rightful responsibility of public agencies? What is the role of government regulation of FES providers? Should there be national standards for emergency services? Or is each local jurisdiction so unique that it should have its own tailor-made standards and customer service expectations?

Questions about the services governments can successfully and effectively provide for citizens need to be answered. Some Americans assume that the private capitalistic economy is the best place to find products and services. After all, this is what has made the United States great.

Others believe government should provide for many of the needs of its citizens, especially the less fortunate. In general, most Americans prefer to rely on free markets, families, places of worship, and other private-sector associations whenever possible. Governments are called on mainly when private-sector institutions prove to be severely inefficient and when they fail to satisfy certain minimum standards of care. Americans have built a strong economy based on competition and the free enterprise system.

## Justification for Government Intervention

When can government legitimately intercede to provide services, goods, or safety to promote its citizens' (the customers') well-being? One type of justification occurs when there is a market failure, which provides validation for regulations or services by government. For example, if education for young children was provided exclusively by private, for-profit institutions, the children whose parents could not afford the tuition would not get an education. In the United States, a high school education is a public service that is provided free to the student and supported by taxes.

Fire departments have a monopoly for services they provide in the community, although EMS might be offered by several suppliers. Because there is no direct competition with the local fire department, improvements, service levels, and fiscal efficiencies are under the control of the department or the municipality that oversees the department. If service is below nationally recognized professional standards, there is no competing department for the customer to compare against or switch to. Unless the public demands better service or there is a regulatory body to oversee the operation, the FES organization sets its own level of service.

Fortunately, most FES organizations are dedicated to community service. However, the administration of these organizations is often at a disadvantage when it would like to make a change that will bring improved customer service. In the business world, if the manufacturer does not make the best product at the best price for the customer, it loses market share and profit, or even goes out of business. However, in the FES organization there is typically no outward indication that a change is needed. Therefore, in many cases it is difficult to gain the additional funding or cooperation of members that is needed to make the change.

Government regulation and elected officials help to provide the balance needed for this monopolistic public service. They provide the FES administration with the independent oversight necessary to ensure the public they are getting professional quality service.

In addition, for overall guidance, a survey of customers may provide some insight into their expectations. If their expectations are supported by professional standards or expert advice, the administration can look for ways to implement the suggestions. If the public's understanding is inaccurate, the agency can fashion an education program to correct this erroneous understanding of the department and its service to the community.

**Public Preference Surveys** One useful technique to assess the needs of the customer and support budget requests is the public survey. These types of surveys, which require prior approval by the chief administrative officer and elected officials, can be used to recognize how the public perceives the department's present service, identify areas in which they would like improvement or expansion of service, and determine how much the public is willing to pay for improvements in service.

For example, budget requests for additional personnel are always hard to gain approval for because personnel costs are always the most expensive items in the budget. After estimating the total costs for the proposal, including all personnel salary and benefit costs, the administrator would express the unit cost per person, per taxpayer, or per household for the community. Then, the survey question that could be asked would be: "Would you be willing to pay X dollars more per year to bring all engine companies up to the staffing levels recommended by national professional standards?" An effective survey would also provide detailed justification as part of the question or as an addendum.

Sharing the results of a public survey is one method of convincing the appointed and elected officials to buy into the proposal. This survey should be the ultimate test of whether the public is willing to spend tax dollars on the FES budget request. Although the results of this survey still do not guarantee approval, the chief officer can argue that this is a fair indication of the public's (customer's) wishes.

Fire and EMS organizations should look for opportunities throughout the year to communicate with citizen groups, the press, and elected officials to gain support for existing programs or enhancement programs that the administration would like to propose in the future. When these opportunities occur, they are often very short-lived. Organizations need to be prepared with all the information for justification and an accurate estimate of all costs.

In most cases, the public, press, and officials have little knowledge or background when it comes to fire, EMS, and emergency preparedness services. In a survey commissioned by the National Fire Protection Association (NFPA), it was noted that most people believe they are more at risk of dying in a fire in a hotel room than in their own homes. Over half of the respondents wrongly assumed they would have plenty of time to escape a typical living room fire. Most had no idea how fast smoke, toxic gases, and heat can fill a room, blocking escape routes and reducing visibility to zero (NFPA, 1996). For surveys to be meaningful, the persons completing them must have some idea of what constitutes quality service.

The first step to resolve this issue may be to educate the public. This should include fire safety education and education on how to evaluate the public FES. Although the latter tends to be complex, giving the public the knowledge of how to assess the public fire services and EMS is necessary if they are to be able to have an informed opinion about the department and its services.

## Keeping Pace with Community Demographic Change

It is necessary that fire and emergency organizations monitor the community for demographic changes that may affect the services requested and tax revenues available. For example, older communities may see a decrease in population and an increase in the number of senior citizens. Some suburban communities are experiencing growth with strong support for improved services in areas such as education. Typically growth areas see an appreciable number of new families with school-aged children.

These demographic statistics are published yearly as estimates and can be found at state or local planning agencies. They are not 100% accurate on a yearly basis but are very close in most cases.

## Expanded Services

The types of services a department can offer affect its relationship with its customers. Many departments look at the addition of services outside of firefighting as a way to market the department to its customers. These organizations are quick to point out to citizens and elected officials that they provide extra value for the tax dollars spent.

The following is a list of services that an FES organization can provide to its customers. These are the primary areas on which to focus when planning customer service efforts. Several of these may be the primary responsibility of other agencies as dictated by local ordinances or historical evolution, but all are provided in one fashion or another.

- Fire suppression
- EMS
- Technical rescue service
- Major disaster mitigation and recovery
- Prevention and education (fire, medical, and major disasters)

## Fire Suppression

As demographics and communities grow and change, fire service and EMS organizations need to revisit the issue of response time. The capability of the fire department to respond with sufficient competent personnel and equipment in time to rescue any trapped occupants and confine the fire to the room of origin is a primary customer service goal. Until 2001 there was no consensus on measuring response times in the American fire service; however, it is now defined in NFPA 1710, *Standard for the Organization and Deployment of Fire Suppression Operations, Emergency Medical Operations, and Special Operations to the Public by Career Fire Departments.*

Another measure that can be used by fire service and EMS organizations is to look at existing average response times for all areas of the jurisdiction to see if their organization's times fall within this average. A fire or EMS department may want to benchmark its response times against a national survey that reports response times. Although not very common, there have been some national studies in this area, such as the 1999 Phoenix Fire Department National Survey on Fire Department Operations. The rationale behind this type of evaluation is that all citizens (the customers) deserve equal fire protection as measured by response time. This type of analysis can point out areas that need new stations and areas that could be served by fewer stations and still meet the jurisdiction's average response time.

However, response times are typically longer in rural areas that are sparsely populated. Response time becomes a cost-benefit calculation. Areas with few residents are not able to collect the taxes needed to support shorter response times. For example, in built-up areas, the Insurance Service Office (ISO) Grading Schedule generally equates a 3-minute response time to a response distance of 1.5 road miles. In prominently single-family dwelling developments, ISO uses a five-mile distance as a maximum for adequate protection (ISO, 2003).

According to the NFPA *Fire Protection Handbook* (NFPA, 1997), evaluation parameters for response times should be based on flashover. Although these times vary based on a number of variables, NFPA notes, "[F]ire departments should operate on the assumption that prevention of flashover requires a response in less than 10 minutes" (NFPA, 1997, p. 7-311). However, this general response time changes if the fire department also provides EMS. In that case, "the widely recommended 4-minute response for non-breathing or trauma victims is very important" (NFPA, 1997, p. 7-238).

The NFPA 1710 standard, the ISO Grading Schedule, and the *Fire Protection Handbook* give some guidance to the planner or administrator trying to determine an appropriate response time for a jurisdiction. The number and location of fire and rescue stations in a given area are controlled solely by the response time chosen as a service indicator for the local jurisdiction. In general, longer response times equate to fewer fire and rescue stations, and the opposite occurs when response times are shorter. For example, assume that a jurisdiction plans for a maximum response time of 4 minutes (NFPA 1710). If the response time is increased by 1 minute to 5 minutes, 36% fewer stations will be needed to cover the same area. If the response time is lowered to 3 minutes, then 66% more stations are needed. These are large variances.

When reviewing a department's equipment for compliance, NFPA provides comprehensive standards for design and operation. Most departments meet these standards and, unless limited by age, it is rare for this item to affect fire suppression.

The other key item that can critically affect fire suppression is competent personnel. In fire suppression and rescue service, competence has been equated to personnel who meet nationally recognized professional standards, such as NFPA 1001, *Standard for Firefighter Professional Qualifications*, through a training system that certifies competencies. In addition, larger metropolitan fire departments have developed "recruit" training programs that generally meet or exceed the national standards.

However, in the United States, there is no consensus on the minimum time needed for a student to become proficient in the job performance items listed in the NFPA standards. There are some statewide certification systems, but no national programs. Generally, the length of the training program is directly proportional to the skills of the students. EMS, however, has a national curriculum and a set number of hours needed for completion of its several levels of competency.

For customers to be able to make a rational judgment about the quality of their emergency services, they should be able to understand the previous discussion on response times, competent personnel (including minimum staffing), and equipment. It is very important that citizens understand that uncontrolled fires grow exponentially, meaning that every minute a fire grows faster than the previous minute. The NFPA has several excellent videos that visually illustrate this rapid growth. It is a good idea to purchase these videos and use them in public fire safety education programs.

## Emergency Medical Services

The provision of EMS is held in high regards by the public and elected officials. At the most basic level, the fire engine responds to 911 calls and renders care and support to patients and their families. The engine may be staffed with emergency medical technicians and provide initial care until advanced life support (ALS) arrives; they may have paramedics that render ALS-level care, then turn over the patient to a private ambulance company for transport; or they could be an ALS fire department that also provides patient transport.

As reported by NFPA, more than 97% of all fire departments that serve more than 100,000 customers provide EMS service (Karter, 2010). It is alarming that this percentage is not 100% given the good public relations EMS can bring to the fire department.

Fire departments that provide EMS transport have an additional challenge over first responder–only or non-EMS fire departments: potential competition with other agencies for patient transport. In many cases, transport fees are paid by medical health insurance or through state or federal funds, such as Medicaid or Medicare. Recently, private ambulance companies have proposed taking over patient transport services in several large cities. The main argument for this change is that the private companies can provide service cheaper than a municipal fire service.

**Disparity of Duties** In departments that respond to every call for medical help, the percentage of a fire unit's calls can approach 80%. For departments that respond only to critical emergency medical calls, the EMS fire responses hover at 45–50%. More and more often firefighters are serving as primary care providers.

This issue is essentially a political decision: whether to send the closest emergency unit (normally a fire company) to all medical calls, or just to true emergencies. Many elected and appointed officials prefer to be on the politically safe side, despite the additional costs and increased risk of mental or physical harm that may be placed on the fire company.

The dilemma is best handled through tiered response, because it allows the fire service to make the most efficient use of their human and material resources. Rather than immediately sending an engine for every call, response decisions are made after dispatch call screening and medical evaluation have occurred. This is also known as "priority dispatching." Questions that must be considered include the following: Does every call warrant an emergency response? Will a basic life support unit suffice rather than an ALS unit? Is an engine necessary at all when the call can be managed by an ALS ambulance?

The only real advantage to running every call as an emergency is that it raises call volume and awareness, giving the community a sense that the fire department is serving their needs rather than sitting around the station waiting for the next fire. This awareness might also increase the likelihood of justifying funds for additional personnel and equipment. Visibility to the public equates to survivability for many departments.

However, increased call volume taxes resources, wears out expensive fire apparatus, and can take a physical and psychological toll on crews. If a department is forced to cut resources but continues to receive the same call volume, time to the scene will be impacted or mutual aid resources will need to be called in. More importantly, scarce ALS resources may be tied up on a minor trauma while the patient with cardiac chest pain is made to wait. If a fire department does not seem to be able to handle the load, a private ambulance company might be able to convince local government that it can.

Another issue involves the risk of hazards present to the public and the fire department when making emergency responses. If only 20% or less of the calls require ALS, the risk seems unjustified. Studies have shown that time saved by emergency responses is minimal, and would not have much effect on a non-emergent patient.

Controversy over the issue of firefighters as paramedics also rages within the fire service itself. Norris W. Croom, Division Chief of Operations for Colorado's Castle Rock Fire and Rescue Department, notes his experience with members of the fire service who believe that firefighters should not provide EMS. Croom goes on to state, "A more disturbing trend, though, is that some new firefighters are adopting the attitude that 'I joined to fight fires, not be a paramedic'" (Croom, 2010, p. 1).

The International Association of Fire Chiefs (IAFC) and the International Association of Firefighters support fire service–based EMS and urge public officials to support all efforts to broaden emergency medical care and ambulance transport by the fire service. In a 2009 position paper, the IAFC states, "The American fire service is strategically and geographically well-positioned to deliver time-critical response and effective patient care rapidly" (IAFC, 2009).

# EMS Challenges

EMS organizations continue to face challenges of their own. The following issues need to be addressed if an EMS organization is to provide the public with the best EMS:

- Overcommitment of resources
- Reduction of the volume of medical calls
- Issues surrounding healthcare reform
- Overloading of system
- Integration with the healthcare system

## Overcommitment of Resources

There is a trend in communities to overcommit EMS resources. As noted by Heightman, editor-in-chief of the *Journal of Emergency Medical Services*, overcommitment can lead to responder morale problems. Heightman notes that this problem comes from "throwing excessive and expensive resources at basic life support (alpha-level) responses and burning out the responders forced to 'mount up' and respond en masse to calls that don't really need them" (Heightman, 2009, p. 6). With the stress that comes from working a 24-hour shift, down time can be critical.

In addition, overcommitting resources causes extensive wear and tear on expensive fire apparatus—apparatus that can increase the life expectancy by 50% or more.

## Reduction of the Volume of Medical Calls

In 2010, former US Fire Administrator Kelvin Cochran called for the creation of public education programs to help reduce the number of medical emergencies in the United States. Speaking at the opening of Fire-Rescue Med in Las Vegas, Chief Cochran noted, "The increasing volume of medical calls cannot be addressed simply by continually adding paramedics and medic units"

## Case Study

### Some Thoughts from the Field

In some agencies, firefighters can be heard saying "I'm here to fight fire, not take care of sick people." This attitude is dependent on the culture of the department and may be more common in older departments that have not traditionally had EMS responsibility. Although firefighting and EMS duties both require time-sensitive lifesaving skills, the skills themselves are very different. Some members may find they have a strong preference for one of the two; still, others enjoy both.

Jane, a paramedic and assistant fire chief said, "I have always felt that I am both equally. That is not true for everyone, but it is for me. I love the medical part, and love fighting fires. I have always thought that it's the best job in the world. Also, I was never very good at working in the for-profit environment. I was always drawn to public service, so that was another reason to move away from private ambulance service."

Michael, firefighter paramedic: "I wanted to do both. The meat and potatoes, or what makes me feel good inside, is helping people (the paramedic stuff). The dessert, the fun adrenaline rush stuff is the fire service areas of expertise. So by far and away being a firefighter medic is much, much more rewarding both personally and professionally than being either a medic or a firefighter."

(Thompson, 2010). Instead, the model of fire prevention programs could be replicated to address non–fire-related injury and death in communities.

### Issues Surrounding Healthcare Reform

In 2010, a healthcare reform bill was adopted to improve access and quality of medical care in the United States. Fire and EMS expert Chief Gary Ludwig cites a number of issues that failed to be addressed with this reform, including the following (Ludwig, 2010):

- The failure to depart from the current model of treating and transporting patients to an emergency room.
- The failure to create injury and illness prevention programs that would change the current model of EMS delivery.
- The failure to fund "treat and release" programs that would save money by not transporting patients that do not need it.
- The failure to implement programs that would reimburse EMS for evaluating individuals on the scene and transporting them to clinics and urgent care centers when more intensive care is not warranted.

### Overloading of System

Nationwide, EMS units have become tasked with far more duties than they have the resources and leadership to handle—far more, in fact, than they were ever intended to handle. This has given rise to a whole host of risks, including sleep-deprived EMS crews; long patient wait times; and emergency workers who lack the training to deal with large-scale catastrophes, such as chemical attacks or another Hurricane Katrina scenario.

The exact depth of America's EMS problem is unknowable: federal patient confidentiality laws limit the amount of information made public, and immunity statutes minimize the number of lawsuits initiated by families. Several years ago, however, a study commissioned by *USA Today* reached a startling conclusion: an estimated 1,000 cardiac-arrest victims die in the United States each year, in part because of slow EMS response (*USA Today*, 2005).

People also leave the profession because of the long hours—another result of budget shortfalls. A recent survey of EMS providers published in the *Journal of Emergency Medical Services* found that 24-hour shifts remain normal, raising concerns about crew and patient safety. A 2008 study in *Sleep Medicine* found that moderate, repeated sleep deprivation is comparable to a blood alcohol concentration of 0.08% (Caruso, 2012). "That's the legal limit for noncommercial drivers," says Brian Maguire, a clinical associate professor of EMS at the University of Maryland at Baltimore County. "That means if the man or woman who comes to treat you is nearing the end of a shift, he or she might as well be drunk" (National Association of EMS Officers, 2011).

### Integration with the Healthcare System

In 1996, the 30th anniversary of the EMS industry, the National Highway Traffic Safety Administration (NHTSA) looked at the status of EMS nationwide, developed a vision for the future of EMS, and published the findings in the paper "EMS Agenda for the Future." NHTSA predicted that EMS systems of the future will be community-based and fully integrated with other healthcare providers and public health and public safety agencies. NHTSA also predicted that future systems will "reduce illness and injury risks through prevention, provide acute illness and injury care and follow-up, assist in the treatment of chronic conditions outside hospitals, and provide community health monitoring" (NHTSA, 1996). Yet, even in light of predicted changes, NHTSA's "EMS Agenda for the Future" recognizes that EMS providers will remain the public's emergency medical safety net.

## What Is Fire and Emergency Service's Customer Service Duty?

On February 20, 2003, 100 people were killed in a tragic fire at The Station nightclub in West Warwick, Rhode Island. The fire was started when pyrotechnics used during a Great White concert ignited flammable foam placed on the walls for soundproofing. There were many lessons to be learned from this fire, primarily regarding the effects from the absence of fire prevention. Patrick C. Lynch, the Rhode Island Attorney General at the time, was "criticized for not bringing charges against fire marshals [fire inspectors] or building inspectors for failing to cite the nightclub for code violations, or against members of Great White" (Laviue, 2006). Besides failing to prevent these deaths, the FES agency and its officials could be personally liable or face criminal charges. In 2008, the State of Rhode Island and the town of West Warwick agreed to pay $10 million as settlement.

To make this tragic fire even more appalling was the realization that a similar fire and life loss had occurred in Boston, Massachusetts, in 1942. The Coconut Grove nightclub fire resulted in 491 fire deaths, and the building shared the following fire prevention deficiencies with the Station nightclub: restrictive exits, flammable interior finish, overcrowding, and no automatic sprinklers.

Fire prevention by FES agencies can be restricted by state and local regulations; however, some states allow the fire department to enforce the fire safety aspects of building and fire codes in new and existing construction. At a minimum, chief officers should determine who has the authority to enforce fire prevention codes in existing buildings. Through state fire organizations, officers should push for changes in the state law that allow this life-saving customer service to be provided by the fire department.

> **Case Study**
>
> **Customer Service Considerations**
> A fire department responds to a house fire and very quickly extinguishes the fire. The officer at the site notices that the grass around the house is extremely high. After assessing that the owner of the house, an elderly woman, has limited financial and physical resources, the officer offers to cut the grass.
>
> The firefighters drive back to their station and load the station lawnmower onto the engine. They then return to the woman's house and cut the grass. At all times they are available by radio for any emergency calls. The woman is extremely pleased with the firefighters, and the firefighters feel they have done a good deed in the spirit of customer service, which has been the subject of several recent training sessions. However, in the following days, headquarters receives several calls from other owners in the same neighborhood asking to have their grass cut by the fire department.
>
> This story points out one pitfall of customer service that must be considered. Before volunteering a service, determine whether it is something that can be provided to anyone who requests the assistance.

## Fire Prevention

Fire prevention is a major customer service area for the public, specifically for building owners. The *NFPA Fire Protection Handbook* states, "One basic aspect of a comprehensive public fire protection plan is the concept that it is infinitely better for a community to prevent fires altogether…than to depend solely on the fire suppression capabilities of the community's fire department" (NFPA, 1997, p. 7-31).

The customer may not have a good understanding of fire prevention and may not be as inclined to support funding of prevention programs. There is often a conflict between funding emergency operations and funding prevention programs, with advocates for funding and supporting emergency operations generally outweighing those that support prevention programs. The lack of support may be caused by the fact that in many cases the beneficiaries of a comprehensive prevention program do not even know that their life or property has been spared. Although this is customer service at its most efficient level, in most cases the outcome is invisible and therefore difficult to document life and property loss prevention.

In The International Society of Fire Service Instructors' publication *The Voice,* Vina Drennan further explains the difficulty of documenting the results of prevention (1998, p. 2):

> We do not know the name of a child that is spared the scars of fire because one of you went into a school to teach the importance of developing an exit plan for their family. We do not know which elderly person you saved when you visited the senior citizen center in your neighborhood and gave a cooking safety demonstration. We do not know the names, but we do know the fatalities are decreasing.

There are a lot of customers who are alive today thanks to the efforts of modern fire prevention. This is the ultimate goal in customer service.

## Home Fire Sprinkler Systems

Home fire sprinkler systems may be the ultimate customer service for fire safety in the future. They are so effective that they are more recommended for life and property protection than smoke alarms.

Still, the code changes needed to mandate fire sprinklers in homes have faced sustained and organized opposition on a national level, specifically from the National Association of Home Builders. When these safety devices were first proposed, the cost of installation per square foot was expensive; but now, partly because the devices are gradually being required by building and fire codes, installation in new homes can cost $1 or less per square foot. Additionally, some of the installation costs can be recovered by reductions in fire insurance.

It is also common for the housing construction industry to argue that adding the cost of sprinklers to a home will put them at a price disadvantage compared with adjoining jurisdictions. This supposed disadvantage would hurt home sales in the jurisdiction adopting the fire safety requirement, resulting in loss of real estate taxes. However, a 2009 NFPA-sponsored study concluded that a sprinkler ordinance has no impact on the number of homes built. This study was conducted using the jurisdictions of Montgomery County and Prince George's County, Maryland, which have sprinkler requirements; and Fairfax County, Virginia, and Anne Arundel County, Maryland, which do not have these requirements. "The selected areas, all developmentally mature, cover a wide geographic area and contain a variety of housing stock and income levels, making them prime for comparing municipalities" (NFPA, 2009, p. 2). The results showed that such issues as schools and proximity to work have more influence on the decision to buy a home than the 1–2% cost differential represented by fire sprinklers.

> **Case Study**
>
> **Fire Sprinkler Requirements**
> In 1992, Prince George's County, Maryland, became the first county in the nation to require fire sprinklers in all newly constructed residential dwellings. During the previous years, Prince George's County studied the development of sprinkler technology, appointing a task force to review fire injury and death
> *(continues)*

## Case Study (Continued)

data during a 10-year period. Participants included representatives from the home building industry, multifamily developers, members from the Board of Trade, zoning attorneys, representatives from water and sewer authorities, local planning and zoning and subdivision commissioners, insurance industry, representatives from the International Association of Fire Fighters, County Volunteer Fire and Rescue Association, county executive's office, and city governments.

During the course of the task force's investigation, it reviewed fire statistics, scenarios, and visual evidence of actual tests. The task force agreed that residential sprinklers save lives and property. Discussion then centered on resolving to what degree they should be mandated and what construction alternatives would be sufficient to offset costs while maintaining adequate fire protection.

During this time, the building industry voiced concerns regarding mandatory sprinkler installation in single-family dwellings, citing high costs and limited trade-offs. To counter these arguments, the task force noted that developers could apply for "density bonuses," which would allow them to create more building lots (up to 10%), thereby increasing property value by 10%.

With some potential cost savings identified and mutual agreement that the cost of residential sprinkler systems would fall as demand increased, a compromise was worked out to implement the requirements.

Fifteen years after the adoption, a research paper from the Home Fire Sprinkler Coalition reported that since the ordinance was enacted, a reported 101 fire deaths occurred in single-family or townhouse structures that were not protected with fire sprinkler systems, whereas no fire deaths occurred in structures that had sprinkler systems (Weatherby, 2009).

For structures that did not have a sprinkler system installed, fire damages averaged $9,983 per incident, with that average going up to $49,503 when the fire resulted in a fatality. However, damages to structures with sprinklers averaged only $4,883. With 13,494 single-family or townhouse fires occurring during this time period, the County's fire damage totaled about $134 million. In contrast, installation of a sprinkler system costs under $2 per square foot (Weatherby, 2009).

The results of a 2011 study of Scottsdale, Arizona's 15-year experience with residential home sprinklers furthers the justification for installation:

- The average fire damage to homes with sprinklers was $2,166; the average for homes without sprinklers was $45,019.
- 90% of the fires were contained with one sprinkler.
- The cost of sprinkler installation was between 1% and 1.5% of the cost of the building (Home Fire Sprinkler Coalition, 2011).

Today, there are many sources of information and statistics to help justify this type of code change. For example, in 2006 NFPA approved a provision supporting fire sprinklers in its Life Safety Code. Additionally, in 2008 the International Code Council began requiring fire sprinklers in new residential construction. Although national building and fire codes now require installation of residential sprinklers in all dwellings, including one- and two-family houses, the next step is for states and local governments to also adopt this requirement.

## Fire Safety Inspections

Many FES departments use line personnel to provide fire inspections of existing buildings. Although this is a good use of standby time, time conflicts seem to be more and more common for station personnel. Such duties as fire calls, EMS responses, training, physical fitness, station and apparatus maintenance, and fire prevention can result in some critical nonemergency functions not being done or being done marginally.

Even though all of these objectives are very important and more than justified, there is not time for everything in a busy fire company. Therefore, tasks must be prioritized. The number one priority has to be response to emergency calls and the training (including physical fitness) that prepares these companies to do an efficient and effective job. Along with the skills training, prefire plans and familiarization inspections are very useful and can save firefighters' lives. All of these functions are necessary if the FES are going to be able to give 100% effort when responding to a structural fire.

There is no better time to improve (or damage) the professional reputation of an agency than during fire inspections or prefire planning visits. Therefore, inspectors and firefighters should be trained and educated in public relations.

If a correction order must be issued, it is important that it is issued professionally, because not doing so can leave a bad impression of the agency. Inspectors should ensure that the violations are documented and validated. It is vital that inspectors cite the correct code while issuing violations to maintain public trust and respect.

Inspectors should maintain an amicable, yet professional, demeanor when presenting the findings of the inspection to the owner or occupant **FIGURE 7-1**. A professional fire safety program has a favorable effect on the ability of the fire department to obtain funding for needed equipment, personnel, salary raises, and new stations.

Fire inspections also provide knowledge that is helpful in firefighting. These fire inspections and prefire planning surveys present an opportunity to learn about conditions and construction features of buildings in the community before a fire occurs.

**FIGURE 7-1** Presenting inspection results to homeowners.

**Home Inspection Programs** Although some fire codes contain fire safety requirements for existing homes, most inspections are done on a voluntary basis because of political realities and expectations of privacy. For example, NFPA 1, *Uniform Fire Code*, contains a requirement for existing homes to have a smoke alarm outside of sleeping areas and on each level of the house. For those jurisdictions that have adopted this fire code or another with a similar requirement, the code is more encouraged than strictly enforced.

Most home inspection programs are primarily devoted to fire safety education. As noted by the NFPA, 80% of fire deaths occur in the home (Home Fire Sprinkler Coalition, 2011). Therefore, it is extremely helpful if smoke alarms can be installed by fire inspectors or firefighters during these inspections. One idea is to contact local civic organizations and request donations to cover the cost of a give-away smoke alarm program. If necessary, some can be bought with public funds. Programs such as these are great public service opportunities and demonstrate the concept of leading by example.

There are two ways to schedule home inspections. One is to advertise the service by cable and regular television, local newspapers, and mailings. The advertisement would ask occupants to call a central number to schedule the inspection. The second, which can be in addition to notifications, is to have fire companies go into the field and knock on doors for entry. Home inspections should be conducted in pairs, because this helps protect against any potential accusations of misconduct. A four-person engine can split up into two separate teams. If a paramedic unit is also in the field, the two-person unit is the right crew size for these inspections. It is a good idea to place signs on the fire and rescue apparatus indicating that the firefighters are performing home safety inspections.

Timing is very important in these visits. Many more people are home in the evening than during the day. Therefore, the best time to do home inspections is between 7 and 9 PM. It is not a good idea to go too late in the evening, such as at dinnertime or when people may be sleeping. Be prepared to return on a different day or time. If a resident is not home, leave a brochure on the door.

For serious violations of fire safety code, entry can be gained either voluntarily or by a search warrant. For example, after interviewing several teenagers who were found with illegal fireworks, a fire department discovers that an individual is selling fireworks out of their home. With this information, the department's investigator can go to the state's attorney and ask for a search warrant issued by a judge. In situations such as this, police should assist in gaining access and enforcement.

## Enforcement of Fire Safety Codes

An effective enforcement program is made up of the following steps:

1. **Code adoption.** New or updated fire safety code ordinances need to be publicized when enacted. This should include an effort to highlight any potential controversial issues. Departments may consider giving a presentation outlining the major proposed changes and invite architects, engineers, developers, owners, and the media. It is counterproductive for a department to attempt to hide something in code language that may become obvious after adoption.

2. **Plan review.** For new construction, it is imperative that a building be designed to comply with all fire safety codes and standards. In the typical building code, about one-half of all requirements are fire safety–based. Because the FES agency has to respond to any fire emergency and put their firefighters in harm's way, it has a greater motivation to make sure the building is constructed in accordance with the fire safety aspects of the building and fire codes. If the jurisdiction uses the NFPA Life Safety Code, it is good practice to have the FES enforce this code. In other jurisdictions that use only a building code, the FES plans examiner can be tasked with reviewing fire hydrants, access to the building for fire apparatus, fire protection systems, means of egress, fire-resistive construction, and smoke control systems.

3. **Fire inspections.** The inspector should be tasked with enforcing those items in the building code that are fire safety–related. If some code items are missed during construction, it may be impossible to gain compliance at a later date. The success of these inspections is directly related to the correctness and thoroughness of the plan review.

4. **Maintenance fire inspections.** These are needed to ensure that fire protection systems, means of egress, and other fire safety items are properly maintained. To do this task properly, it is mandatory that inspectors know the fire code

requirements and the fire safety aspects of the building code. In addition, housekeeping items that, for example, could mean the accumulation of excessive amounts of combustible trash in the structure must be enforced. These inspections can be combined with prefire plan visits.

## Preventive Medicine

Similar to fire prevention, emergency medical problems can also be prevented in many cases. It is always preferable for a person to lose weight, quit smoking cigarettes, and start exercising before a serious illness occurs. Although fire safety education is used to help with the fire problem in existing and nonsprinklered structures, preventive medicine education and treatment are the method of choice to prevent diseases and injuries.

The FES organization can provide the customer with some preventive medical services, such as high blood pressure and cholesterol screenings. These types of customer services can be effective at identifying symptoms that may lead to serious illnesses, thereby securing many friends for the agency.

## Customer Service at Emergencies: The Extra Mile

Many experienced fire protection experts estimate that each person will experience a hostile fire in his or her home once in a lifetime. These relatively rare occurrences can earn the department a lifelong friend. The same can be said for the more common, but equally traumatic, medical emergencies. If a death has occurred, these painful events become even more distressing for those close to the victims. Department members can be very helpful as a source of comfort and in locating resources to help a grieving family.

FES personnel should be provided with information and techniques for handling the grief process. For example, careless comments such as "I understand exactly how you feel" or "Calm down; everything is all right" can be hurtful to the bereaved. Short seminar courses can provide members with the knowledge and awareness they need to provide compassionate support in distressing situations.

The following is a list of services that may be provided to the victims of fire or other emergencies:

- Explain all FES operations
- Separately talk to owners and occupants
- Salvage furnishings, especially personal items
- Provide temporary lodging
- Provide meals and clothing
- Provide transportation
- Notify relatives and friends
- Notify religious services, if appropriate
- Contact the Red Cross and Salvation Army
- Provide temporary pet services
- Contact the insurance company
- Contact the police for security patrol
- Secure the building
- Install smoke alarms, if needed

### CHAPTER ACTIVITY #1: Tragedy Strikes

At 2 AM on a cold winter night, a department is dispatched to an apartment fire. The caller, a resident who lives on the third floor, said she can smell a strong odor of smoke and the fire alarm is ringing.

When the fire company arrives, they find a large fire on the second and third floors of a four-story garden-style apartment building. The initial dispatch is three engines, two ladders, and a battalion chief. The first engine to arrive reports a working fire and requests the assignment be upgraded. This calls for the dispatch of two more engines, one ladder, a heavy rescue, a second battalion chief, the safety chief, and a deputy chief.

After the fire is extinguished, all of the occupants escape without any deaths. However, several occupants are burned and require hospitalization. In addition, 10 apartment units are so heavily damaged by fire that the occupants are not allowed to return, becoming homeless.

### Discussion Questions

1. Make a list of the services that the FES personnel could either provide or direct these victims to for help, and identify a contact person and telephone number (24-hour access) for each.

2. If an elected official or the press shows up, describe how they should be treated and what information should be gathered for their knowledge.

## CHAPTER ACTIVITY #2: A New Model of Customer Service

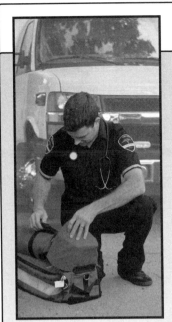

The emerging trend known as "Community Paramedicine" began with EMS providers in Colorado, Texas, and Minnesota providing primary and preventative healthcare through house calls. The program addresses the issue of shortages in primary care, especially in rural communities, but it is also being considered in urban communities to reduce the dependence on emergency rooms for primary care.

Through this program, paramedics are given specialized training and use down time between emergency calls to visit patients. They check blood sugar and blood pressures; confirm prescription compliance; change bandages; and occasionally run diagnostic checks, such as an electrocardiogram. They can also evaluate an older person's home for fall hazards and advise on other safety issues.

Community paramedicine allows many patients to obtain the routine care that would otherwise not be available. It also gives the paramedic the opportunity to recognize a more serious problem before it becomes an emergency. The paramedic's report is evaluated by a physician, who can recommend further care options for that patient.

The Western Eagle County Health Services District has created the "Community Paramedic Program Handbook" to explain its program in detail. The handbook is available through the following link: http://www.communityparamedic.org

### Discussion Questions

1. Do you consider community paramedicine to be a viable option in your community? Why or why not?

2. Do you believe that this service would prevent illness, injury, and unnecessary emergency room visits will make a difference? Why or why not?

3. Do you consider this to be a valuable asset to FES or EMS in terms of garnering support or funding? Why or why not?

## References

Caruso, C. C. (2012). *Running on empty: Fatigue and healthcare Professionals.* Retrieved from http://www.medscape.com/viewarticle/768414_2

Croom, N., III. (2010, May). "I didn't join to be a paramedic": Challenging bad attitudes toward fire service-based EMS." *IAFC OnScene, 2*, 8.

Drennan, V. (1998 November/December). Forty more are alive today. *The Voice, International Society of Fire Service Instructors.*

Giebel, G. Program Manager, Office of the Fire Chief, Montgomery County, Maryland, and 45 year member and past President of BCCRS, personal communication, April 13, 2012.

Heightman, A. J. (2009). Resource overkill: We can do more for less. *Journal of Emergency Medical Services, 34*(3).

Home Fire Sprinkler Coalition. (2011). *Scottsdale report 15-year data.* Retrieved from http://www.homefiresprinkler.org/fire-department-15-year-data

Insurance Service Office. (2003). *Fire suppression rating schedule.* Jersey City, NJ: Copyright Insurance Service Office Properties, Inc.

International Association of Fire Chiefs. (2009). *Position statement: Fire-based emergency medical services.* Retrieved from http://www.iafc.org/files/IAFCposition_Sinclair_Fire-Based%20EMS_Position%20Statement.pdf

Karter, M. J., Jr. (2010). *U.S. Fire Department profile through 2009.* Quincy, MA: National Fire Protection Association.

Laviue, D. (2006). R.I. attorney general says he opposed deal, sentences in nightclub fire case. *The Associated Press.*

Ludwig, G. (2010). EMS misses chance for health-care. *FireRescue magazine.*

National Association of EMS Officers. (2011). *National EMS assessment.* Retrieved from http://ems.gov/pdf/2011/National_EMS_Assessment_Final_Draft_12202011.pdf

National Fire Protection Association. (1996, May/June). Americans underestimate fire risk. *NFPA Fire 77.*

National Fire Protection Association. (1997). *NFPA fire protection handbook* (18th ed.). Quincy, MA: National Fire Protection Association.

National Fire Protection Association. (2009). Comparative analysis of housing cost and supply impacts of sprinkler ordinances at the community level. Retrieved from http://www.firesprinklerinitiative.org/~/media/Fire%20Sprinkler%20Initiative/Files/Reports/FSI_Comparison_Analysis_Final_Report.pdf

National Highway Traffic Safety Association. (1996). *EMS agenda for the future.* Retrieved from http://www.ems.gov/emssystem/agenda.html.

Phoenix Fire Department. (1999). Phoenix Fire Department national survey on fire department operations.

Thompson, J. (2010). *Chief Kelvin Cochran on injury prevention programs. FireRescue1.* Retrieved from http://flashovertv.firerescue1.com/Media/2604-Chief-Kelvin-Cochran-on-injury-prevention-programs/

USA Today. (2005). *The price of just a few seconds lost: People die.* Retrieved from http://www.usatoday.com/news/nation/ems-day2-cover.htm

Weatherby, S. (2009). *Benefits of residential fire sprinklers: Prince George's County 15-year history.* Retrieved from http://www.homefiresprinkler.org

# CHAPTER 8

# Training and Education

## Fire and Emergency Services Higher Education (FESHE) Objectives

**Module IV: Leading Change**

The students will:

1. Describe the importance of accepting and managing change within the fire and emergency service department. (pp 114–124)
4. Assess ways to create a positive climate for change and introduce new ideas within the organization. (pp 115, 116–120, 121–124)
5. Describe how an organization can respond to current or emerging events or trends. (pp 116–117, 118–120)
8. Describe ways to increase and reward professional development efforts. (pp 121–124)

## Learning Objectives

After studying the chapter, the student will be able to:

1. Understand the necessity for continuous reinforcement of basic emergency services skills and knowledge. (pp 114–115)
2. Identify training needs for FES personnel. (pp 114–116)
3. Recognize the best training practices that lead to safe, professional emergency operations. (pp 114–117)
4. Understand the requirements for emergency medical technician (EMT) and paramedic certification. (p 116)
5. Examine local and national developments that require new training and education efforts. (pp 117–120)
6. Recognize ways to determine acceptable competencies for training. (pp 120–121)
7. Identify advanced educational opportunities for FES professionals and chief officers. (pp 121–122)
8. Understand the National Professional Development Model. (pp 123–124)

## Introduction to Training and Education

Administrators are responsible for the efficient and effective emergency operations of the department. They should be aware that there are no shortcuts to having members who are well trained, equipment that is properly maintained, or standard operating procedures (SOPs) that are followed for emergency operations. Success at the scene of an emergency is the result of a well-trained team, meaning the providers have the ability to complete manual skills with little or no conscious effort. This level of ability is achieved only after practicing the skills many times, beginning with an intensive recruit school.

In 1873 (ironically the same year that the International Association of Fire Chiefs was founded), Captain Eyre Shaw, a chief officer in the London Metropolitan Fire Brigade, made the following comment after visiting several fire departments in the United States: "The day will come when your fellow countrymen will be obliged to open their eyes to the fact if a man learns the business of a fireman only by attending fires, he must of necessity learn it badly...I am convinced that where study and training are omitted, the fire department will never be capable of dealing satisfactorily with great emergencies" (Parow, 2011). Jack Parow, former president of the International Association of Fire Chiefs, poses the following question regarding this account, "Did we listen to his words of wisdom or did they fall on deaf ears?" (Parow, 2011). Parow reminds us that there are still many issues regarding contemporary FES training and education that need to be resolved.

## Firefighting Training and Certification

To be proficient at emergency operations, members have to be thoroughly trained before participating, and must receive updated training regularly. Because essential skills, such as those required for structural fires, are used infrequently, some officials make the argument that there is less value to training for firefighting. Actually, the opposite is true. This profession requires a higher level of training because of the emergency nature of the service and the infrequency of the use of these skills. In emergency operations the job must be done right the first time; otherwise, the result is the potential for increased property loss, deaths, and injuries. For this reason, training must be an ongoing and never-ending process.

Take the example of the New York City Fire Department. This department is the largest and busiest fire department in the United States. In 1995 it instituted a back-to-basics training program designed to ensure that firefighters maintain their skill level in the most important aspects of fire suppression operations while also receiving training in new areas **FIGURE 8-1**. Even with a large number of experienced firefighters (as a result of the high number of working structural fires in the area), this department saw a need to reinforce basic skills by using a structured mandatory in-service training program (US Fire Administration, 1997, p. 13). If the largest, busiest, and most experienced fire department in the country sees a need for in-service basic training, all fire departments could benefit from this type of training.

Although training can sometimes present a problem to smaller departments in which finding the time and resources for exercises can be difficult, training (including live fire exercises) is essential to retain skills and ensure safety. Smaller departments will find that partnering with another department can often provide the necessary resources. Larger departments might also need outside sources to provide training in specialized areas such as hazardous materials incidents, acts of terrorism, technical rescues, wildfires, catastrophic disasters, high-rise fires, multialarm mutual-aid incidents, large-scale emergency medical incidents, and other infrequent events. An instructor experienced in these types of incidents can enhance the training program and the retention of information by the students.

The fewer working fires a department experiences or the less frequently a skill is used, the greater the need for training. It is common for members to have a better memory for the skill than their ability to actually perform the skill. Skills that require muscle coordination can degrade rather quickly. This varies for each individual, but because firefighters work in teams, training must be directed to the least able member. All members, including those assigned to staff positions, should receive this refresher training. It is not uncommon to have firefighters working in staff functions, such as fire inspectors, not experience actual firefighting for years at a time.

At a minimum, the FES organization should schedule drills for incumbents that cover all the basic essential job functions throughout each year. Technical competencies are most successfully taught to the adult learner through hands-on training. Many higher-level skills, such as emergency command, hazardous materials evaluation and operations, fire prevention, and emergency medical services (EMS), can be best learned through simulation drills. Although many of these skills require massive amounts of special knowledge, the highest level of knowledge retention comes from applying the knowledge to an actual incident or training exercise and practicing the skill or administrative task. This type of learning exercise also tends to keep the interest of the adult learner better than just memorization of facts and regurgitation on an examination. Although some multimedia presentations and reading assignments are necessary in the familiarization process, an exercise should be included in the program of instruction to ensure long-term retention and immediate recall at the emergency incident.

Most of the in-service training drills can take place at the fire station, ensuring a convenient location without requiring the company to go to a training facility. Still, there are also times, such as for live-fire training, when fire companies must go to a central training facility and the department may have to make provisions for a stand-by crew to be available at the station. For example, a metropolitan department may choose to provide a special shift consisting of four 10-hour days per week that fills in

**FIGURE 8-1** Firefighters stretching a hose line.
© Glen E. Ellman

for departments that need to attend multicompany drills or paramedic recertification classes.

For station drills, one effective technique is to have the rookie administer basic drills. This new firefighter will have recently learned the most current methods at the training academy, because these agencies are often the first to adopt new skills and equipment. In addition, having the rookie administer the drills also helps reinforce the skills in new members. Often, the best way to learn something is to teach it.

A firefighter's job is complex; there is a need to provide challenging, relevant training. A department with a negative training culture will find lack of motivation, inattention, poor retention, and nonadherence to expectations in the field. The culture of the department needs to set the tone that training is taken seriously and is demonstrated in practice. Unfortunately, nonmandatory training programs are often the first thing to be cut during a budget crisis.

Therefore, time, opportunity, and budget for firefighter training have to be carefully considered. For training to be effective, it needs to be timely and relevant. Curriculum needs to be meaningful, and easily and immediately adapted to actual practice. The instructor not only needs to be knowledgeable but also relatable and someone to whom the employees can confer respect.

Motivation is one factor that determines successful training, but another important consideration is the delivery method. Training programs need to have a variety of delivery methods, a challenging and relevant curriculum, and a direct application to the field setting. There is added value when firefighters from different departments, with a different set of experiences to share, can train together. Integrating multiple companies in the training exercises, exchanging instructors, and promoting other regional efforts to share training allows the departments to more easily be able to cooperate when a large-scale emergency calls for them to work together.

## Basic Firefighter Training

As a guideline, most firefighter basic training uses NFPA professional qualifications standards, such as NFPA 1001, *Standard for Fire Fighter Professional Qualifications*. There are a few states that have a mandatory statewide basic fire training or certification system; generally, however, each state and even larger city fire departments have their own system. In addition, many states and cities have different certification requirements for volunteers. Therefore, although they follow the same training and education outlines from the NFPA, no two systems are identical and they often vary substantially in length. This variance is addressed by the fifth goal of the National Fallen Firefighters Foundation's Everyone Goes Home Firefighter Safety Initiatives: "Develop and implement national standards for training, qualifications, and certification (including regular recertification) that are equally applicable to all firefighters based on the duties they are expected to perform" (National Fallen Firefighters Foundation, 2013).

As noted by the National Fallen Firefighters Foundation, firefighting training and education should be standardized. Because of an increasing amount of manmade and natural large-scale incidents, there are going to be far more situations in the future that require responses that cross local, county, and state lines. This kind of

### Case Study

**Collaborative Training**

In 2005, the County of Contra Costa, California, EMS Agency piloted a unique collaborative training model called the Contra Costa County Fire-EMS Training Consortium. This model pools the resources of nine fire departments, a large ambulance company, a trauma center, and two air ambulance providers. The curriculum for this training is created quarterly by different system partners. Time, energy, and money are saved by eliminating redundancy in curriculum creation and sharing resources and materials. Relevancy is ensured by using quality assurance and quality improvement data. Easy access to standardized curriculum provides a forum for fire and transport providers to train together in a real world setting, thus creating better cooperation, understanding, and improved patient care in the field.

The Contra Costa County Fire-EMS Training Consortium Advisory Board uses quality improvement data collected by the ambulance provider, county fire departments, and the EMS agency to determine the subjects that are targeted for training. This not only ensures that training is pertinent but also addresses identified system deficiencies. It also provides the opportunity for the EMS agency to provide input and ensure standardization of the message and the materials.

The program has its own quality improvement process, which includes instructor education and development, an annual needs analysis from field providers, feedback and evaluation on each training module, and regular updates on project design and implementation.

Experienced instructors from multiple departments provide variety and expertise. Educators are respected people to whom the students can relate. Training needs are easily and immediately adapted to actual practice. This program has proved to provide challenging, satisfying training with relevant curricula and realistic skills training. The end result is improved motivation and self-esteem on the part of Contra Costa County firefighters and EMS providers.

cooperation might also be precipitated by reduced resources from a struggling economy, prompting more departments to ask for help from their neighbors. FES agencies have to continue to figure out how to work together seamlessly and be confident in their neighbors' abilities. Standards and credentialing for firefighters and officers need to be national and portable.

Traditional in-service training programs rely on company officers and their knowledge, experience, and individual approach to training. Even with the most competent officers, many of these programs lack consistency because of the numerous fire station locations and different shifts. The training given in one station on a particular day is not the same as that given in another. Therefore, any in-service training program should start with department-wide planning and implementation.

For consistency, NFPA 1001, *Standard for Fire Fighter Professional Qualifications*, can be used as a basis for a departmental in-service training program. This program should be in addition to a new member's basic skills training. The key to providing the best fire and rescue service is a comprehensive, exhaustive recruit school for new firefighters and an in-service training schedule that stresses and reinforces basic skills. NFPA 1500, *Standard on Fire Department Occupational Safety and Health Program*, calls for monthly structural firefighting training, totaling a minimum of 24 hours per year. Again, this is a minimum number of hours and is only appropriate for departments that protect a small number of single-family dwellings.

## Live-Fire Training

One of the best training drills that results in higher levels of competency and morale is the live-fire drill. It can also be one of the most dangerous and should not be done without extensive planning and research, and the supervision of certified, experienced fire instructors. A great place to start in planning a live-fire drill is NFPA 1403, *Standard on Live Fire Training Evolutions*.

This document has a thorough process that, if followed, provides a safe but valuable experience for the firefighter. Two categories of structures are defined for use in a live-fire drill: acquired and live-fire training structures. Each has its own pros and cons; however, the acquired structures are more realistic. The biggest drawback is in gaining permission to burn, especially in urban areas, because this often generates large quantities of smoke.

Because NFPA 1403 is heavy on safety aspects, some skills are not tested. For example, in a real fire it is not possible to perform a walk through before arriving on the scene. Therefore, it is valuable to also include some other confidence building drills, such as navigating through a maze with a self-contained breathing apparatus with the face mask covered to simulate heavy smoke conditions.

This training is also a good time to put members through other seldom-used but critical skills, such as climbing to the top of the department's ladder truck, forcible entry, and ventilation skills. Generally, a whole day is set aside to complete this training and several companies are required to participate. Because many of these drills are physically demanding, members should be checked for minimum medical and physical benchmarks at the start of training and periodically throughout the day.

## EMS Training and Certification

Departments that have EMS-trained firefighters have to accommodate EMS and firefighter training. EMS has a national standard for training and testing. However, some minor differences can be found in a few states. Formal training and certification are required to become an EMT or paramedic in all states. Training requirements and minimum hours are recommended in a national curriculum created by the US Department of Transportation and the National Highway Traffic Safety Administration.

A certified EMT needs to accumulate 24 continuing education hours in each 2-year certification cycle, or he or she can take a 24-hour refresher course for recertification. A paramedic, however, needs to accumulate 48 continuing education hours. These hours are generally acquired through merit badge refreshers, such as advanced trauma or pediatric courses.

In addition, many states also require testing through the National Registry of EMTs before final certification or licensure. In some states it is a first-time certification requirement; in other states it is required as a basic part of continued certification or licensure.

It does not make a difference if an EMT is in a city, county, or state agency, or if the area served is urban or rural; each person has essentially the same training and certification testing. The only variable in practice may be optional skills as appropriate to an EMT or paramedic scope of practice, but those skills must be requested and approved by a local medical director.

## Technology-Based Training and Certifications

Interactive multimedia training is a very useful technology to FES. This means of training is especially useful for required yearly recertification training, such as that needed for hazardous materials operations. It is even being used in basic training in lieu of classroom lectures and to reinforce self-study material. The big advantage to this technology is the ability to obtain the certified training at the local fire station or at a member's own home.

The National Fire Academy and the Emergency Management Institute of the Federal Emergency Management Agency (FEMA) offer numerous online courses that when completed successfully will result in the student receiving a FEMA certificate. Some of these courses can even be applied to college credit and may help document the student's knowledge for a promotion.

Podcasts are another technique for distributing training to many FES members separated by distance. A 2009 study of podcast-based learning conducted by the State

University of New York at Fredonia discovered that students who listened to a podcast lecture outperformed those who attended in-person lectures. Furthermore, these same students performed better when they took notes, which was indicated by their use of the pause and rewind buttons (SUNY Fredonia News Services, 2009).

Although written, supervised examinations may still have to be administered in person, many certification tests can also be taken on the Internet. Web technology has the greatest potential for reducing the off-duty time needed for recertification. In addition, some computer-based training packages have the ability to administer a test and record the results in a tamper-proof computer file. Although these training packages may seem pricey, if costs are compared to classes that require a certified instructor to provide the same training, the interactive program would win. Plus, costs can be divided if several small departments share the use of the program. As with any new trend, changing the institutionalized certifying agencies is difficult but not impossible.

## Seminars and Training Sessions

Many FES organizations send their members to seminars or training sessions. Because there are more than 30,000 fire departments in the United States, it is impossible for one organization to have all the knowledge and experience needed for every potential emergency situation. Seminars are a great method for increasing knowledge and gaining valuable experience from national experts on advanced or specialized topics.

There are numerous state, federal, and national training programs and conferences that have excellent content. These training sessions are sometimes provided at a local training academy by outside experts. In some cases, a local fire department may be fortunate enough to have a nationally recognized expert to use as a resource. The training bureau may choose to record a webcast of the session for those who cannot physically attend, for circulation, or for future use. Another alternative may be to have the attendees teach their new knowledge to all members when they return home.

## Recertification

A weakness in most certification systems and in the NFPA professional standards is that there are no recertification or continuing education requirements. In the future this may change and existing members may be required to update their knowledge and skills on a yearly basis. For example, the State of Florida already requires people holding fire officer, fire inspector, and fire instructor certifications to complete 40 hours of continuing education every 3 years.

For departments located in areas where there is no requirement for recertification, however, administration can include items in the yearly performance appraisals that require self-study or the attendance of seminars and formal courses to update knowledge and skills. Over a 20- to 30-year career, many things change and it is very difficult to retain information that is used infrequently.

Therefore, each department should create a standard training program and schedule. This approach helps to ensure that all members have equal abilities, knowledge, and skills, and can easily move from assignment to assignment. It also ensures that an engine company that services a rural farm district is able to operate with efficiency and skill when dispatched or transferred to an area with high-rise buildings.

## Training to Fit a Need: Standard Operating Procedures

SOPs are rules that clarify consistent actions that result in proven outcomes and safe practices. There is a reliance on SOPs in the FES organizations because of the nature of emergency incidents. In many cases, split-second decisions must be made by the first arriving company officer, and these decisions can have a huge effect on the outcome of the incident. The likelihood of deaths and property damage increases when there are no SOPs in place; therefore, SOPs should be followed with few exceptions.

To illustrate the complexity of the initial actions at an emergency, the following list includes major questions that could confront the officer of a first-arriving engine company at a single-family dwelling fire:

- Should a supply line be laid for water, or should tank water on the engine be used?
- From the first observations, can it be determined where the fire might be in the structure and if there are any potential victims inside?
- What size hose line should be used?
- What type of nozzle is appropriate?
- Are there enough firefighters to attack the fire safely?
- If the building is locked, how do the firefighters gain access?
- Is there any information or observations that need to be communicated by radio to other responding companies or the incident commander?
- Which door should be used to enter the structure for fire attack?
- Does the building show any indications of a backdraft or flashover condition?
- Are there signs of structural weaknesses?
- What method will be used to account for the progress and location of all firefighters?
- Can it be determined if more resources will be needed than are presently on the scene or responding?

Answering questions such as these can be difficult, and many of the variables are typically unknown when the first responding company arrives. Therefore, after SOPs are adopted, training sessions must be required to ensure that all members are proficient in following procedures. Any actions or situations that might require variance need to be addressed and incorporated into the SOPs and training sessions. For example, even though many structural fires can be extinguished with a small amount of water, most departments require the use of a minimum-size hose line because the size of the fire might not be known until the company has spent a few minutes on the scene and has stretched its initial handline. SOPs such as these should be reviewed frequently and updated for their technical correctness and their relevance.

## Consistency and Reliability

SOPs are the basis for consistent and reliable levels of service. They should be followed with few exceptions, and must be studied, reviewed, and practiced frequently. The public expects a high degree of reliability in the service provided by the FES agency. SOPs that are rigorously enforced—similar to the military's commitment to obey orders—will provide the consistency the public expects.

## Variances in SOPs

There have been numerous arguments and discussions over the years involving the use of different fire tactics and equipment, such as the type of nozzle to be used (solid stream or fog). For controversial subjects, the chief officer should create an advisory committee that can analyze and make recommendations on SOPs. In addition, the chief officer should contact other FES agencies and request copies of their SOPs, finding out if they are actually followed or if there are enforcement issues.

The advisory committee would then be instructed to find the best solution to the issue, with the understanding that exceptions may be handled in the field. For example, a company has a fireground SOP that states that the first due engine take action to extinguish the fire. On one rare occasion, however, when approaching an apartment building, an officer of the first due engine spots a woman holding a baby at a third-story window. The officer orders the crew to remove the ground ladder from the engine and use it to rescue the victims. The officer should immediately radio to other companies and any responding chief officers that his or her company varied from the SOP and another company must accomplish the original assignment. Any time a variance such as this occurs, it needs to be communicated immediately to the command officer and the other companies that are responding to the same incident.

## Critique of SOPs

In the FES business, a critique of operations can point out deficiencies that, if not corrected, may cause poor firefighting practices, civilian or firefighter deaths, or other negative outcomes. An example of an illustrative, honest, and comprehensive critique of a firefighting operation was performed by the Prince George's County, Maryland, Fire Department after a single-family dwelling fire response required the rescue of a downed and unconscious firefighter.

The critique required many hours of investigation, interviews, and research to produce. It pointed out actions that were not in conformance with generally accepted safety practices and the department's own SOPs. The department then made these discrepancies known to all members.

Although only a large department with extensive resources is capable of a comparable effort, small departments also benefit from a brief critique. Comparing the actions taken with the department's SOPs is very valuable. A checklist of actions required by each company with an evaluation of yes/no is helpful. Even if this critique indicates that some companies and their officers do not comply with departmental SOPs, the critique can reduce the chance of a reoccurrence by the officer or other departmental officers. A critique should be primarily a learning experience, although its results can also be useful during the annual performance evaluation of members.

## Training to Fit a Need: Hazardous Materials

In the past two decades, there has been an increase in the number of hazardous materials teams created in the public sector. For example, because of the high density of oil and chemical industries in the area, the Houston (Texas) Fire Department has one of the busiest hazmat units in the world, with 956 runs in 2004 (*Firehouse*, 2005).

Emergencies involving hazardous materials have required a rethinking of many of the traditional approaches the fire service has relied on in past firefighting and rescue work. The speed with which an incident is approached, the level of preincident planning required, and the dependence on specialists from other fields including the private sector are all radically different from the SOPs that have been developed for firefighting. Although the experience obtained from an actual incident is a team's most valuable resource, it is still very important to practice hands-on drills and field training with team members and teams likely to be called in for mutual aid.

Still, many communities have found that since formation, these specialized teams—which are generally under the jurisdiction and command of a local fire department—have rarely been used. According to incident records collected from state and local jurisdictions, and discussions with members of various operational teams, hazardous materials incidents account for only 1% of the fire calls made by local departments. Approximately 50% of these incidents involve very small quantities (less than 55 gallons) of the hazardous materials, and 75% of these incidents occur at fixed facilities where the material is routinely stored and used.

In the initial rush to create hazardous materials teams everywhere, many experts originally supported the creation of teams. However, in the face of infrequent calls, limited chances to test their training, and an increased need to broaden their experience and knowledge in the field, hazardous materials team members often become discouraged and burned out. In addition, because a hazardous materials response team should consist of approximately 20 members to provide coverage on a 24-hour basis (including holidays and weekends), costs of staffing, training, and equipment are a major expense.

The increase in demand and costs for this type of training often reduces the team head count to less than optimum levels. Department chiefs who experience few hazardous materials–related runs have difficulty justifying the continued cost of training and investment in new technologies and equipment that keep the team at the highest state of readiness. This is one of the primary reasons that many communities have consolidated their teams into regional or district hazardous materials response units. A great example of a regionalized approach (several counties in each region) was created in Massachusetts FIGURE 8-2.

This same type of analysis can be used for any special service that a FES agency considers offering or presently provides. Many special services can be better provided through the cooperation of several departments to share the costs and staffing. Such services might include ladder or rescue services that are difficult to staff and whose personnel are difficult to train adequately in a small department with few calls.

## Training to Fit a Need: NIMS

After several national emergency incidents resulted in poor coordination of resources, the federal government created the National Incident Management System (NIMS). NIMS establishes a uniform set of processes, protocols, and procedures for which all responders, including EMS personnel, must be compliant. Primarily, the function of this new management system is to command and control operations at catastrophic emergency incidents.

NIMS works with the National Response Framework (NRF). NIMS provides the template for incident management and is the command system used to control and allocate the resources provided by the NRF. The NRF provides structure and mechanisms for national-level policies, allowing emergency response partners to prepare for and provide a unified national response to disasters.

Incident Command System (ICS) is a subcomponent of NIMS and the initial model of command for local incidents. With the exception of large disasters, all

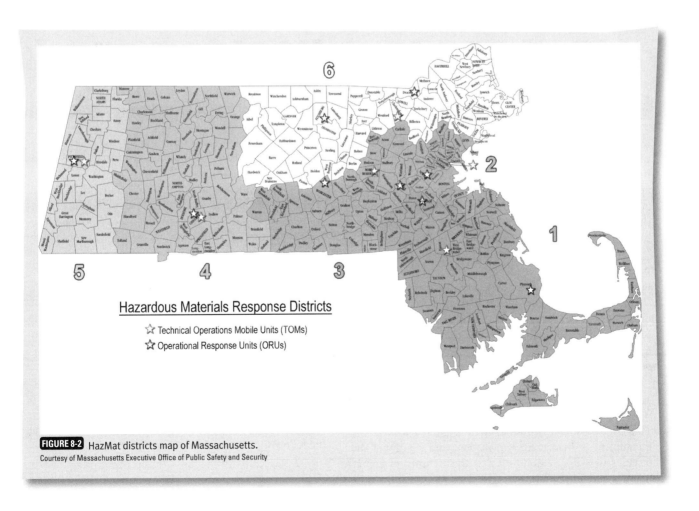

FIGURE 8-2 HazMat districts map of Massachusetts.
Courtesy of Massachusetts Executive Office of Public Safety and Security

emergencies start as local emergencies. When the situation gets more complex and starts to involve multiple local, state, or federal jurisdictions, then an expanded model of management is necessary. Fire departments are encouraged to use this system for everyday incidents so that when large incidents occur, the use of NIMS is second nature. As is true for all emergency operations, practice ensures emergency skills and procedures are accomplished successfully and safely. Making anything routine more likely results in it being done correctly.

Key incident command positions at an emergency medical incident may include the following:

- Triage unit leader
- Treatment unit leader
- Medical transportation officer
- Staging supervisor
- Safety officer

All public safety and EMS agencies must preplan and become competent in the use of tools, techniques, practices, and the vocabulary of the ICS. According to FEMA, each agency should do the following (Kirkwood, 2008):

1. Develop policies and procedures for training, implementation, and evaluation of NIMS.
2. Have a cache of tools necessary to function in an ICS/NIMS environment. These tools include radios programmed with common channels, reflective vests, and other incident management tools, such as ICS forms and tracking boards.
3. Develop ICS/NIMS competence by getting involved with local planning, exercises, and after-action reports.
4. Be prepared to act as the incident commander in cases where they may be the first arriving unit on an incident.

NIMS-compliant ICS should be used for every incident involving more than one resource. The first arriving unit—whether a police vehicle, private ambulance, or fire engine—initiates initial command. As other agencies appear on scene, command should be passed to the most appropriate and qualified agency. Which agency takes that role depends on the type of incident, the local structure, and the resources available.

For example, in situations involving multiple-patient injury or illness scenarios, the initial command or operations section chief would most appropriately be an EMS expert. Another option in these scenarios might be to have an experienced and skilled EMS personnel serve as a medical group supervisor to help with patient care–related decisions. How these roles are determined depends on the level of the agencies' ICS expertise.

In some cases, a private or nonfire EMS unit may arrive on the scene and set up a parallel command, which may or may not coordinate with the fire service. This could result in redundancy of requested resources, incomplete or inaccurate information, and a number of other issues. For reasons such as these, unity of command is very important.

> **Facts and Figures**
>
> As a result of the mutual-aid agreements in place by Metropolitan Washington Council of Governments in the Washington Metropolitan Area, a standard numbering system was put into place for area fire departments. All units have a three-digit designator, with the first digit denoting what jurisdiction the unit is from. The following are examples:
>
> - 000 series Washington, D.C.
> - 100 series Arlington County, Virginia (includes City of Falls Church)
> - 200 series City of Alexandria, Virginia
>
> In this system, Engine 101 would be an engine from Station 1 in Arlington County, Virginia. This number-identifier system is a great help for mutual aid calls across jurisdictions that can be city, county, or state.

EMS supervisors and senior staff should be encouraged to get higher-level ICS training and experience, and to participate in state and regional Incident Management Teams. Not only will this teach the EMS agencies practical application of the system, but it will also show other public safety agencies that they are experts and can function in an NIMS/ICS incident with a unified command.

## Training Goals: Initial Fire Attack

This section provides an example of one portion of the successful extinguishment of a structural fire and is a good example of how to determine acceptable competencies for firefighting. Similar approaches can be derived for other emergency services operations. Many other skills and knowledge are needed to manage the varied emergencies a department is called to handle.

To determine a department's capabilities for initial fire attack performance at structural fires, NFPA 1410, *Standard on Training for Initial Emergency Scene Operations*, can be used. This standard describes methods for evaluating the adequacy of a department's training program, including typical fireground evolutions broken down into three separate items (handlines, master streams, and automatic sprinklers) and recommended completion times for each. The fire company or department can evaluate the effectiveness of its training program by comparing the times that its crews attain performing the evolutions.

Experience has shown that with appropriate training, SOPs, and adequate staffing, most fire departments can complete these evolutions within the suggested times. If the times are slower than those recommended in the standard, the administration may want to evaluate the following:

- Frequency of training sessions or actual fireground experience. If firefighters perform

these basic evolutions frequently at actual fires, their skill level is higher. If the fire experience is low, the frequency of in-service training needs to be increased.
- SOPs. These may contain evolutions that slow the emergency operations. The critical importance of SOPs has been discussed previously.
- Staffing. Staffing can affect the times of each evolution. Staffing is described in NFPA 1410 as "the average number of personnel that ordinarily respond" (NFPA, 2000a). However, the staffing provisions listed in NFPA 1500, *Standard on Fire Department Occupational Safety and Health Program*, must be followed to ensure the safety of firefighters participating in the evolutions and on the fireground. For handline evolutions, NFPA 1410 explains, "A minimum of two firefighters shall be used on each hose line to keep interior attack lines under control" (NFPA, 2000a). Similar wording is found in NFPA 1500, which states, in part: "Members operating in hazardous areas at emergency incidents shall operate in teams of two or more" (NFPA, 2000b).

All three of these areas (training frequency, SOPs, and staffing) should be considered individually and in combination when analyzing a department's problems with meeting competencies.

Other training goals can be formulated from standards, such as NFPA 1001, *Standard for Firefighter Professional Qualifications*, which specifies the minimum competencies for a firefighter, and NFPA 1500, *Standard on Fire Department Occupational Safety and Health Program*, which provides the framework for a safety and health program.

# Higher Education

There are many forms of incentives to encourage members to take advantage of a formal college-level education. For example, some departments use educational incentive pay or policies regarding credit or minimum educational requirements for promotional opportunities.

Most higher education programs in FES are offered at community colleges, with a few universities offering related bachelor's degrees. Many community colleges have a committee made up of representatives of FES organizations to provide curriculum advice. In addition, the National Fire Academy's Higher Education program has developed a recommended core curriculum for fire science programs. The model curriculum can be found at the US Fire Administration Web site (http://www.usfa.fema.gov/nfa/higher_ed/index.shtm).

When considering a college-level education, it is important that individuals are sure the particular program meets their needs. To check whether the college is accredited, visit the websites of the US Department of Education (http://www.ope.ed.gov/accreditation/) or the Council for Higher Education Accreditation (http://www.chea.org/search/). Each contains a list of accredited colleges. In addition, individual courses that are recommended by The American Council on Education may not be accepted at many of the regionally accredited schools. Individuals should ensure that course work will transfer if there are potential plans to go to another institution of higher learning.

Every college and employer has the right to set standards and refuse to accept transfer credits and college degrees from institutions they deem substandard. Students that have gone to a nationally accredited school may find it especially difficult to transfer credits. A 2005 study by the US Government Accountability Office (GAO) found that, in making decisions on college credit and degree acceptance, about 84 percent of the country's higher education institutions considered the accreditation of the institution; many had policies stating that they would accept college credits and degrees only from regionally accredited institutions. One reason given for regional institutions' reluctance to accept credits from nationally accredited institutions is that national accreditors have less stringent standards for criteria, such as faculty qualifications and library resources (US Government Accountability Office, 2005, p. 3).

In addition, examine the courses offered by the institution. Many institutions offer only fire science courses but do not offer courses on EMS or emergency preparedness management and operations, which may be of interest to those who would like a more inclusive education in emergency services.

Finally, look at the official degree title. Some institutions add a small number of courses to an existing program, such as business or public management. On a transcript, a degree with an official title of Bachelor of Science in FES means more to employers than a Bachelor of Science in Public Administration with a concentration in fire science.

Bachelor's degrees and some associate's degrees can be achieved by distance learning over the Internet **FIGURE 8-3**. This is the most common method of instruction

**FIGURE 8-3** A distance learning student.

for FES education at the bachelor's degree level because most students are spread throughout the country.

There are some drawbacks to Internet-based programs. They require a great deal of self-discipline in setting aside the time for studying and completing homework. However, those who have adapted to the rigid structure of firefighting and EMS generally do well. In fact, many of them complete entire degree programs through the online medium.

## Officer Education and Training

The methods for selecting FES officers range from popular election to sophisticated testing, education, and experience criteria. Currently, there is no national consensus for the minimum training, education, and experience for a fire officer. The closest thing that FES has for professional qualifications is the comprehensive guide to education, training, and self-development discussed in the International Association of Fire Chiefs' *Officer Development Handbook*, second edition (2010). This book is an excellent resource for a career fire officer, because it outlines the training, experience, and education needed to successfully gain a promotion and be effective in an officer position.

The following is a checklist of training and characteristics that are helpful in preparing to function as a professional fire officer:

- NFPA professional qualification certifications
- College degrees in fire science or administration
- Promotional examinations
- Officer candidate schools
- In-service training
- National Fire Academy courses
- FES textbooks (self-study)
- Professional periodicals
- Conference and seminar attendance
- Experience and common sense
- Knowledge from the experiences of other departments of all sizes and types (there are good ideas everywhere)

Chief officers need to build on the experience and training they gained during recruit and in-service training, company officer training and education, and formal higher education (bachelor's level is recommended) to be effective **FIGURE 8-4**. As a prerequisite, the general educational (i.e., reading, writing, math, history, humanities, and so forth) knowledge for the company officer is relevant to the chief officer and is offered at many community colleges. A chief officer should look at formal education as another tool to help be effective at their job. Many of the other agency heads, appointed officials, and elected representatives will have at least a bachelor's degree. A good educational background puts the officer and any staff on equal ground with contemporaries, and that equality gives the chief many opportunities, such as to acquire a fair share of the tax revenues. In most cases, education gives the chief the tools needed to be a change agent.

**FIGURE 8-4** Fire Administration class at the Florida State Fire College.

Conferences and seminars have programs for advanced and infrequently used skills and sessions for management and leadership. These are great opportunities to gain contemporary knowledge and education, and to reinforce basic management and leadership skills. However, there is no peer review of the content contained in these educational and training sessions. Therefore, it is good to confirm what is heard with others' knowledge and one's own experience. Have an open mind and be prepared to accept new ideas that may be different than what was previously accepted as the truth. Many administrative procedures and techniques evolve and change over time.

Another great opportunity for an administrator is to be a presenter at one of these sessions. With the appropriate knowledge and experience, the seasoned administrator has a lot of wisdom that can be passed on to others. Ideas become clearer as they are justified and explained to an audience. During the presentation, there may be questions or feedback from the audience. A good lecturer tries to learn something from every audience.

For emergency management competency, the Emergency Management Institute of FEMA provides a nationwide training program of resident courses, nonresident courses, and interactive Internet courses to enhance US emergency management practices.

In addition, training and education are crucial for quality customer service. It takes a strong commitment to train at the level needed to keep all basic skills sharp over 20–30 years of service. It also takes a disciplined and rigorous promotional and in-service training program to keep officers at their peak of effectiveness. It is more common to replace an inefficient fire engine than retrain a seasoned firefighter to the skill level achieved earlier in his or her career. Just as water pumps are tested once each year, testing firefighters once each year is good practice.

# National Professional Development Model

High-quality, professional public FES is primarily the result of two basic factors. The first of these, equipment and facilities, is the most obvious and typically gets the most attention. The other factor is the human resources members who must be in good physical condition, well trained, and well educated. If any one of these elements is weak, critical emergency service suffers.

**FIGURE 8-5** is a representation of the balance of training and education necessary in the FES. The model was developed by the US Fire Administration/National Fire Academy's Fire and Emergency Services Higher Education program.

The right side of the graphic represents training, the middle represents the experience and self-development attained over many years, and the left side represents education. Training is typically attained in firefighter basic training, fire officer courses, seminars, and company drills. The basis for most of this training is contained in the NFPA fire-service professional qualifications standards, a series of distinct criteria, starting with those that pertain to entry-level firefighters and going up through four levels of officers.

For instance, Fire Officer I and II are generally company officers, or officers who may command up to three companies. These positions normally carry the job title of lieutenant or captain, although in smaller one- or two-station departments, the position title may be chief. On the education side of the pyramid, the banner notes that an associate's degree should parallel the positions of Fire Officer I and II.

The best practice is to offer combined fire officer training and education. Many community colleges offer fire officer courses that also meet the requirements of the state certification system. This is beneficial to the students since they do not have to take two identical courses: one for college and the other towards certification.

The most common associate's degree is in fire science. Fire science degrees primarily contain coursework that focuses on fire prevention, management, and incident command skills. The modern, progressive fire department realizes it can play a leading role in reducing loss of life and property by preventing fires and lowering their risk. Achieving this goal requires officers who are professionally educated to offer fire safety education and code enforcement, and modern emergency incident skills. As is true in all professions, it is not realistic to expect fire officers, especially newly promoted ones, to be proficient in all situations as a result of only their experience. In addition, it is always better to learn from the mistakes of others, because these mistakes have the potential to result in death and property loss. For these reasons—along with the new likelihood of being first to arrive at a terrorist incident—the knowledge and skills needed to respond successfully to an emergency can be gained only through a formal education.

As the complexity of administrative roles increases, so does the need for officers within the roles of battalion, division, deputy, or assistant chief. The National Professional Development Model recommends that

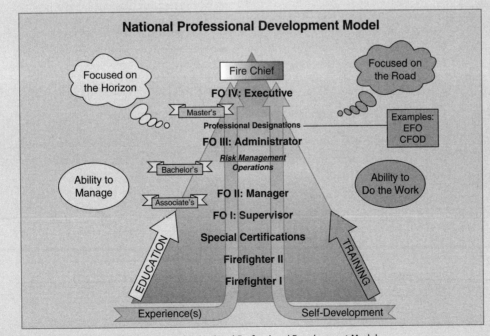

**FIGURE 8-5** Fire and Emergency Services Higher Education National Professional Development Model.
Courtesy of FEMA

officers with these types of titles have a bachelor's degree. Although implementing educational requirements can be challenging, discussions with department members and their representative employee organization are always helpful. Such incentives as educational pay and tuition reimbursement may be used. Two additional techniques may also be helpful.

The use of the first technique was exemplified by the New York City Fire Department. In 2001, days before the World Trade Center terrorist attack, the department announced that for an officer to be eligible to take the battalion chief and deputy chief examinations, the officer needed to have a bachelor's degree. However, it was not until 2008 that the requirement began taking effect. This type of phase-in is common for education requirements to allow existing department members time to acquire the education.

Another method used to implement educational requirements is to award bonus points to written test scores depending on different levels of higher education. These bonuses may be modest at the beginning of the policy's implementation and become substantial after several years. An appropriate time frame would allow a member the opportunity to gain the necessary education. In addition, as the need for higher education increases with higher officer ranks, point level can be increased accordingly.

Awarding additional points on the promotional list score offers an advantage, especially in smaller departments. This practice is more flexible than the New York City Fire Department plan during the implementation phase, but it does not guarantee that the people promoted have a minimum level of education. For example, a department that has only a few individuals with the minimum educational requirements might still find that no one on the promotion list is suitable. This department's only option might be to hire from outside. This option usually has merit only at the fire chief's level because of the need for junior officers to be familiar with local conditions.

The National Professional Development Model effectively takes the fire service to the final level of professionalism. Obtaining the knowledge and skills associated with this model results in a highly professional fire service that delivers better service to its customers. Still, training and education are only one part of the effort to become a professional; experience is also needed. Gaining knowledge from past experience and seeking to understand new ideas should be a never-ending effort for the FES officer.

## CHAPTER ACTIVITY #1: Would Training Have Made a Difference?

After a firefighter within a small community was killed fighting a house fire, the community was quick to blame the department's poorly organized training program. Many believed that the department ought to refocus on the basics of structural firefighting rather than spend time training for EMS, terrorist and hazmat incidents, and numerous other services. Although these critiques may be valid, deaths such as this open the door to a very sensitive subject. Often, a person or organization cannot make needed changes without facing uncomfortable situations.

### Discussion Questions

1. Would you support mandatory firefighter training including on-the-job refresher training by either the state or federal government?

2. Make a list of those training programs needed and the frequency with which they should be held.

3. Review a local fire department's training program and make an analysis of its training needs.

## CHAPTER ACTIVITY #2: Infrequently Used Skills

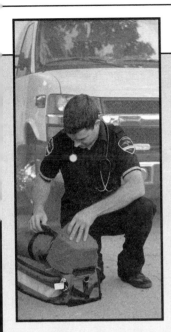

A suburban fire department gets a call for a trench rescue. The first due crew finds a contractor buried to his waist in an 8-ft deep trench. He is conscious and alert and explains that he was hired to replace some pipes. Trench rescues are not common for this area, and the battalion chief and firefighters feel unprepared.

A few firefighters jump into the trench to approach and assess the patient, but the battalion chief orders them to retreat until the trench can be stabilized. A paramedic then approaches the patient from the edge to assess and treat, but he does not remember specific treatments other than fluid infusion. It is quickly determined that the team does not have the equipment or expertise to mitigate the incident safely. A trench rescue team from a department across the county is requested, but it takes 2 hours for them to assemble and transport personnel and equipment.

Four hours later the patient is extricated and transported. Later discussion reveals that most members of the department had not had trench rescue training for 5–7 years. The program was considered less important than others and was cut because of budget woes.

### Discussion Questions

1. What is the issue in terms of safety for the firefighters and the patient?
2. Could this incident have created political issues for the fire district?
3. What actions should be taken by the department after the fact to ensure the situation is not repeated in the future?

## References

Firehouse. (2005, June). 2004 national run survey. *Firehouse*, 54.

Firehouse. (2005, August). 2004 national run survey. *Firehouse*, 42.

International Association of Fire Chiefs. (2010). *Officer development handbook* (2nd ed.). Fairfax, VA: International Association of Fire Chiefs.

Kirkwood, S. (2008). *NIMS and ICS: From compliance to competence*. Retrieved from http://www.emsworld.com/article/10321278/nims-and-ics-from-compliance-to-competence

National Fallen Firefighters Foundation. (2013). *16 firefighter life safety initiatives*. Retrieved from http://www.everyonegoeshome.com/initiatives.html

National Fire Protection Association. (1997). Evaluation and planning of public fire protection: Planning fire station locations. In *NFPA fire protection handbook* (18th ed., chap. 10–20). Quincy, MA: National Fire Protection Association.

National Fire Protection Association. (2000a). NFPA 1410, *Standard on Training for Initial Emergency Scene Operations*. Quincy, MA: National Fire Protection Association.

National Fire Protection Association. (2000b). NFPA 1500, *Standard on Fire Department Occupational Safety and Health Program*. Quincy, MA: National Fire Protection Association.

Parow, J. (2011, April 15). Higher education: The key to professional fire/EMS recognition. *IAFC On Scene*. Retrieved from http://www.iafc.org/Operations/LegacyArticleDetail.cfm?ItemNumber=4279

SUNY Fredonia News Services. (2009). *Study tests effectiveness of podcasts vs. lectures*. Retrieved from http://ww2.fredonia.edu/news/AllNewsReleases/tabid/1101/ctl/ArticleView/mid/1878/articleId/1449/Study_tests_effectiveness_of_podcasts_vs_lectures.aspx

US Fire Administration. (1997). *The aftermath of firefighter fatality incidents, preparing for the worst*. Retrieved from http://www.usfa.fema.gov/applications/publications/

US Government Accountability Office. (2005). Transfer Students. Retrieved from www.gao.gov/new.items/do622.pdf

# CHAPTER 9

# Health and Safety

## Fire and Emergency Services Higher Education (FESHE) Course Objectives

### Module IV: Leading Change

The students will:

1. Describe the importance of accepting and managing change with the fire and emergency service department. (pp 126–127, 129–138)
5. Describe how an organization can respond to current or emerging events or trends. (pp 131–138)
6. Explain the benefits of employee involvement in departmental decisions. (pp 133–134)
8. Describe ways to increase and reward professional development efforts. (pp 132–134)

## Knowledge Objectives

After studying this chapter, the student will be able to:

1. Understand the importance of safety for fire and emergency services (FES) and its members. (pp 126–127)
2. Recognize the influence that the National Fire Protection Association (NFPA), US Occupational Safety and Health Administration (OSHA), Everyone Goes Home, and other safety organizations have on FES. (pp 127–130)
3. Examine progressive health and safety trends in FES. (pp 127–138)
4. Realize the impact that work-related injuries and deaths have on an organization and its members. (pp 129–138)

## An Introduction to Health and Safety for Emergency Personnel

Any organization's most valuable resource is the people that work for the organization, and the well-being of these people is vital to the organization's success. In a profession that is so dependent on physical skills—as is the case in the FES—it is especially important to ensure the health and safety of its members.

### Firefighter Safety

In the past, firefighter safety has been at times a controversial topic. When the first safety standard for the fire service was issued by the NFPA in 1987, many FES leaders, along with city and county managers, predicted that it would lead to departments going out of business, or that huge sums of money would be required to comply with the safety standard. Neither of these predictions has come true; instead, many FES departments have used safety standards to justify additional funding for important upgrades, such as more modern protective clothing or new hires for inadequate staffs.

Nevertheless, there still seems to be selective compliance by FES agencies. Recently, the NFPA reported, "While the number of structure fires and deaths at structure fires has dropped, the rate of firefighter deaths at structure fires has not followed the same pattern [2000-2009]" (Fahy, 2010, p. 7). The National Firefighter Burn Study, which was conducted in 2002 in association with the Washington Hospital Burn Center and the Maryland Fire and Rescue Institute, provides even more insight into this issue of noncompliance. The data for this study were voluntarily and confidentially collected from firefighters treated at 20 burn centers throughout the United States. The following are some of the results (National Firefighter Burn Study, 2002):

- 22% were required to wear glasses but were not wearing them
- 16% had not achieved Firefighter I competency
- 22% received burns at a training fire
- 6% were alone when burned
- 29% were standing in the fire area
- 6% had consumed alcohol within 12 hours
- 64% were advancing a hose line
- 59% described the heat buildup as "instant"
- 24% were in a flashover
- 20% were not wearing a hood
- In 18% of the hand burns, the firefighters were not wearing gloves
- 88% reported that they were "reluctant to return to duty"
- 45% believed the injury was preventable

In addition, the results of a study conducted by the University of Minnesota's Carlson School of Management/City of Minneapolis indicated that firefighters who tended to ignore safety rules and regulations had accidents more frequently and suffered more severe injuries than firefighters who conscientiously followed safety rules (Liao, Arvey, & Butler, 2001). All of these troubling statistics indicate that it is time for FES agencies everywhere to review their compliance with standards and recommendations set out by the NFPA.

## Emergency Medical Services Personnel Safety

Although there are fewer sources of data identifying threats to and injuries of emergency medical services (EMS) personnel, a few commonly used sources include the Federal Census of Fatal Occupational Injuries, National Highway Traffic Safety Administration's Fatality Analysis Reporting System, the National Emergency Medical Services Memorial Service, and the Bureau of Labor Statistics. The US Bureau of Labor Statistics (2012) lists a number of reasons why EMS personnel face higher risks of illness or injury than most other professions, such as working outdoors in all types of weather; kneeling, bending, and performing heavy lifting; and being exposed to communicable diseases or violent patients.

Long-term effects of multiple injuries to knees, backs, and shoulders could affect the quality of life, and certainly the job longevity, of EMS personnel. These types of non–vehicle-related injuries pose a challenge for researchers, because there are fewer established data definitions and repositories for collection and analysis. However, one study published in the *American Journal of Emergency Medicine* stated that, in July 1990, researchers in Pittsburgh examined 254 injured EMS personnel. Low back strain was the most common injury found, accounting for 36% of the total injuries (Chapeau, 2006).

Unfortunately, back injury is only one type of serious injury that can be career ending to personnel. FES administrators should also recommend the cessation of damaging behaviors, such as smoking; encourage surveillance and management of weight, blood pressure, and cholesterol; and reward healthy behaviors, such as stretching muscles, staying hydrated, maintaining good bone health, getting enough sleep, and keeping positive eating habits.

## NFPA Safety and Health Standard

NFPA 1500, *Standard on Fire Department Occupational Safety and Health Program*, was established in 1987 with the goal of making the FES profession less dangerous by reducing the risk of accidents, injuries, and fatalities in the course of conducting emergency operations and routine duties at the station. NFPA 1500 is also effective at helping personnel become healthier and more physically able to undertake their duties. This combination directly benefits the individual, the organization, and the public they serve.

Although many FES organizations are in the process of adopting the requirements contained in this standard, very few actually comply with it in its entirety. This partial adoption is not an unusual situation, and NFPA 1500 allows a phase-in schedule for compliance.

Following the standards set in NFPA 1500 should be one of the chief officer's highest priorities for several reasons. The first is a moral obligation to provide a safe working environment for the members of this very hazardous profession **FIGURE 9-1**.

Second, caring for personnel injured on the job is expensive. In the case of burns, the costs can be exorbitant and complete recovery can take years. Many FES personnel are better motivated if they know that their administration is looking out for their welfare.

Finally, many NFPA 1500 requirements ensure that members of the FES organizations will be competent professionals. This standard outlines the minimum training, education, and physical fitness requirements needed to provide the skills and knowledge that are necessary to be an effective firefighter.

## NFPA 1500 Major Topics

The following is a list of major topics found in NFPA 1500, along with a brief explanation of each:

- Health and wellness. Health problems, including heart attacks, are consistently noted as the number one cause of firefighter deaths. FES has made little progress in addressing this issue, and the health and wellness requirements in NFPA

**FIGURE 9-1** A hazardous profession.
© Glen E. Ellman

1500 are sometimes poorly defined. This topic is discussed in greater detail later in this chapter.
- Vehicle accident prevention. Next to heart attacks, vehicle crashes are the most common cause of duty-related firefighter deaths.
- Equipment and vehicles. Issues regarding equipment and vehicles have been addressed by many FES organizations. The federal government's Assistance to Firefighters Fire Grant program helps needy departments purchase newer and safer apparatus and equipment. Unfortunately, many rural and other small departments are still struggling to find enough money to buy safety equipment.
- National Incident Management System. Progress in this area is ongoing but common in most larger fire departments. Online and in-person National Incident Management System certification courses are offered throughout the United States.
- Safety committee. These committees are a major part of implementing a comprehensive safety program and provide the avenue to plan and implement changes needed to comply with the standard. Safety committees also provide a real-time discussion of current, potentially unsafe equipment; employee behavior; and operations. Members of this committee should be selected from management, labor, and mid-level officers. Multilevel recruitment helps ensure that discussions focus on considering and assessing everyone's point of view. It is also useful to invite a representative from the municipality's risk management office. Besides providing the opinion of a safety expert, the representative has extensive knowledge of the fiscal impact of proposed changes, including insurance costs and worker's compensation issues.
- Standard operating procedures (SOPs). SOPs set out safe and efficient methods of managing specific emergency operations, such as attacking structural fires. The safety committee provides input that can prevent deaths and injuries. Previously developed SOPs are available from multiple sources, but what works for one organization may not be effective for another. Therefore, blindly copying another organization's work is strongly discouraged. Instead, it may be useful to review the Federal Emergency Management Agency's (FEMA) publication *Guide to Developing Effective Standard Operating Procedures for Fire and EMS: Federal Emergency Management Agency United States Fire Administration* (1999).
- Training and education. These issues are covered in more depth in the chapter *Training and Education*.

## Federal Safety Regulations

Because of language within the US Constitution that prevents the federal government from having too much power over the states, many federal regulations—including NFPA safety standards and OSHA regulations—are not enforceable at the state or local level. However, when a state adopts the federal OSHA regulations (and receives federal grants in return), the regulations then become enforceable for state and local government employers and employees. Many departments in nonparticipating states use the federal and NFPA safety standards, with the understanding that they are providing a safe working environment for their members and reducing the chance of liability.

### Facts and Figures

**OSHA Regulated States**
The following states have OSHA-approved safety regulations:
- Alaska
- Arizona
- California
- Connecticut
- Hawaii
- Indiana
- Iowa

*(continues)*

## Facts and Figures (Continued)

- Kentucky
- Maryland
- Michigan
- Minnesota
- Nevada
- New Mexico
- New York
- North Carolina
- Oregon
- South Carolina
- Tennessee
- Utah
- Vermont
- Virginia
- Washington
- Wyoming

In addition, Florida, Illinois, and Oklahoma have typically adopted OSHA regulations for use by fire departments.

Courtesy of National Fallen Firefighters Foundation's Everyone Goes Home® Program

## Firefighter Injury and Fatality Prevention

Every year, approximately 100 firefighters in the United States lose their lives in the line of duty. Recent experience shows a slightly lower number of 83 for 2011 and 2012. Over the past few years there has been a realization that, although firefighting is very dangerous, many of these deaths and injuries are preventable. Although various national leaders and organizations have attempted to convince the fire service to follow safety standards, it is clear that for some organizations, the culture of risk remains unchanged. For example, consider live fire-training exercises for recruit firefighters, in which instructors deliberately create unsafe, extremely high-temperature evolutions. Think about the firefighter with heart disease who continues performing regular firefighter duties, performing strenuous physical work even when it is medically unsafe.

Examples such as this point to a FES safety paradox. Every effort is made to recruit members who are courageous and take risks. When faced with a dangerous situation, this type of person looks for ways to rescue victims and save property from destruction, even at their own peril. Although these brave members are needed, the challenge—through discipline and training—is to make these members safety-conscious and have them practice safe behavior at emergency incidents, ensuring their own well-being and taking risks only when necessary.

The National Fallen Firefighters Foundation (NFFF) is working to do just that. In 2004, with the cooperation of the US Fire Administration, the NFFF gathered about 200 prominent national fire service leaders together to focus on prevention of line-of-duty deaths and reduction of firefighter injuries. The NFFF described the meeting by stating the following: "The Summit marks a significant milestone, because it is the first time that a major gathering has been organized to unite all segments of the fire service behind the common goal of reducing firefighter deaths. It provided an opportunity for all of the participants to focus on the problems, jointly identify the most important issues, agree upon a set of key initiatives, and develop the commitments and coalitions that are essential to move forward with their implementation" (National Fallen Firefighters Foundation, 2004).

The result of the meeting was the NFFF's Everyone Goes Home program, which cites the following 16 Firefighter Life Safety Initiatives:

1. Define and advocate the need for a cultural change within the fire service relating to safety; incorporating leadership, management, supervision, accountability, and personal responsibility.
2. Enhance the personal and organizational accountability for health and safety throughout the fire service.
3. Focus greater attention on the integration of risk management with incident management at all levels, including strategic, tactical, and planning responsibilities.
4. All firefighters must be empowered to stop unsafe practices.
5. Develop and implement national standards for training, qualifications, and certification (including regular recertification) that are equally applicable to all firefighters based on the duties they are expected to perform.
6. Develop and implement national medical and physical fitness standards that are equally applicable to all firefighters, based on the duties they are expected to perform.
7. Create a national research agenda and data collection system that relates to the initiatives.
8. Utilize available technology wherever it can produce higher levels of health and safety.
9. Thoroughly investigate all firefighter fatalities, injuries, and near misses.
10. Grant programs should support the implementation of safe practices and/or mandate safe practices as an eligibility requirement.
11. National standards for emergency response policies and procedures should be developed and championed.
12. National protocols for response to violent incidents should be developed and championed.
13. Firefighters and their families must have access to counseling and psychological support.
14. Public education must receive more resources and be championed as a critical fire and life safety program.

15. Advocacy must be strengthened for the enforcement of codes and the installation of home fire sprinklers.
16. Safety must be a primary consideration in the design of apparatus and equipment.

Initiatives five and six call for national standards for training and certification, and national standards for medical and physical fitness. Respectively, these specific standards have the potential to save the greatest number of lives and reduce injuries among firefighters. However, they both start with the words "develop and implement," indicating that a lot of work still needs to be done and action must be taken by a standards-making organization, such as the NFPA. Although the NFPA does already have standards that include these topics, they are more of an outline than a specific standard that would result in firefighters having equal abilities, health, and fitness (refer to the chapter *Training and Education*). At this time, there does not seem to be any movement for the appropriate NFPA committees to more specifically confront these issues, and there is always the potential that influential special interest groups might resist efforts to develop national criteria for training, education, and medical and physical fitness requirements for firefighters. In most cases, this leaves the responsibility of development and implementation to the local or state FES agencies.

## EMS Injury and Fatality Prevention

The *EMS Workforce Agenda for the Future*, prepared by the University of California, San Francisco, with funding from the National Highway Traffic Safety Administration and the Health Resources and Services Administration, investigates the issue of developing and retaining a professional, well-trained EMS workforce. This document, which was released in 2011, follows and compliments the 2008 *EMS Workforce for the 21st Century: A National Assessment*. It identifies four key research components that are considered critical to developing a thriving EMS workforce (National Highway Traffic Safety Administration, 2011):

- Education and certification. Variations of EMS certification and licensing create a lack of reciprocity—an issue that is especially important for EMS personnel responding to disasters outside of their immediate area or state. The *Agenda* recommends a national EMS education process to eliminate this issue and establish a high standard of quality across the country.
- EMS data. It is necessary to research and evaluate workforce issues; however, until accurate data are available, attempts to improve issues for the EMS workforce are difficult to justify. The challenge seems to be centered on two issues: the lack of trained EMS researchers and the lack of reliable funding sources. One solution is for EMS agencies to contribute to the nationwide collection of EMS patient and system data, the National Emergency Medical Services Information Systems.
- Health, safety, and wellness. The *Agenda* recognizes that EMS can be dangerous, because employees are vulnerable to infectious diseases, patient assaults, and vehicle crashes. It recommends the establishment of a national EMS Workforce Injury and Illness Surveillance Program to study these issues and look for solutions.
- Planning and development. To predict future demands for workforce needs, planning and development of resources are essential. There are currently few data on which to predict future needs. It is recommended that state agencies help collect information on what drives EMS requests, workforce supply and demand, and compensation issues.

## Ambulance and Fire Apparatus Crash Prevention

States have provisions in their vehicle driving regulations that allow emergency vehicles to exceed posted speed limits and proceed through controlled intersections without coming to a complete stop. These emergency driving regulations contain some language that assumes the emergency vehicle driver will use all due caution in being granted this privilege. If the emergency vehicle hits another vehicle while exceeding the speed limit or proceeding through a stop signal or sign, the driver was likely driving with some carelessness. Many vehicle crashes are preventable. Safety professionals dislike calling them accidents and prefer to call them crashes because they generally are not accidents. Instead, most serious vehicle crashes are the result of excessive speed, failure to stop at traffic signals and stop signs, or general lack of caution at intersections.

Provisions to control the most common causes of FES serious accidents are outlined in NFPA 1500. For example, when drivers encounter a stop signal or sign, a stopped school bus, or an unguarded railroad crossing, they should come to a complete stop and not proceed until it is confirmed that it is safe to do so (NFPA 1500 6-2.8). Very specific written policies (SOPs) governing vehicle speeds and limitations during inclement weather are needed to reduce accidents and any potential liabilities **FIGURE 9-2**.

In the 1997 Pennsylvania Supreme Court case *Schwarzbek v. City of Wauseon*, the City of Wauseon was sued by the estate of Blaine and Mildred Sanders, who were killed when their car was struck by an EMS vehicle reporting to a scene. The Court Justice issued the following statement, which touches on the issue of speed and emergency vehicles: "while the law—in view of the vital and urgent missions of fire department vehicles—wisely allows them certain privileges over other vehicles, it

**FIGURE 9-2** SOPs governing vehicle speeds reduce accidents and potential liabilities.
© Peter Willott, St. Augustine Record/AP Photos

does not assign to them absolute dominion of the road… speed, which is uncontrolled, is capable of wreaking as much havoc and causing as much sorrow as fire itself" (Brannigan, 1997, p. 28).

In addition to standards regarding speed, NFPA 1500 contains other standards that help prevent FES crashes and injuries. For example, when operating at a vehicle crash on a highway, members should wear a garment with fluorescent reflective material for visibility, use fire apparatus as a shield, and place appropriate warning devices including highway caution signs to warn oncoming traffic. Furthermore, the inspection, maintenance, and repair of vehicles have also been identified as methods of preventing serious vehicle accidents. Requirements for creating a program to eliminate these causes are outlined in NFPA 1500. Finally, accident investigation and analysis have also been incorporated into the NFPA standard. Without feedback from analyses, safety prevention programs would not be able to adjust to new hazardous situations, information, and equipment.

## Crashes Involving Ambulances

Ambulance crashes are very common, especially in rural areas. The report "Toward Zero Deaths: A National Strategy on Highway Safety," prepared by the National Association of State Emergency Medical Services Officials, notes the following statistics: "EMS providers are at greater risk of death on the job than their police and firefighter counterparts, with 74% of EMS fatalities being transportation related… The estimated crash rates of ambulances are 7 to 10 times greater than heavy trucks, and as many as 9,000 crashes per year among 50,000 vehicles" (National Association of State Emergency Medical Services Officials, 2010, p. 4).

Factors surrounding the vehicle itself, such as maintenance, operator error and training, and safety restraints, play a significant role in these crashes (Sanddal, Sanddal, Ward, & Stanley, 2010). Another important factor is the use of lights and sirens. Studies comparing code-3 and non–code-3 trips have shown that time saved by a code-3 response is insignificant to the outcome of the patient in nearly all cases (Sanddal et al.).

The National Institute for Occupational Safety and Health (NIOSH) conducts research to improve safety for EMS responders. As part of the National Occupational Research Agenda, researchers at NIOSH have been looking for ways to reduce ambulance crash-related injuries and deaths. One major issue is that EMS employees often ride on the squad bench without wearing restraints, because restraint systems do not allow easy access to the patient or equipment. Without the use of restraints, personnel are at higher risk of head injuries from striking their heads on bulkheads or cabinets (Centers for Disease Control and Prevention, 2011).

In January 2012, the NFPA published a draft of NFPA 1917, *Ambulance Vehicle Safety*. This standard defines requirements for new ambulances to set the highest standard of safety specifications. It addresses the issue of head injuries in the patient care compartment and the need to secure equipment that has the potential to seriously injure the provider or patient. Ambulance manufacturers should develop restraint systems designed to increase crash survivability of EMS responders and patients while accommodating accessibility to patients and equipment.

Further considerations to reduce injuries include the layout and structural integrity of ambulance compartments, hardware design, occupant restraints, and additional factors that can contribute to employee injuries and death (Centers for Disease Control and Prevention, 2011). To improve ambulance safety, federal funding is needed for ongoing research and testing of ambulance vehicle design. Furthermore, EMS systems need to develop and implement safe driving policies and procedures, monitor driving behaviors, and require proficiency training and oversight of continuous vehicle operation.

### Case Study

**EMS Ambulance Safety**

In April, 1989, Michael Montecalvo, a firefighter-paramedic and EMS educator, became the first EMS professional to be sentenced to prison for an ambulance driving accident. He had driven his ambulance through an intersection, killing a woman and seriously injuring her 6-year-old son. He was sentenced to 2 years in the Ohio Mansfield State Penitentiary. Michael made use of his time by teaching the penitentiary employees first responder skills and was released in 7 months, but as a convicted felon he was never licensed as a paramedic again. This case had a huge impact on the EMS community, because many had previously assumed that an accident (crash) in response to an emergency would not be criminally prosecuted.

## Crashes Involving Fire Apparatus

According to NFPA statistics, from 1987 to 1996, 272 US firefighters were killed in motor vehicle–related incidents. Of these 272 deaths, 127 were the result of crashes, with most occurring while the firefighter was responding to or returning from an alarm. Forty-nine of the crash deaths involved fire department apparatus, 60 involved firefighters' personal vehicles, and 15 involved aircraft (NFPA, 2005).

A vehicle accident can cause a financial and operational disaster for the fire department, because these crashes may result in firefighter injuries and deaths, liability issues, and out-of-service time for fire apparatus. There is also the concern for civilian deaths, injuries, and property damage. Firefighters who risk their own lives to rescue a victim from a fire may unknowingly place themselves, fellow firefighters, and civilians at risk while aggressively responding to that same emergency call. When an emergency vehicle driver drives too fast for conditions, fails to fully stop at a controlled intersection, or places too much dependence on their emergency lights and sirens, they are risking the lives of unaware civilian drivers and their passengers.

The NFPA has created key resources for reducing these types of crashes. NFPA 1451, *Standard for a Fire and Emergency Service Vehicle Operations Training Program*, contains training requirements for defensive driving techniques. NFPA 1002, *Standard for Fire Apparatus Driver/Operator Professional Qualifications*, and NFPA 1500, *Standard on Fire Department Occupational Safety and Health Program*, Section 4-2, Drivers/Operators of Fire Department Vehicles, contain guidelines for the basic training new drivers should receive before being permitted to operate fire department vehicles.

It is always easier to learn a new skill correctly than to change existing habits. Nevertheless, existing drivers should also be encouraged to complete a driver-training course designed to comply with these safety standards. In-service training for existing drivers is required at least twice a year and is necessary to change the culture and tradition that encourages unsafe driving, such as speeding and not stopping at stop signs or red traffic lights. In-service training provides valuable hands-on experience and reinforces defensive driving techniques and emergency response SOPs.

## Physical Fitness

The FES is a physically demanding profession that involves heavy lifting of equipment and exposure to toxic chemicals. Recruits are generally healthy and physically fit, but their health can decline over time because many agencies do not require regular exercise or yearly medical examinations. Also, the health requirements are usually less stringent for volunteers, who tend to continue firefighting as they age, a time when most heart problems occur.

The main causes of firefighter deaths are not a secret, nor are they new. The firefighter fatality report for 2012 showed that heart disease accounts for 49% of firefighter fatalities. For example, in an interview about his experience as the first due battalion chief at the World Trade Center on September 11, 2001, Battalion Chief Joseph Pfiefer, New York Fire Department, described the health problems caused by the extensive stair climbing: "There were messages, urgent messages of firefighters having chest pains as they went up" (*Firehouse*, 2002, p. 42). Another New York Fire Department officer at the scene reported similar findings, "As we got higher into the building, we heard numerous 'Urgents' and 'Maydays' from firefighters with chest pains, in need of oxygen or worse" (*Fire Engineering*, 2002, p. 97). Although all the other items in the safety standards are important and do save lives and prevent injuries, preventing heart disease has the potential to save a great number of firefighter lives.

## The Need to Stay in Shape

A large US study conducted by the Harvard School of Public Health suggests that firefighters face a far greater risk of dying from heart problems while battling a blaze than was thought. In the study, researchers examined a federal registry of 1,144 on-duty firefighter deaths between 1994 and 2004 (excluded were the 343 firefighters who perished in the September 11, 2001, terrorist attacks). Nearly 40% of the on-duty deaths during that period were caused by heart disease. Thirty-two percent of those deaths occurred while fighting blazes. Researchers found that the risk of suffering a heart-related death while putting out a fire was up to 100 times higher than the risk during down time, even though fighting fires accounts for only a small percentage of these workers' time.

A 2010 FEMA-sponsored study conducted at Saint Joseph's Hospital in Atlanta, using the medical examinations of 300 firefighters aged 40 and older, reported that

---

**Words of Wisdom**

"Let me start by saying that there seem to be a lot of people dropping over because of heart attacks. This does not surprise me, as I have lived the lifestyle of the brave and cholesterol ridden for many years. The causes of this malady are well known. You eat too much, you smoke too much and you don't exercise at all. And then suddenly you are racing to the scene of a blaze, raising ladders, dragging hose and rescuing victims."

– Dr. Harry Carter, municipal fire protection consultant

*Source*: Carter, H. R. (2001). Why are we dying? *Firehouse*, *26*(8), p. 21.

one-third of the firefighters had heart disease that was unrelated to traditional risk factors, such as high cholesterol. According to Dr. Superko (principal investigator), the risk of heart attack was tremendously impacted by such factors as stress and psychological pressures, diet, exercise, inherent personality, and genetic predispositions (Saint Joseph's Hospital, 2009).

Diet and exercise should be priorities in all FES organizations **FIGURE 9-3**. As noted by Dr. Linda Rosenstock, dean of the UCLA School of Public Health, "You may not be able to prevent all these deaths, but to the degree you can prevent some deaths by paying attention to underlying risk factors and better fitness programs, that's the goal" (Chang, 2010). These organizations could do more to improve health by requiring annual physicals and fitness tests, and wellness and fitness programs geared toward reducing heart disease risk factors, such as obesity and high blood pressure.

As part of a Firefighter Wellness class in 2009, Memphis Fire Department training instructors calculated body mass index values for 670 firefighters ranging in age from 22 to 66 years. Based on the World Health Organization body mass index scale, 33% of these firefighters were considered obese and 5% were considered morbidly obese. More than two-thirds of the firefighters in this assessment were found to be overweight. "Since obesity is considered a major cardiovascular disease risk factor, many overweight and obese firefighters are predisposed to a higher incidence of adverse cardiac events than their normal-weight peers" (Spratlin, 2011).

> **Words of Wisdom**
>
> "Imagine being awakened from a dead sleep by a loud, shrieking siren several times during the night, responding through the rush of adrenaline, carrying a hundred pounds of equipment on your back, and meeting people at the very worst possible moments in their lives every day, and you can begin to understand the toll it takes on the first responders."
>
> – Saint Joseph's Hospital
>
> *Source*: Saint Joseph's Hospital. (2009). Study: Heart disease: An epidemic for firefighters. Retrieved from: http://www.prnewswire.com/news-releases/heart-disease--an-epidemic-for-firefighters-61857577.html

In addition, excess body weight adversely affects firefighters' abilities to perform their duties. In 2011, researchers at Auburn University's Human Performance Laboratory studied Montgomery, Alabama, firefighters and found that there is a negative correlation between weight and physical performance by firefighters (Spratlin, 2011). Obese firefighters are also more likely to get sick or injured more often. There is concern that obesity in the firefighter demographic might only get worse.

A two-pronged approach works best for reducing the number of stress- and heart-related firefighter deaths. First, medical examinations, as outlined in NFPA 1500 and 1582, should be scheduled for all members. Any indication of heart disease should be treated without delay. In most cases, the member should be completely evaluated before being allowed to participate in emergency operations duties.

Second, an ongoing physical fitness program has many beneficial outcomes. This type of training reduces the chance of heart attack, lowers cholesterol, reduces body fat, and strengthens the heart. Individuals with stronger hearts as a result of aerobic exercise are more likely to survive a heart attack.

Voluntary fitness programs have reportedly reduced injuries by 30% (Rawlyk, 2010). This benefit affects more than just the individual, who has a decreased chance of suffering a heart attack and a better chance of surviving if one does occur. Team members also benefit, becoming safer working with other team members who are physically fit. Furthermore, firefighters in good physical condition consume less air from their self-contained breathing apparatus than do firefighters who are in poorer physical condition. This allows them to work for longer periods of time in a hostile atmosphere, which makes a tremendous difference, because the low air alarm activation on

**FIGURE 9-3** Diet and exercise should be priorities in all FES organizations.

> **Words of Wisdom**
>
> "[Y]ou have contributed in many ways to many people in a positive manner. You are the most important person in the world to your family. You are a valuable asset to your fellow firefighters...And you are an integral part of the fabric of your company and department. Do yourself, your fellow firefighters, and your family a favor. When you wake up tomorrow, stand in front of the mirror and answer this question honestly: 'Are you fit?'"
>
> - John J. Salka, Jr., Veteran Battalion Chief of the New York Fire Department
>
> *Source*: Salka, J. J. (2009). Are you fit? Are you really up to the physical demands of firefighting? *Firehouse, 34*(3), 121.

a firefighter with poor aerobic capacity requires the entire team to withdraw from the scene.

The entire department also receives a better reputation from the public and elected officials who derive their impression of the department based on what they see of the members. Furthermore, the customer receives better service, because the results of the team are always limited by the weakest member. The public has a right to expect that all firefighters who are responding to their emergency are trained, organized, staffed, equipped, and in top physical shape. Physical fitness should be directly related to the job, not gender or age. Predictably, the human body loses strength and endurance as it ages; however, with proper physical training a person can function at high levels of strength and endurance into his or her 60s. Discussing the Candidate Physical Ability Test (CPAT), the International Association of Fire Fighters (IAFF) website comments, "It is the position of the IAFF/IAFC [International Association of Fire Chiefs] Joint Labor Management Wellness-Fitness Initiative Task Force that fire departments should increase the diversity of their workforce by actively recruiting candidates from throughout their communities rather than lowering candidate physical ability standards" (IAFF, 2011). In addition, as a result of fewer injuries from members, the city and county end up paying less for worker's compensation claims, sick leave, and medical insurance.

Many fire departments do not have a physical fitness program that is effective and creditable. Any program that is not mandatory will have a percentage of nonparticipants. Although resistance can be found among individuals (typically the individuals who need it the most) and labor groups, firefighter groups are increasingly taking notice of heart risk. In 2003, the National Volunteer Fire Council began an awareness program promoting fitness and nutrition. This program offers free health screenings, healthy cooking demonstrations, and fitness techniques (Chang, 2007).

## Physical Fitness Program

Physical fitness has long been recognized as an essential trait for members of the FES. The profession requires a high level of strength and endurance, and without it many tasks cannot be done properly. As previously exemplified, a physical fitness program serves the public and the FES member. The following recommendations can help guide the implementation, evaluation, and execution of a department's physical fitness program.

- When deriving physical fitness minimum levels, the needs of the public (customers) and the safety of the firefighter must come first.
- Validating physical fitness by testing incumbents is sometimes inaccurate. There are incumbents who perform poorly because they do not participate in a physical fitness program. Also, if there was no physical fitness standard when they were hired or accepted as a member, incumbents may be unable to perform all of the essential job functions. Most members have the potential of doing better with training.
- A comprehensive wellness program that includes medical evaluations must be used in combination with an exercise program, and the costs should be borne by the department. The program might allow firefighters up to 2 hours to exercise on duty each day.
- When first implemented, a physical fitness standard for incumbents should recognize that some firefighters have not exercised for many years, so a phase-in period should be used. Most firefighters who find themselves identified as below standards for a physical fitness evaluation are able to meet standards with the appropriate training and time.
- A firefighter who fails to meet the minimum levels of physical fitness (different levels for incumbents during the phase-in period) should be prevented from participating in physically demanding work. As noted by Stephen Foley in *Fire Department Occupational Health and Safety Standards Handbook*, "Members must not be allowed to begin or resume suppression duties until they pass the physical performance requirements" (Foley, 1998, p. 87).
- A voluntary or nonpunitive physical fitness program guarantees that there are members in poor physical fitness. Imagine if a traffic intersection in a town had stop signs posted to prevent accidents, but the local ordinance was voluntary or nonpunitive. Some drivers would disregard the stop signs, guaranteeing some serious accidents. Safety and health requirements must be mandatory to work.

**Joint Labor-Management Wellness and Fitness Initiative** The IAFF and the IAFC have launched a national initiative to establish a program for medical and fitness evaluations, rehabilitation, behavioral health, and data collection. The basic design premise of this effort is to not be punitive. For example, in a punitive program, a member who is not fit for duty based on either medical or physical fitness criteria may be fined or placed on nonpaid leave. In a nonpunitive program, unfit members are placed on light duty, injured-on-the-job leave, or some other placement with no reduction in wages.

The next phase for a full salaried, unfit firefighter is placing him or her in rehabilitation. However, several questions need to be considered in these cases. For example: How many salaried firefighters can a department afford to place in noncombat positions before it becomes a hardship? How many volunteers can be taken off combat status before it adversely affects responses? This is a complex and highly individualized problem that can have a substantial impact on the budget and operation of a department. If a reasonable implementation plan is used, however, the number of firefighters needing rehabilitation will probably be small.

When a new program is initiated, it is reasonable to allow existing firefighters who have medical and physical fitness problems an equitable time frame in which to participate in a rehabilitation program. Also, NFPA 1500 allows phased implementation of safety requirements. For example, if a fitness test was chosen that has a pass-fail time of 7 minutes, the first year could start out with a time of 14 minutes for incumbents. Then, each year, the pass-fail time would be reduced by 1 minute until all members meet the 7-minute standard.

## Physical Fitness Testing

Any physical fitness test should be based on validated job performance. The IAFC and the IAFF developed the CPAT in 1999 to address this issue in the FES. This test was developed to provide fire departments with a tool that would enable them to select physically capable individuals to train as firefighters. The CPAT includes the following eight sequential events:

- Stair climb
- Hose drag
- Equipment carry
- Ladder raise and extension
- Forcible entry
- Search
- Rescue
- Ceiling breach and pull

Although the CPAT is considered to be the best evaluation to date, the test does have limitations. For example, the test was devised with data provided by 10 metropolitan fire departments, some of which had crews of more than four firefighters. Therefore, the data may or may not accurately represent minimum physical fitness required to accomplish each task. Also, some of the tests or the background data were gathered from firefighters who were already on the job, and it can be argued that just because these individuals were already firefighters they were not necessarily physically competent to perform all critical tasks. Instead, valid studies that focus on job performance should be used.

Another limitation of the CPAT is that certain critical tasks were underrepresented in the test. One example is the dummy drag, officially known as Event 7. This task, which was reported as critical by 97% of those surveyed, simulates removing a victim or injured partner from a fire scene. According to the IAFC, "The candidate grasps a 165-pound mannequin by handles on a shoulder harness, drags it 35 feet to a drum, makes a 180-degree turn around the drum, and continues an additional 35 feet across the finish line." The established weight used within this test was based on the average weight of adult patients and firefighters. The established distance was based on the average distance from the front door to the living room in a single-family home. The final determination of pass-fail was made after a committee evaluated this data, using the lowest tenth percentile as a basis for passing. This means that a person passing the test can rescue only 10% of incumbent firefighters (without the weight of personal protective equipment and self-contained breathing apparatus). This brings up two questions about this particular evaluation: (1) Why did the committee choose such a low pass-fail? (2) Why did the committee not use the NFPA standards-making process and committees for devising this standard?

### Case Study

**Physical Fitness Fairness**

In the late 1970s, the first female firefighter graduated from a rookie school of a metro county fire department. At that time, this department had no physical fitness test for new employees. As her first duty station, she was assigned to an engine company. The officer at this station liked to think outside the box and planned to have the rookie teach a series of basic drills to the crew. He was aware that teaching the skills necessary to be a safe and effective member would help the rookie commit them to memory.

The first drill was to take the extension ladder off the engine, place it so the firehouse roof can be accessed, and climb it with a roof ladder. The officer observed the drill and immediately realized that the rookie was not strong enough to complete the evolution.

Subsequently, the officer made an appointment to talk to his Battalion Chief and related his concern for the rookie's safety and the effectiveness of the crew. The BC stated, "This is the first woman firefighter." He went on to advise the officer to "just make it work."

Seven years later, the same firefighter was assigned to a truck company. It was at this time that she reported a back injury as a result of lifting ground ladders and started medical treatment. Eventually, she was awarded a job-connected disability of

*(continues)*

## Case Study (Continued)

70% of her pay tax-free. This account points to three issues regarding safety:

1. The firefighter now has a permanent painful back injury that may have been prevented if she had successfully completed an initial physical fitness test.
2. As part of the company crew, the firefighter handicapped the team's ability to perform all of the necessary job functions. This affects the safety of fellow firefighters and the service the public receives.
3. If the back injury was contrived, the county and its taxpayers are paying a very generous disability benefit to someone who does not deserve it.

This case is not meant to point to issues of gender; instead, it illustrates the danger for departments that do not test new members. The issues listed above could have been resolved if job performance physical fitness testing had been correctly implemented. Giving firefighters who cannot pass a fitness test a pass is not fair to the individual, the crew, the department, or the public.

needed, quarantine suits to avoid exposure to these toxic substances FIGURE 9-4.

Some cancers, such as colon cancer and brain cancer, are being reported in a higher percentage of firefighters than in the general public. In fact, some jurisdictions have now accepted some cancers as job connected and provide the appropriate benefits (Gagliano, 2009, p. 89).

In addition, the Center for Research on Emergency Medical Services' Emergency Responder Human Performance Lab has examined protective programs that are common to the fire service but seem to be lacking in EMS. The Center contends that NIOSH provides firefighters with resources specific to heat- and work-related injuries but does not provide programs specific to EMS. As the role of the traditional EMS responder expands to include response to disasters or weapons of mass destruction, the effects of personal protective equipment will become even more prevalent.

## Violence

There is also a psychological toll taken when, in the attempt to render aid, EMS personnel become victims of violence. All too often, personnel are threatened by anxious family members who feel the response was too slow, people who feel threatened by the uniform, those with

## Other Occupational Hazards

There are many other safety items identified in NFPA 1500 that departmental safety committees should address. In addition, safety committees should review the annual US Firefighter Fatalities and Injuries report issued by the NFPA. Also, the US Fire Administration Firefighter Fatality Retrospective Study: 1990–2000 identifies the major causes of deaths and injuries from a multiyear perspective. These reports contain many clues to areas where deaths and injuries can be prevented.

Whether within a fire-based or private organization, FES personnel are exposed to a wide variety of occupational hazards including infectious disease and bloodborne pathogens, hearing loss from exposure to sirens and loud environments, temperature extremes, a higher incidence of vehicle collisions, musculoskeletal injuries, long work hours, sleep deprivation, and inordinate physiologic and psychological stress.

### Personal Protective Equipment

OSHA has created standards to protect FES personnel from some of the hazards previously mentioned. For example, OSHA's 1910.120 requires that employers provide all employees with the personal protective equipment necessary to safely do their jobs. This may include air respiratory masks, firefighting protective clothing, filter masks, radiation suits, and, when

**FIGURE 9-4** Employers must provide all employees with the personal protective equipment necessary to safely do their jobs.

metabolic impairments or trauma that affects their behavior, or those seeking drugs. Violence against EMS personnel is a real and pervasive problem. During a presentation at the 2009 Mid-Year Conference of the National Association of State EMS Officials, EMS Scholar Brian Maguire stated, "EMS personnel are seven times more likely to be killed by workplace violence than other healthcare professionals and 22 times more likely to experience injury from assault than the population at large" (Maguire, 2009, p. 6).

EMS personnel may become so focused on the patient and the task that they become blindsided by events occurring around them. Force protection and self-defense training are not typically covered in an emergency medical technician or paramedic school curriculum. An agency or company that invests time in training personnel on how to be aware and safe can save themselves grief, the cost of a worker's compensation claim, and disability and death for their employees. One such class is a collaborative effort between EMS-1.com, an online education resource, and Calibre Press, a provider of law enforcement training: the EMS-1 Street Survival Seminar. The class covers the following topics:

- Gang threats
- Weapons search
- Intoxicated, psychotic, and combative patients
- Specific conditions, such as excited delirium and autism
- Ruling out medical complaints

Employers have a duty to provide their employees with a safe workplace, and prepare them for instances in which hazards are not known.

## Stress

FES personnel are constantly exposed to stress and can easily develop stress-related disorders if situations are not mitigated properly or in a timely manner. For example, chronic stress disorder can result from exposure to stress that occurs over time because of such issues as salary; work schedules; an unsupportive company or partner; a lack of confidence in their own skills; or general fears of exposure to communicable diseases, violence, or vehicular accidents. These issues can be exacerbated by 24-hour work schedules, lack of sleep, general burnout, or home and family issues.

Critical incidents are also likely to have a psychological impact on a responder. These include such scenarios as injury or death of a team member, a situation in which an error caused an injury or death of a patient, child abuse, an unsuccessful rescue, or a scene where the responder felt threatened. There are several questions that need to be asked when determining whether stress from a critical incident can be a more serious problem for a first responder:

- Can this incident be mitigated by normal coping mechanisms? Often a responder just needs time to process and adapt to the events of an incident. They may be able to reframe the issue from negative to positive or evaluate the problem from an internal perspective rather than blaming it on an external force.
- Does the person have a good social support network? Coworkers and partners are often caring and understand the issue. It is even better if the family is a part of the support chain. Responders without sufficient support may adopt negative coping mechanisms.
- Does the company or agency have good Critical Incident Stress Management (CISM) resources? CISM is a comprehensive crisis intervention system that includes components to manage all phases of a crisis. It starts with precrisis training so employees know what to expect and understand how to access available resources. The acute crisis phase may be managed by a team of peer counselors and mental health professionals and provides interventions to individuals, crews, or larger groups. In the postcrisis phase, debriefings may occur or references may be provided for ongoing care.

If an issue is not mitigated properly, posttraumatic stress disorder (PTSD) can occur. This is considered by some to be a silent epidemic for FES personnel. PTSD develops from situations in which the FES member "has experienced, witnessed or was confronted with an event or events that involved actual or threatened death or serious injury, or a threat to the physical integrity of self or others and the person's response involved fear, helplessness, or horror" (Alexander & Klein, 2001, p. 80). The symptoms of PTSD can include re-experiencing of the event or flashbacks, avoidance, or persistent symptoms of increased arousal. Similar calls, sights, smells, or other triggers can propel the person back to the original incident and could create a dangerous situation. Companies and agencies are wise to create their own CISM teams or partner with other agencies for resources. The agency needs to ensure that the right resources with the right training are available, and that a mental health professional can be accessed when necessary. For most employees who suffer critical incidents, timely CISM can make a difference in whether the employee continues to be psychologically healthy.

## Sleep Deprivation

Sleep disruption is another type of hazard for FES personnel who work a 24- or 48-hour shift. "[A]t any time during the night, that person may be awakened suddenly and abruptly by a bunkroom light coming on and some form of a bell indicating there is an alarm" (Gagliano, 2009, p. 90). A 2009 study of firefighters noted that their heart rates "typically rise to 80 percent of maximum when the alarm is received" (Gagliano, 2009, p. 91). Furthermore, the inability to sleep easily and deeply may result in sleep deprivation, which may physically predispose an individual to becoming a stress casualty.

Recent studies have suggested possible policies that would reduce the negative outcomes of a 24-hour shift. These include the following:

- Restrict members from working more than 24 hours without a 12-hour break.
- Because most departments require approval of a second job, add the provision that they cannot work the second job 12 hours before or after their 24-hour shift.
- Encourage better than average physical fitness levels, which would result from a firefighter physical fitness program.

## CHAPTER ACTIVITY #1: Injuries on the Job

Approximately 14.5% of all fireground injuries are categorized as wounds, cuts, bleeding, or bruises (Karter, 2012, p. 1). Because there are a large number of injuries in this category, a program could be initiated to prevent these injuries. Training and education would be a part of the overall plan, but the department's officers would also want to be sure that all members are wearing all their safety equipment and following SOPs.

One important item for the prevention of cuts, wounds, burns, and other injuries of the hand is the use of issued firefighter safety gloves. If a safety officer receives an injury report of a cut to the palm of the hand from a firefighter, he or she would want to examine the glove of the hand that was cut. It should have an identical cut and blood around the cut in the glove. Without the matching cut in the glove, it could be inferred that the firefighter was not wearing gloves when the injury occurred, and corrective action should be taken. There may be times when there is a suspicion that cannot be proved. Even in these cases, if the members know the administration is checking, they are more likely to wear all their personal protective clothing in the future.

### Discussion Questions

1. Is it fair to check up on the firefighters?
2. Would it be appropriate to have the firefighter use his or her own health insurance rather than the jurisdiction's worker's compensation if the firefighter was careless or neglected to follow safety SOPs or wear safety equipment?
3. Is this an example of managing by walking around? If yes, are there other actions that could be taken using managing by walking around that would prevent injuries?
4. How could the administration be assured that all firefighters know the safety SOPs, including the use of safety equipment?

## CHAPTER ACTIVITY #2: Critical Incident Stress

As a paramedic at a busy suburban ambulance company, you have worked well alongside one of your coworkers for over a year, when you suddenly begin to notice changes in his behavior. He always seems tired, gets angry about calls he feels are unnecessary, and seems to take advantage of long transports. You know he has experienced a number of pediatric calls with negative outcomes in the past few months and have heard him mention that he is having issues with his wife. You mention your concerns to a supervisor, but your coworker's behavior is brushed off as simply a "bad attitude." Then one day after a particularly bad call, you notice him take the station's supply of narcotics and leave. You report him to the company supervisor, and he is eventually caught by police and arrested for theft of controlled substances. The company fires him.

### Discussion Questions

1. This coworker hadn't always been an angry person. What may have happened to him over time?

2. Was the company remiss in not listening to your initial concerns?

3. How could all of the negative outcomes have been prevented?

## References

Alexander, D. A., & Klein, S. (2001). Ambulance personnel and critical incidents: Impact of accident and emergency work on mental health and emotional well-being. *British Journal of Psychiatry, 178*, 76–81.

Brannigan, V. (1997). The real hazard is driving the fire truck. *Fire Chief, 41*(7).

Carter, H. R. (2001). Why are we dying? *Firehouse, 26*(8), 21.

Centers for Disease Control and Prevention. (2011). *NIOSH continues research to improve safety for ambulance service workers and EMS responders.* Retrieved from http://www.cdc.gov/niosh/docs/2011-190/

Chang, A. (2007). Firefighters face heart risks in a blaze. Associated Press. Retrieved from: http://www.washingtonpost.com/wp-dyn/content/article/2007/03/21/AR2007032101600.html.

Chapeau, W. (2006). Preventing ambulance personnel injury. *Fire Apparatus and Emergency Equipment, 11*(11). Retrieved from http://www.fireapparatusmagazine.com/articles/print/volume-11/issue-11/departments/ems-equipment/preventing-ambulance-personnel-injury.html

Fahy, R. F. (2010). *U.S. Fire Service fatalities in structure fires, 1977–2009.* Quincy, MA: National Fire Protection Association.

Federal Emergency Management Agency. (1999). *Guide to developing effective standard operating procedures for fire and EMS: Federal Emergency Management Agency United States Fire Administration.* Retrieved from http://www.usfa.fema.gov/downloads/pdf/publications/fa-197.pdf

Fire Engineering. (2002). Anonymous letter. *Fire Engineering, 155*(9), 97.

Firehouse. (2002, July). WTC—This is their story. *Firehouse, 27*(7), 42–44.

Foley, S. N. (Ed.). (1998). *Fire department occupational health and safety standard handbook.* Quincy, MA: National Fire Protection Association.

Gagliano, A. (2009). What every firefighter's spouse should know. *Fire Engineering, 162*(12), 89–92.

International Association of Firefighters. (2011). *Candidate physical ability test program summary.* Retrieved from http://www.iaff.org/hs/CPAT/cpat_index.html

Karter, M. J. (2012). *U.S. Firefighter Injuries – 2011.* Quincy, MA: National Fire Protection Association.

Liao, H., Arvey, R. D., & Butler, R. J. (2001). Correlates of work injury frequency and duration among firefighters. *Journal of Occupational Health Psychology, 6*(3), 229–242.

Maguire, B. J. (2009, June 8–9). Occupational risks among EMS personnel. Presented at the 2009 Mid-Year Conference of the National Association of State EMS Officials, Hilton Alexandria Mark Center, Alexandria, VA.

National Association of State Emergency Medical Services Officials. (2010). EMS occupational risks. Retrieved from http://www.nasemso.org/meetings/midyear/documents/maguire-ems-occ-risks-jun09.pdf

National Fallen FireFighters Foundation. (2004). Everyone Goes Home: Firefighter Life Safety Initiatives. Retrieved from http://www.everyonegoeshome.org/summit.html

National Fire Protection Association. (2005). *Selected special analyses of firefighter fatalities*. Retrieved from http://www.nfpa.org

National Firefighters Burn Study. (2002). Retrieved from http://www.firefighterburnstudy.com

National Highway Traffic Safety Administration. (2011). *EMS workforce agenda for the future.* Retrieved from www.ems.gov/pdf/2011/EMS_Workforce_Agenda_052011.pdf

Rawlyk, H. (2010). Fitter firefighters saving taxpayers money: Health program reducing injuries. *The Capital Gazette*, 10.

Saint Joseph's Hospital. (2009). Heart disease: An epidemic for firefighters. Retrieved from http://www.prnewswire.com/news-releases/heart-disease--an-epidemic-for-firefighters-61857577.html

Salka, J. J. (2009). "Are you fit? Are you really up to the physical demands of firefighting?" *Firehouse*, 34(3)

Sanddal, T. L., Sanddal, N. D., Ward, N., & Stanley, L. (2010). *Ambulance crash characteristics in the US defined by the popular press: A retrospective analysis. Emergency Medicine International.* Retrieved from http://www.hindawi.com/journals/emi/2010/525979/

Spratlin, K. (2011). *Firefighter obesity: A public safety risk.* Retrieved from http://www.fireengineering.com/articles/print/volume-164/issue-1/departments/fire-service_ems/firefighter-obesity-a-public-safety-risk.html

US Bureau of Labor Statistics. (2012). *Occupational outlook handbook.* Retrieved from http://www.bls.gov/oco/ocos101.htm

US Fire Administration. (2002). *Heart attack leading cause of death for firefighters.* Retrieved from http://www.usfa.fema.gov/about/media/2002releases/02-193.shtm

# CHAPTER 10

# Government Regulation, Laws, and the Courts

## Fire and Emergency Services Higher Education (FESHE) Course Objectives

**Module I: Leading and Managing Purposefully with a Community Approach**

The students will:

2. Explain the importance of a good working relationship with public officials and the community as a whole. (pp 144, 145–146, 152)

**Module II: Core Administrative Skills**

The students will:

4. Describe the key elements of successful communication. (p 144)

**Module III: Planning and Implementation**

The students will:

7. Describe the purpose, function, and current and future security concerns of working document publication, storage, and integrity. (pp 143–144)

**Module IV: Leading Change**

The students will:

7. Demonstrate innovative ways to address traditional problems with the organization. (pp 143–146, 151–152)

## Knowledge Objectives

After studying the chapter, the student will be able to:

1. Identify and understand the moral and ethical justifications for government oversight of the public fire and emergency services (FES). (pp 141–143)
2. Understand the creation process and the importance of government regulations and laws. (pp 143–146, 151)
3. Examine building codes and regulations at the state and local level, and understand their effect on FES. (pp 144–145)
4. Examine progressive laws and regulation trends in FES. (pp 146–150)
5. Recognize the influence that the Occupational Safety and Health Administration (OSHA) has on FES. (pp 146–148)
6. Recognize the influence that the National Fire Protection Association (NFPA) has on FES. (pp 148–150)
7. Understand the impact and influence that the legal system and the courts have on FES. (pp 150–152)

## Government Regulation

Most FES providers are either a part of local or state governments, or independent volunteer fire companies or emergency medical services (EMS) organizations. Often the elected officials and appointed administrators of these organizations argue that there is no need for federal or state government intervention in their affairs. They protest that standards of service are best determined at the local level of government.

This is especially true when a government regulation requires the expenditure of tax dollars at the state or local level but is unaccompanied by funding. For example, in 2001, Robert Johnson, writing in the Wall Street Journal, predicted that the implementation of NFPA 1710, *Standard for the Organization and Deployment of Fire Suppression Operations, Emergency Medical Operations, and Special Operations to the Public by Career Fire Departments*, would "prompt fire departments to hire 30,000 firemen nationwide, an 11 percent increase" (Johnson, 2001). While this particular prediction wound up being unfounded, it would have had a substantial fiscal impact on the nation's FES and taxpayers. For regulations such as these, it is not uncommon for city or county administrators and elected officials, along with FES officials, to protest the mandates. Most mandates are never fully funded; however, some may at times be partially funded. For example, when the US Department of Transportation issues new safety design features for roads, the federal government often provides matching funds to encourage states to comply.

Regardless of this issue and the fact that regulations and standards for local FES are relatively new (starting in the 1980s) in the United States, there seems to be adequate justification for the public to expect and encourage government oversight. In general, there are two factors influencing the justification for government regulation: (1) the role of government to protect its citizens and (2) the concern over a failure in the market preventing a public good or service from becoming available or widely used.

## Role of Government to Protect Citizens

It is the role of government to protect the consumer (citizens) in cases where it can be demonstrated that the consumer lacks the knowledge, background, or education to make a conscious judgment of the safety, competency, or adequacy of the service or product.

For example, governments typically regulate the contents and labeling of food products. Without a label, it is impossible for the consumer to know the health risks or contents of a can of vegetables without first sending it to a laboratory to be tested. The consumer would have no readily available knowledge about how the vegetables were planted, grown, harvested, processed, and canned. Because of government health regulations, a consumer can be assured that the contents are accurately described on the label and are acceptable for consumption. Without some assurance of the safety of this food product, few consumers would continue to buy the product. Therefore, this regulation benefits the consumer and the producer.

Similarly, the state or federal government regulates most professional services, such as doctors, lawyers, and engineers. It is impossible for most consumers of these services to accurately assess the competency of the professionals that provide them. The customer would have to know the university they attended and if the academic standards were sufficient to meet a minimum standard of education. Currently, these professions require a competency examination to prove mastery of the knowledge needed for the profession before they are granted a license to practice in their state. When consulting a doctor or seeking advice from a lawyer, the client must exercise a certain level of trust. Minimum license requirements ensure the public that the provider they are dealing with has basic knowledge and competency regarding his or her profession.

Can the same be said about FES? When a fire engine shows up at a house fire, can the typical customer determine the competency of the firefighters or the adequacy of the fire equipment? If a person suffers a serious injury, can he or she tell if the responding ambulance crew is proficient in providing advanced life support? Becoming a professional, competent firefighter or EMS provider requires extensive time in a specialized training program to master necessary skills and knowledge, but the typical citizen has no way to make an educated assessment of the competency of these providers. In a few larger states, professional certification is required for these emergency providers; however, this is not universal throughout the country.

## Market Failure Consequences

A market failure may prevent a safety product or service from becoming available to the general public. Typically, market failures are more common in products or services that can prevent injury, sickness, or death. For example, fire safety experts recognized a market failure shortly after the first smoke alarms came on the market. These alarms were somewhat expensive due to the low volume of purchases. Although it was recognized that this safety device was extremely beneficial, experts realized it would not necessarily be purchased if left up to individuals or building owners who did not see the benefit of spending money for these detectors.

This could result in troubling consequences for even the fire-conscious individual. For example, if a fire in a multifamily apartment building starts in an apartment without a detector, the fire will endanger the other occupants of the building, regardless of whether those occupants have a smoke alarm in their own apartment. An undetected fire in an apartment can become very sizable and hard to extinguish. In a worse-case scenario, it may trap victims by blocking the exits from other apartments. Due to this market failure, the smoke alarm became mandated by building and fire codes in existing buildings.

Seat belts are another example of a safety product that is now required by government regulation in all vehicles sold in the United States. At the time the regulation went into effect, seat belts were offered as an option, but very few car buyers chose them because they were uncomfortable and the likelihood of being involved in a car crash seemed low to the car purchaser. Thanks to the perseverance of safety groups, such as the American Automobile Association and insurance associations, the federal government issued a regulation requiring seat belts in all newly manufactured cars.

After seat belts were mandated in cars, it was assumed that drivers and passengers would use them. However, there was resistance; so states—encouraged by insurance companies, the federal government, and safety organizations—passed laws requiring the usage of seat belts. In most cases, regulations that make safety equipment a requirement must be paired with rules that require its use and provide some type of disciplinary action for noncompliance. Even with vigorous enforcement efforts, it is not uncommon to read about a death as a result of a vehicle crash where the occupants were not wearing seat belts.

## Monopolies

When a product or service becomes a monopoly, the consumer loses the ability to choose a product based on a preferred provider or price. Some typical consumer monopoly products are electricity, water, and sewer services. Private companies or the government can provide these, but they all work most efficiently when granted a monopoly regulated by government.

For example, imagine that local cable television service has four separate providers competing for the public's business. Every telephone pole in the neighborhood would have four separate wires in addition to the electrical and telephone wires. Because of the heavy costs of the infrastructure, the government grants a monopoly to one provider and controls quality and price.

Monopolies also exist in the realm of emergency services. Emergency services would be very expensive if there was more than one provider in an area. In addition, because response time is such a critical part of emergency service, a given system can provide reduced response time by having all stations be part of the same company rather than having two or more providers servicing the same neighborhoods. Therefore, when an individual dials 911, they do not have a choice of fire departments or EMS; whoever is the local provider responds to the call. This inability to choose by itself justifies government oversight.

## Federal Regulations

At the federal level of government, Congress passes laws and the president approves them; however, Congress generally does not include specific guidance for implementing regulatory legislation. It usually defers to the government bureaucracy to formulate the rules of compliance with the new or revised law. Congress refrains from setting these "rules" for two practical reasons.

First, Congress normally lacks the expertise of the professional bureaucracy. Because of this reasoning, Congress passed a public law approved by President Nixon in 1970 that created the Occupational Safety and Health Administration (OSHA). The law contained guidelines and a process for adopting regulations to "help employers and employees reduce injuries, illnesses and deaths on the job in America" (OSHA, 2006). The adoption process required OSHA to seek out and heavily rely on subject matter experts from the industry being regulated. As exemplified by the following extract, the expertise provided by this process had a great impact on employees and employers in all occupations (OSHA, 2006):

> Since then, workplace fatalities have been cut by more than 60 percent and occupational injury and illness rates have declined 40 percent. At the same time, U.S. employment has more than doubled and now includes over 115 million workers at 7.2 million worksites.

Second, lawmakers typically do not like to place themselves in the middle of controversial regulations. Instead, there are two groups that typically have more of a stake in the outcome of the regulatory process: employers and employees. In the case of an OSHA safety regulation, employees are interested in safety at any cost, whereas their employers are more likely to be interested in making a profit by keeping costs as low as possible. Because of the conflicts that can arise with proposed regulations, a long period of time is often required for review and public comment before final adoption of a regulation.

The review process begins with the federal agency publishing notice of their proposed rule in the *Federal Register*, which can be thought of as the "government's legal newspaper" (US Government Printing Office, 2005). These documents can also be found on regulations.gov. This publication gives the public and other organizations the ability to comment on the proposal and participate in the federal decision-making process. Once a regulation is reviewed, the public can submit their comments either in writing or at a hearing. After this period, which usually lasts between 30 and 90 days, agencies then publish their final regulations and address any significant issues that were raised and discuss any changes made in response.

Although this review can take a long period of time, a new process may help shorten this period. Issued on February 10, 1998, the Executive Office of the President, Office of Management and Budget, Circular No. A-119 directs federal agencies to use voluntary consensus standards, such as the ones promulgated by the NFPA, instead of waiting for government-issued standards (Office of Management and Budget, 1998). The voluntary consensus standards are described as:

- Being open process
- Supporting a balance of interest
- Maintaining due process
- Providing an appeals process

- Working toward consensus that is substantial agreement, but not necessarily unanimous support

Because this order is relatively new, many existing regulations are still based on requirements created by federal bureaucrats. However, moving forward, federal agencies should use voluntary consensus standards in their procurement and regulatory activities, except where inconsistent with law or otherwise impractical.

## Offering Effective Input on Codes, Standards, and Regulations

Many national FES organizations and special interest groups routinely review the *Federal Register* and circulate items that may be of interest to the organization. Besides reviewing the information internally, members of these organizations are typically notified of proposed regulations of interest, and notices are placed on these organizations' Web sites.

Controversial policy topics tend to attract the attention of strong advocacy groups who have political allegiances and commitments. These groups may focus on issues that are of benefit to them and their members. When faced with political pressure from a special interest group and a lack of input from citizens, chief officers need to rely on their own ability to argue for policy and regulation changes. Although doing so might take a lot of courage, honesty, perseverance, and knowledge, the officer must remember that the primary advocate for the customer (citizen) is the FES administrator.

Chief administrative officials of a FES organization should represent the public in any discussion of policies that affect service level and quality. These officials can represent the public's interest impartially. It is in these times and situations that the chief officer must use the best negotiating and consensus-building techniques possible to come to a fair and equitable solution.

If the chief officer decides to submit a proposal or comment concerning public policy, it is very important that he or she researches and justifies his or her position thoroughly and accurately. The federal government and the NFPA are looking for consensus from participants, advocacy groups, committee members, and public officials. The submitter of the proposal or comment must be able to convince the regulators that their input has merit. Proposals that are supported by hard data and backed by well-thought-out lines of reasoning are more likely to succeed.

In most cases, the submitter is not physically present when the committee or public officials review the proposal or comment. These members and officials only have the information that is submitted in writing. However, submitters who can attend the meeting are allowed to present their case personally (although permission to testify usually must be approved before the meeting). An alternative for submitters who have contacts with national organizations, committee members, or other participants in the rule-making process is to contact these members and voice opinions personally on the subject. Participants in the rule-making process have more influence in the debate than the individual submitter.

## State Regulations

When the federal government adopts a regulation, it may or may not be enforceable by state or local governments and their employees. Many FES departments do not need to comply with federal regulations because they are binding only on private sector employers; however, as explained in the chapter, *Health and Safety*, when a state does adopt the federal OSHA regulations, the regulations then become enforceable for state and local government employers and employees. Many departments in nonparticipating states use the federal and NFPA safety standards, with the understanding that they are providing a safe working environment for their members and reducing the chance of liability. (Refer to the chapter, *Health and Safety*, for a complete list of all states that are federal partners.)

For the purposes of compliance with state OSHA regulations, volunteer firefighters are considered to be agents or contractors, not employees, and so they are exempt. Again, this situation can differ from state to state. For example, in one state, volunteers are considered employees when it comes to workers' compensation protection. However, one OSHA regulation is an important exception: OSHA 29 CFR 1910.120, *Hazardous Waste Operations and Emergency Response*, was adopted in 1993 by the Environmental Protection Agency, therefore making the regulation a national mandate.

## Building and Fire Codes

Most states and many local governments adopt building and fire codes that regulate the construction and maintenance of structures. These codes are adopted and enforced in a variety of ways. For example, in some states adopted codes cannot be changed locally, but in others local amendments to the codes are allowed.

Two national organizations currently promulgate building and fire codes: the NFPA and the International Code Council (ICC). The ICC is an organization that was created in 1994 by the three leading building code groups in the United States: (1) Building Officials and Code Administrators International, (2) International Conference of Building Officials, and (3) Southern Building Code Congress International. The ICC's goal is to produce one building code to be used throughout the United States.

The NFPA has also created a building code to complement NFPA standards and codes that previously prescribed only fire and electrical safety features for structures. Most states have opted to adopt the ICC codes, using the NFPA standards for reference standards. These codes can have a major impact on a fire department's ability to protect newer structures. For example, it is common for local fire protection regulations to require fire sprinkler protection for structures over a specified height or area, or that are in a certain use-group; however, one code may require this regulation and the other may not.

In addition, a few cities have their own separate building and fire codes; however, there has been a steady trend away from this practice. For example, in 2008, New York City decided to replace its own local codes with the ICC's International Building Code. In 2005, prior to the implementation, New York Buildings Commissioner Patricia Lancaster explained, "As it stands now, our Building Code is the most stringent set of construction regulations in the nation, yet its complexity is seen by many as an impediment to progress. The International Building Code will allow us to streamline the construction process while not sacrificing the effectiveness of these regulations in keeping our city a safe place to live, work, and build" (Lancaster, 2005).

Building and fire codes are very complex and are best studied through a formal education process, such as by taking fire prevention courses at a community college. In many states, inspectors and plans reviewers have to pass a rigorous certification process.

## Zoning Regulations

Zoning regulations can be difficult to generalize based on their origin. One simple example of a zoning regulation would be prohibiting a major manufacturing building in a residential area. Local governments generally use zoning to prevent noncompatible uses of buildings and land; however, there are many unincorporated areas and a few cities, such as Houston, Texas, that have no zoning regulations.

Some zoning regulations have their roots in fire prevention and were encouraged by insurance companies, and some are for esthetics, such as the Washington, DC, regulation that no structure can be built higher than the Washington Monument or the US Capitol building. This zoning ordinance has an unintended benefit for the local fire department—there are no buildings over 130 feet to contend with in the city.

The best way to gain some understanding of zoning regulations is to review examples. The list below cites examples of zoning regulations that the FES administrator may come across:

- In residential neighborhoods, small convenience stores can be built only at major intersections.
- Before a major commercial development, such as a covered mall, is approved, potential locations must be detailed on a map of the municipality, and requirements must be specified for adequate transportation to the property by the adjacent road network. The developer might have to pay for road improvements and traffic signals to gain zoning approval from the local government.
- In a congested downtown area, a masonry firewall is required on the property line of each building (as a result of a major conflagration in years past).
- If a new building is more than three miles from a ladder truck station, that building has to be protected by fire sprinklers.
- Streets in new developments, such as gated retirement communities, must be built to a minimum width.

Local governments might also control the type of development that can occur in each geographic area by using land-use zoning maps, which are maps approved by the local zoning authority that visually identify different use areas. For example, zoning might be approved for a large commercial development, but with the condition that the developer donates a parcel of the property for a new school, library, or fire rescue station.

## Union Contracts

In addition to federal and state regulations, union contracts have become vehicles to adopt standards and local regulations for FES employees. By mutual consent, and with the formal approval of the legislature and executive branch, these standards and regulations become legally binding for both parties. In essence, they are virtually the same as a law or ordinance.

Unions deserve special attention because they are an important part of the political landscape. In right-to-work states, unions may attempt to develop clout with elected officials (and those who influence them) because they cannot mandate members to join or pay dues, or mandate management to negotiate a binding contract for wages and benefits. (The chapter, *Introduction to Administration*, includes a complete listing of the right-to work states.) In other states, unions might have mandatory bargaining rights FIGURE 10-1. A union contract acts just like a regulation and is binding on all parties.

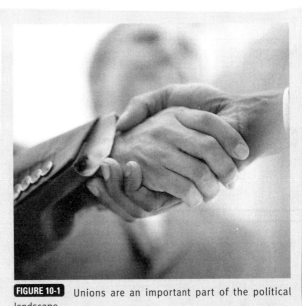

FIGURE 10-1 Unions are an important part of the political landscape.
© Yuri Arcurs/ShutterStock, Inc.

The relationship between union members and the administration can be adversarial or confrontational. The main goals of a union are to maximize pay and benefits while ensuring the health and safety of their members. FES administrators, however, often strive to improve services at the lowest practical cost to the taxpayers. These goals can conflict.

For example, an administration might propose an 8-hour duty limit in response to data showing that shifts longer than 8 hours have an adverse effect on a worker's mental acuity. Theoretically, this proposal would ensure well-rested FES providers who would make fewer mistakes during response. However, firefighters who work 24-hour shifts often have a second job and, in many cases, they spend almost as many hours at the part-time job as at their primary job. Therefore, a union might oppose this change because of the hardships it places on its members.

To facilitate change to existing services, the administrator should consider the following advice:

- Determine which decisions and policies are required by law or ordinance to be negotiated with the union.
- Delegate policies that do not require negotiation to a nonbinding task group made up of representatives from all interested parties.
- Understand the language in the existing labor contract, especially the concept of past practice.

If a decision is not a mandatory subject of bargaining, the chief does not technically have to secure the union's permission before implementation. However, the bargaining process requires good faith discussion with the union and a willingness to include the union's ideas and concerns in the final outcome. The union should be given a full opportunity to express all its viewpoints.

Some jurisdictions might have a formal impasse process. If both parties cannot come to agreement, an arbitrator might be brought in to decide the issue. This process is called binding arbitration. Even if mandatory negotiations are not required by a jurisdiction, formal meetings between the administration and members are always a good idea. Regular contact improves member morale and is useful in implementing change, helping the organization provide better services to the public.

# Tax Regulation

The ability to charge taxes and fees is a powerful tool; with it, governments can encourage or discourage behaviors. For example, home ownership is encouraged with the exemption of real-estate taxes and interest on property. Products like cigarettes, on the other hand, may have raised taxes on them to discourage consumption. Both of these are examples of the government using taxes to regulate behavior for what it deems as good outcomes.

Taxes are also used to regulate behaviors that affect the FES. For example, after a number of failed attempts to gain approval for a mandatory single-family dwelling sprinkler ordinance, one county adopted a reduction in local real estate taxes on new single-family dwellings with fire sprinklers. When a sprinkler system is installed in a new home, the cost is reflected in the total cost, which is then used to set the real estate tax. This temporary reduction offsets most of the cost of sprinkler installation.

At the federal level, the Fire Sprinkler Incentive Act was introduced in the House of Representatives and Senate in 2003. Currently, building owners are discouraged from investing in a sprinkler system due to the depreciation schedule of the 1986 Internal Revenue Tax Code: 39 years for a commercial building and 27.5 years for a residential structure. The Fire Sprinkler Incentive Act, which was reintroduced in 2013, covers small- and medium-sized businesses, allowing property owners to deduct the cost of a sprinkler system from their taxes. In addition, it covers high-rise structures by reducing the depreciation schedule of these structures to 15 years (National Fire Sprinkler Association, 2013).

# Hiring and Personnel Regulations

There are numerous rules, regulations, and laws that pertain to hiring and personnel actions for employees, such as the Fair Labor Standards Act (FLSA), Equal Employment Opportunity Commission (EEOC), Americans with Disabilities Act (ADA), affirmative action, and the Family and Medical Leave Act. These regulations, laws, and acts are discussed in the chapter, *Human Resources Management*.

# OSHA Regulations

In 1970, Congress created OSHA to assure the safety and health of employees in the United States by setting and enforcing safety standards and providing training, outreach, education, and assistance. State and local employees are only covered in state plan jurisdictions **TABLE 10-1**.

**TABLE 10-1 OSHA State Plan Jurisdictions**

| | |
|---|---|
| Alaska | New Mexico |
| Arizona | New York |
| California | North Carolina |
| Connecticut | Oregon |
| Hawaii | Puerto Rico |
| Illinois | South Carolina |
| Indiana | Tennessee |
| Iowa | Utah |
| Kentucky | Vermont |
| Maryland | Virgin Islands |
| Michigan | Virginia |
| Minnesota | Washington |
| Nevada | Wyoming |
| New Jersey | |

## OSHA 29 CFR 1910.146: Permit-Required Confined Spaces

This federal regulation covers employees who enter areas with limited or restricted means for entry and exit, such as trenches, pipes, tanks, and vaults. Confined spaces have the potential to trap a person and may have oxygen-deficient atmospheres. When the fire department receives a 9-1-1 call that someone has been overcome and is unconscious in a confined space, the firefighters must enter the confined space to rescue the victim. Firefighters have died when entering these atmospheres without self-contained breathing apparatus. The OSHA regulation requires training, proper equipment, and written procedures for all employees operating in these hazardous areas.

## OSHA 29 CFR 1910.134: Respiratory Protection

Among other requirements for training and equipment, this OSHA regulation—typically abbreviated as "two in/two out"—requires that four firefighters be on the scene before an interior structural fire can be attacked while using self-contained breathing apparatus. Fire departments generally address this regulation by adopting a standard operating procedure (SOP). Many municipalities are increasing their staffing on engines to four firefighters to comply with this federal regulation.

If a department staffs its engine companies with fewer than four persons, the first crew has to wait until another company (or volunteer/call firefighters) arrives at the structural fire to make up the minimum of four firefighters. In these situations, the only opportunity to start extinguishment is using an outside attack. This practice can be successful if the room where most of the fire is located has a window or door to the outside—a layout common in residential occupancies. For the safety of any occupants who remain inside, the attack should be done with a solid stream, which tends not to push the fire, heat, and smoke into other areas of the structure where victims may still be present. This tactic is practiced infrequently and is frowned on by some individuals in the fire profession.

The following are more specific requirements for this regulation, as noted by the International Association of Fire Chiefs and the International Association of Fire Fighters (1998):

- Personnel must use self-contained breathing apparatus when fighting a structural fire.
- A minimum of two firefighters must work as a team inside the structure, maintaining contact with one another at all times.
- A minimum of two firefighters must be on standby outside the structure. Any tasks performed by these firefighters must not interfere with the responsibility to account for the firefighters inside the structure.

This requirement for four firefighters for safe operations is also found in NFPA 1500 and 1710. In 2010, fireground field experiments conducted by the National Institute of Standards and Technology (NIST) also concluded that four firefighters was the best number of firefighters for safe and effective fireground operations.

## OSHA 29 CFR 1910.1030: Occupational Exposure to Bloodborne Pathogens

This regulation requires the employer to reduce or eliminate employee exposure to diseases that can be transmitted by bloodborne pathogens or other potentially infectious materials. Any agency involved in patient care is required to have a comprehensive education and control program for personnel who may become exposed to bloodborne pathogens or other potentially infectious materials. This includes providing all personal protection equipment necessary to protect personnel from bloodborne pathogens and offering hepatitis B vaccinations at no cost to the employee. These requirements are typically covered in a FES agency SOP.

Standard precautions generally include gloves; approved eye protection; and, in cases where it is warranted, a disposable gown. It is also prudent to have waterless antibacterial hand cleaners available until crews can reach soap and water. Repeated handwashing is the best defense against disease transmission in general.

## OSHA 29 CFR 1910.120: Hazardous Waste Operations and Emergency Response

The Hazardous Waste Operations and Emergency Response regulation is based on a 1986 draft of NFPA 472, *Standard for Competence of Responders to Hazardous Materials/Weapons of Mass Destruction Incidents*. It is generally a good idea to base training and certification on the latest edition of NFPA standards, because OSHA has not updated its hazardous materials regulation since 1989. OSHA recognizes and accepts these more recent standards as equivalent to the original rule.

This regulation was also adopted by the Environmental Protection Agency; therefore, it covers all persons who would have contact with hazardous materials in all states, including those not covered by OSHA state plans. All firefighters and other emergency responders who respond to hazardous materials incidents, such as police and EMS personnel, must be trained to a minimum level of competency.

For firefighters, this minimum competency is the operations level and is usually incorporated in recruit firefighter training programs. First responders, such as police officers and EMS support staff, who need to protect themselves but do not take emergency action during a hazmat incident are required to meet the awareness level, which is the lowest level of training and competency.

To gain a better understanding of this regulation and comply with its mandates, the NFPA *Hazardous Materials Response Handbook* provides an in-depth explanation of this standard and several useful appendix articles dealing with issues of organizing a hazardous materials response team.

## OSHA 29 CFR 1910.156: Fire Brigades

This OSHA regulation, first adopted in 1980, has been largely ignored by the public fire service. Many of its requirements are nonspecific and therefore a challenge to interpret and implement. Also, the title *Fire Brigades* seems to confuse readers, making them believe it is not applicable to public fire departments; however, "fire brigade" is just another name for "fire department," although used more commonly in other countries and industries.

Items required by this safety regulation include:

- An organizational statement
- Training and education similar to fire training offered in such states as Maryland, Georgia, and Washington
- In-service training
- Protective clothing and respiratory protection at no cost to employees

In the appendix, the regulation also goes on to suggest:

- Preplanning
- A physical fitness program
- Fire officer training and education

Most of these same items can be found in NFPA 1500, *Standard on Fire Department Occupational Safety and Health Program*. NFPA 1500 contains more specific language regarding requirements for safety policies, procedures, and equipment, and therefore many find it easier to comply with and enforce its requirements. OSHA 29 CFR 1910.156: Fire Brigades is also sometimes used in conjunction with the General Duty Clause.

## OSHA's General Duty Clause

This clause requires each employer to furnish a place of employment that is safe from hazards that are likely to cause injury, sickness, or death. OSHA and court precedent have shown that in the absence of specific or up-to-date OSHA regulations, a national consensus standard, such as NFPA 1500, can be used to assess compliance. In state plan jurisdictions, numerous violations for failure to comply with nationally recognized safety standards have been issued using the General Duty Clause.

For example, firefighters in a large suburban county filed a complaint with their state OSHA agency to stop the practice of firefighters riding on the back step of fire engines. Even though there is no specific requirement in the federal OSHA Fire Brigades regulation, the state agency cited the General Duty Clause along with NFPA 1500 when it filed a violation against the county. The practice was halted immediately and the proposed fine was eliminated.

Chief officers should always keep in mind the General Duty Clause. At a minimum, departments should consider complete compliance with NFPA 1500 for the FES agency. Otherwise, the chief and the organization could be open to civil liability or state OSHA citations.

## NFPA Codes and Standards

NFPA was formed in 1896 by a group of insurance companies who met to address the inconsistencies in the design and installation of fire sprinkler systems. According to the NFPA, "At that time there were nine different standards for piping size and sprinkler spacing, and these business people realized that unless these discrepancies were resolved, the reliability of these sprinkler systems would be compromised" (NFPA, 2005). The group created a standard for the uniform installation of sprinkler systems, and the NFPA standard-making process began. The organization cites their mission as follows: "to reduce the burden of fire on the quality of life by advocating scientifically based consensus codes and standards, research, and education for fire and related safety issues" (NFPA, 2005).

The NFPA home page contains several links to information under the heading "Codes and Standards." There is also a link entitled "Documents Accepting Public Input" that allows the review of a proposed standard or changes to an existing standard, so comments regarding these proposals can be prepared and submitted. The electronic version of *NFPA News* also contains a list of codes and standards that are in the process of being revised.

## NFPA 1500, *Standard on Fire Department Occupational Safety and Health Program*

NFPA 1500 was first issued as a standard in 1987. At that time, many in the FES industry and city or county administrators proclaimed that it would be too expensive and impractical to implement. Some even believed that the standard would be the end of many smaller departments, which would collapse under the substantial economic burden that came from safe operations and equipment. Although most fire departments have not yet completely complied with NFPA 1500, many have made substantial changes to provide a safer work environment for firefighters. As with any change in the FES field, safety improvements under NFPA 1500 have been evolutionary rather than revolutionary.

NFPA 1500 can be looked at as an umbrella document that adopts by reference many other NFPA standards, such as:

- NFPA 1001, *Standard for Fire Fighter Professional Qualifications*
- NFPA 1021, *Standard for Fire Officer Professional Qualifications*
- NFPA 1403, *Standard on Live Fire Training Evolutions*
- NFPA 1582, *Standard on Medical Requirements for Fire Fighters and Information for Fire Department Physicians*

- NFPA 1901, *Standards for Automotive Fire Apparatus*
- NFPA 1971, *Standard on Protective Ensembles for Structural Fire Fighting and Proximity Fire Fighting*
- NFPA 1981, *Standard on Open-Circuit Self-Contained Breathing Apparatus for Emergency Services*

These standards, in addition to providing for firefighters' safety, also provide firefighters with the training, tools, and protective clothing needed to operate effectively and professionally in mitigating dangerous uncontrolled fires and other emergencies that threaten to destroy property and injure or kill occupants.

One aspect of this standard that still has not been defined clearly is physical fitness. Although firefighting requires high levels of strength and endurance, the NFPA committee that was assigned the responsibility for defining this standard—the Fire Service Occupational Safety and Health Committee—has not been able to come to a consensus on specific levels and definitions of fitness.

## NFPA 1710, *Standard for the Organization and Deployment of Fire Suppression Operations, Emergency Medical Operations, and Special Operations to the Public by Career Fire Departments*

This standard, which was adopted in 2001, requires minimum levels for response times, staffing, and training. This standard will probably have more impact on FES than any other standard, with the exception of NFPA 1500. The word "career," as used in NFPA 1710, does not necessarily mean that a department is fully paid. Career fire departments include departments that rely on firefighters who are on-duty at the fire station. Therefore, this standard can also cover partially and fully volunteer organizations that schedule their firefighters to cover shifts; however, this is not common at this time.

One of the items in NFPA 1710 that often needs further explanation is the definition of career fire departments. Obviously, completely paid departments belong in this definition, but many combination departments can also use this standard, especially if they are in the process of increasing their on-duty personnel.

NFPA 1710's definition of "member" allows volunteers who are on-duty at the station to count toward the minimum number of firefighters on fire companies and the total first alarm assignment. In theory, it is possible for an all-volunteer company that assigns its members to specific duty times to comply with NFPA 1710. The key is that the firefighters and officer of the company have to be on-duty and trained to minimum levels of competency in accordance with the NFPA Professional Qualifications Standards.

There are many legal implications of not complying with NFPA 1710. However, even if a department's liability exposure is limited, it is still a good idea to use the NFPA standard as a benchmark for planning for public fire protection in a community. The citizens of a community deserve to know what level of public fire protection they are receiving.

**Four-Person Companies** The preferred method of complying with NFPA 1710 is to provide on-duty, in-station staffing of three firefighters and one officer per company. The NFPA 1710 document contains many references that support this requirement, and chiefs and elected or appointed officials are encouraged to review them if they do not understand or have a different opinion (International Association of Fire Chiefs, International Association of Fire Fighters, 2002, Section 5, Resources). One of the main reasons why a four-member company works best is that each firefighter is prepared to perform an assigned role. Firefighters that arrive individually have to fill a particular role that might be unfamiliar to them. Even with well-trained firefighters, this can result in an uncoordinated effort.

One staffing method for a company might be to assign four firefighters, two of whom are also trained paramedics. This company would operate an engine and an EMS transport vehicle. To make this company cost-effective, appropriate charges for emergency medical transports could help offset the salaries of the two EMS members. This staffing method works well until the call volume exceeds the point when the company can no longer provide a 4-minute response with 90% reliability, which are the requirements of the standard. However, one drawback of this method becomes apparent when the paramedic unit is transporting a patient to the hospital, or returning from a transport run, and a fire call is received. The first due company would arrive at the scene with only two firefighters, and would have to wait for the firefighter paramedics to arrive to start an interior attack.

**Response Time Requirements** The NFPA 1710 response time requirements for a structural fire state that the first due engine must arrive within 4 minutes (80 seconds are allowed for turnout) or the full first alarm assignment must be made within 8 minutes (80 seconds are again allowed for turnout). For record-keeping purposes, the response can be considered successful when either time target is met. These response times were set with the understanding that engine companies might be out of position when notified to respond, such as during preincident planning, fire prevention education, physical fitness, or training exercises. It is not uncommon for engine companies to be out of the station but in their first due areas. In some jurisdictions with areas that are undeveloped or that have low population densities, the 4-minute response time is not justified.

Response time performance should be measured at the 90% level (i.e., 90% of responses are made in less than or at the maximum time limit). This flexibility allows for exceptions because of the nature of a situation. For example, when a company from a distant station answers a call in the first due area of a company that is already responding to or working at an emergency incident, it is nearly impossible, or at least unreasonably expensive, to provide anything close to 100% compliance.

The number of times that a company is busy when a second call comes in is in direct proportion to call volumes (i.e., the greater the number of calls per year, the greater the amount of busy time and percentage of calls not meeting the maximum response time standard). Circumstances vary but, for the average company, volume in the range of 2,000 to 3,000 calls annually can frequently lead to situations in which the maximum response time is not met. The Insurance Service Office Municipal Fire Grading Schedule once contained the requirement that a second company would be needed if more than 2,500 calls were responded to annually.

As exemplified by the following list, departmental policies and SOPs can help solve—or exacerbate—the increasingly common problem of excessive responses:

- Dispatch policy can sometimes lead to responses to alarms that may not be necessary. For example, some departments send a full first-alarm assignment to an automatic alarm, whereas others send only one engine.
- A vehicle crash may generate the automatic response of a paramedic unit, an engine, and a heavy rescue squad in one jurisdiction; in another jurisdiction, only the paramedic unit responds unless additional information indicates a serious accident with occupants trapped.
- Other than the required companies that must be dispatched to a structural fire (totaling 14 firefighters), chief officers should review all dispatches that send more than one unit and calculate the percentage of times each unit was actually used by call type.
- Adding a second unit in the same station cuts the call volume substantially. For maximum efficiency, only one unit should be dispatched on any fire call, leaving the other unit (engine) to answer a second call in the first due area within the expected 4 minutes.

NFPA 1710 will have a distinct impact on fire and EMS services in the future. It is a planning document that can help identify and quantify weaknesses in fire protection and EMS. Quantifiable comparisons help citizens and both elected and appointed officials learn which parts of their communities have adequate fire and EMS coverage and which parts are beyond response time standards.

## National Highway Traffic Safety Administration

In the 1960s, it seemed clear that the quality of prehospital care was an important determinant of survival from sudden injury and preventable deaths from highway

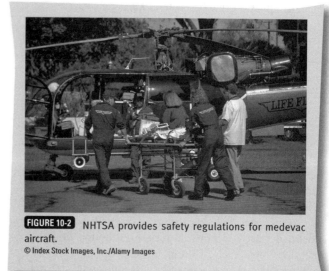

**FIGURE 10-2** NHTSA provides safety regulations for medevac aircraft.
© Index Stock Images, Inc./Alamy Images

trauma. It was for this reasoning that federal funding for standardized training of ambulance attendants was provided through the Highway Safety Act of 1966. In 1969, the Highway Safety Bureau came into existence, funding the development of the National Standard Curricula, which set the precedent for the way EMS education would be standardized for the next 40 years.

In 1970, the Highway Safety Bureau transformed into the National Highway Traffic Safety Administration (NHTSA), with the goal of reducing the number of vehicle crashes, injuries, and deaths. The NHTSA collected data, investigated safety defects, reduced the threat of drunk drivers, promoted the use of safety belts and airbags, and provided consumer information on motor vehicle safety topics. It has also been instrumental in implementing a nationwide 9-1-1 emergency reporting system; setting certification standards for emergency medical responders, emergency medical technicians, and paramedics; and setting standards for medevac helicopters and shock trauma centers **FIGURE 10-2**.

## Legal and Court Issues

Legal issues and opinions are in a continual state of change as a result of new legislation and legal decisions from numerous judges throughout the country. When reviewing decisions from a court, keep in mind the hierarchy of the judicial system in this country. Unless the Supreme Court of the United States has made a determination about a legal issue, the issue can always be appealed for another opinion. For example, a lawyer may explain that a circuit court awarded a large settlement to an employee as a result of some action by a FES organization. If the administration is using this court ruling to determine if a similar policy at another department is illegal or defensible in court, the circuit court's decision should have less of an influence on the administration's actions than a decision from a higher appeals court. Out-of-court settlements might have even less influence.

In addition, each lawyer, attorney, or counselor may have a different opinion about a particular law or regulation. A critical emergency service to the public can be compromised if a department is persuaded by legal advice to reduce professional standards for its members. Legal actions can have major and undesirable consequences for the department. The chief officer should be wary of any attorney who wants to settle quickly and not take the issue to court (Edwards, 2010). Public service is a very serious responsibility that should never be taken lightly by the administration. In some situations, it is wise to seek a second opinion and to do independent research. Contact peers and check with professional organizations. Look for articles that may have been written about the subject. Do not rely on one source. Many individuals have their own agendas.

Chief officers can take several measures to protect the department and its members from legal problems. When applicable, follow the requirements of an appropriate professional standard or regulation (e.g., NFPA or OSHA). Ensure that all the department's policies treat all people equally. Demand that any legal advice be documented by case histories from courts at the appeals level or higher. Ask the lawyer to present all sides of the argument, not just the legal precedent that supports the lawyer's opinion or the department's policy.

## The Court System

The better the FES administration understands the court system and the legal process, the better able it will be to protect the interests of its organization and the public it serves.

Federal and state courts have three separate levels: (1) the district court, (2) the appellate court, and (3) the Supreme Court. Judges in the court systems interpret existing laws and regulations to come to their decisions. If a decision made in a district court is appealed, the case moves on to the appellate court. The decision made in the appellate court then becomes known as a precedent, or common law, and may be used by other judges in similar cases. For the most part, lower appellate court decisions can set precedents only in the district or state in which they are issued; it is possible for appeals courts to make decisions that are in conflict with appeals courts in other districts.

The US Supreme Court, whose decisions are binding nationwide, has final authority **FIGURE 10-3**. Its decisions become precedent for the entire nation. If the Supreme Court refuses to hear a case, its refusal indicates that it supports the lower court's decision. If the Supreme Court makes a decision based on the US Constitution, this decision can be changed only by changing the Constitution, which is difficult, if not impossible, to do. Although there is a process for adding to and changing the Constitution, it is used only in rare cases because it requires a three-quarter majority vote by Congress and US voters.

### Chief Officer Tip

**Legal Crises**

Chief officers should not wait for a legal crisis to meet the lawyers representing their department. By developing and maintaining a relationship with their lawyers, chief officers can get ahead of any legal crises that may occur. Regular conversations allow the chief officer and lawyer to familiarize each other with information on legal issues and trends, court cases, and other organizational developments.

## Attorneys

As is true for many areas of management, when it comes to legal issues, it is a good idea to seek help as soon as possible. The chief officer also needs to make sure to get advice from an attorney who has the appropriate background or best interests of the FES organization in mind.

Attorneys might not understand the internal workings of an emergency organization. Chief officers should educate them, starting with the basics. For example, the FES organization may have a strict policy that any new applicant may not have any record indicating that he or she cannot be trusted, such as a criminal record, excessive driving citations, or poor credit. Chief officers would need to explain to the attorney that there are many occasions when members are in private homes during an emergency call, and the organization cannot tolerate theft of private property or any other criminal activity.

If the case involving the department is lost at the district court level, make sure the department's attorney and

**FIGURE 10-3** US Supreme Court decisions are binding nationwide.

the elected officials support an appeal. A judge may make a decision based on his or her personal beliefs about the law or the specific circumstances. This is known as "legislating from the bench." Do not compromise when the issue is very important to the organization and its ability to provide quality service to the public. Always be prepared to appeal a misguided or inaccurate ruling. Make sure the other side is aware of the department's resolve. Showing this resolve can sometimes keep the department and the chief officer out of court.

If the organization or the municipality is too small to have a staff attorney, or if the attorney does not have enough time to provide quality service, ask the municipal administrator or mayor to select an outside attorney. This is especially helpful if the loss of the case will have a serious adverse impact on the organization. If the present attorney does not seem to be meeting the department's needs, the administrator should insist on changing attorneys.

## Administrative Rule Making

Administrative rule making allows the public administrator to function as a lawmaker. As public administration expert Gerald Garvey states, "The opportunity for administrators to function, in effect, as lawmakers often arises because the legislators themselves may lack the interest, the information, or the political courage to settle every last provision of a new public program" (Garvey, 1996, p. 138). For example, when the Florida legislature adopted a state-wide minimum building and fire prevention code in 2001, it did not specify the exact model building code to adopt but instead delegated that authority to the Florida Building Code Commission, which was made up of government representatives, subject matter experts, and stakeholders. The Florida state fire marshal was then given the authority to adopt, modify, revise, interpret, and maintain the Fire Prevention Code, which was based on NFPA 1, *Fire Code*, and NFPA 101, *Life Safety Code*.

Garvey goes on to explain, "The rule maker's role—one of the most important in public administration—requires administrators to function as lawmakers, or at least in a quasi-legislative mode. That mode casts appointive officials in a role which strict democratic theory once reserved for elected representatives sitting in formal legislative bodies" (Garvey, 1996, p. 139). These administrative rules are sometimes issued under the signature of the head administrative official, such as the secretary of labor or the state fire marshal. However, a special team of experts within an agency or a task force of stakeholders generally is assigned to see rule making through from start to finish. This team then provides their recommendations to the administrative official.

Many meetings and numerous hours are spent in consultation and negotiation with individuals and representatives of groups whose members will be affected by the rule. Public hearings are held, in theory, to solicit the input of all concerned parties. However, special interest groups, such as big business or labor groups, may concentrate their influence during the rule-making process, and their efforts sometimes allow them to have more of a voice than the public. Although appointed officials should be above local politics and should use system-wide planning to approach the problems of interdependent communities, it is not uncommon for vote-conscious politicians to be influenced by influential business leaders, voters, and special interest groups in their election districts.

For this reason, many controversial regulations are delegated to the administrative rule-making process. This process goes a long way toward moderating special interest influence and leveling the playing field because, in most cases, it is open to public input and scrutiny. Still, the chief officer must remain cognizant of the possibility that special interest groups want to advance their own agendas and should, therefore, play the role of advocate for all those affected by the rule. In the FES business, there is a slow and steady movement away from the excessive political influence from special interest groups and elected officials. Instead, progress is driven by national standards, mutual and automatic aid agreements, consolidations, and college-level education for FES administrators.

## Legal Aspects of FES

The following is a general list of advice for the FES administrator:

- First and foremost, do not become paralyzed by fear of liability.
- Get the facts; do your own independent research.
- Check with more than one attorney.
- Check on the department's immunity.
- Comply with all laws, regulations, and safety standards even if they are voluntary.
- Perform all duties imposed by law or regulation.
- Be fair.
- Train, train, and train.
- Consult recognized experts (remember, not just one).
- Have access to a well-informed attorney.
- Document all actions in writing.
- Be proactive in proposing changes or new laws and regulations when necessary.
- Take prompt action on complaints.
- Follow up on discipline and violations of department SOPs.
- Know the professional expectations of the job (experience, training, and education).
- Know the statutes, regulations, and court cases.
- Review legal and occupational publications.
- Treat all employees in the same, fair, evenhanded way.
- Keep customer service first when creating the department's goals.

## CHAPTER ACTIVITY #1: NFPA Compliance

Using the chapters *Training and Education*, and *Health and Safety*, complete the NFPA 1500 Worksheet found in Appendix B of the standard. For this activity, it is not necessary to complete all columns; just complete the columns marked "Compliance," "Partial Compliance," and "Compliance with Administrative Action." Contact an official representative of a department who will help provide copies of written official documents supporting compliance, including SOPs and policy directives.

Remember, for an accurate understanding of a section or paragraph, you may need to do some research, starting with a review of the NFPA handbook. If you are not convinced that a requirement has been explained accurately, contact another knowledgeable person for an opinion. An e-mail to an NFPA staff person is always a good option.

## CHAPTER ACTIVITY #2: Legal Advice

Find an article from an EMS periodical that discusses a recent appeals or Supreme Court decision. Then find two more articles that discuss the same issue either before or after the court's decision. Summarize the articles, comparing and contrasting the sides being taken. After you think you have a good understanding of the issues and the court decision, find an attorney who will provide additional legal advice regarding the issue. (Explain that it is a school project. Many attorneys will give a student a limited amount of time without professional charges.) After writing down a synopsis of the attorney's advice, express your own personal thoughts about the case. Limit the answer to no more than two pages.

## References

Edwards, S. T. (2010). *Fire service personnel management* (3rd ed.). Upper Saddle River, NJ: Brady/Prentice Hall Health.

Garvey, G. (1996). *Public administration: The profession and the practice*. New York, NY: St. Martin's Press.

International Association of Fire Chiefs, International Association of Fire Fighters. (2002). *NFPA 1710: A decision guide*. Fairfax, VA: International Association of Fire Chiefs.

International Association of Fire Fighters, International Association of Fire Chiefs. (1998). *IAFF/IAFC two in/two out questions and answers*. Retrieved from http://www.iaff.org/safe/pdfs/2in2out.pdf

Johnson, R. (2001, February 7). Lines of duty: As blazes get fewer, firefighters take on new emergency roles—Cities complain about costs, but safety group's plans may help fuel the trend. *Wall Street Journal*.

Lancaster, P. J. (2005, June 24). Press release. New York City Department of Buildings.

National Fire Protection Association. (2005). *History of the standards development process*. Retrieved from http://www.nfpa.org/itemDetail.asp?categoryID=2563&itemID=57046&URL=Codes%20&%20Standards/Standards%20development%20process/How%20codes%20and%20standards%20are%20developed/History%20of%20the%20standards%20development%20process&cookie_test=1

National Fire Protection Association. (2001). *1710, Standard for the organization and deployment of fire suppression operations, emergency medical operations, and special operations to the public by career fire departments*. Quincy, MA: National Fire Protection Association.

National Fire Sprinkler Association. (2013). *Legislation*. Retrieved from http://www.nfsa.org/?page=Legislation

Occupational Safety and Health Administration. (2006). *All About OSHA*. Retrieved from http://www.osha.gov/Publications/about-osha/3302-06N-2006-English.html

Office of Management and Budget. (1998). *Circular No. A-119, Federal participation in the development and use of voluntary; consensus standards and in conformity assessment activities*. Retrieved from http://www.whitehouse.gov/omb/circulars_a119_a119fr

US Government Printing Office. (2005). *Your one stop site to comment on federal regulations*. Retrieved from http://www.regulations.gov

# CHAPTER 11

# Ethics

## Fire and Emergency Services Higher Education (FESHE) Course Objectives

There are no objectives in this chapter.

## Knowledge Objectives

After studying this chapter, the student will be able to:
1. Understand the ethical justification for government oversight of the public FES. (p 156)
2. Understand the creation, effect, and importance of a professional organization's code of ethics. (p 156)
3. Identify commonly cited justifications for lying. (pp 157–158)
4. Understand the impact that lying and other ethical issues have on fire and emergency services. (p 159)
5. Examine the role of the administrator as a caretaker of public trust. (pp 159–160)
6. Examine how ethics affect cost-benefit analysis in the FES. (pp 160–161)

## Ethical Behavior

The American Heritage Dictionary defines *ethic* as a "principal of right or good conduct" or "a system of moral principles of values" (American Heritage Dictionary, 2005). Ethical behavior is behavior guided by that system of moral principles. Although using one's ethics to make appropriate moral decisions might seem straightforward, increasingly complex situations and ever-changing policies can complicate decision making. Some believe that only those actions that are illegal, such as robbing a bank, are wrong. However, illegal actions are only a subset of the larger category of potentially unethical actions.

Unethical behavior often involves some form of lying, such as deliberately presenting a false statement or action as true or hiding factual information with the intention of deceiving. This can be a difficult issue to discuss and study because many people hate being called a liar and either rationalize their behavior or are in denial of their actions. No one is perfect and everyone has made mistakes. However, it is hoped that this chapter offers a new way to govern future decisions. The trick to success is to learn from one's mistakes; and to learn, one must admit the mistake and commit to correcting the error in the future.

Administrators of FES agencies and emergency medical services (EMS) organizations make many decisions each day, and every decision has potential ethical consequences. Because many of these decisions are ultimately made by the chief officer, the officer needs to consider whether he or she is making morally correct decisions. This is not always easy to determine, especially for the decision maker.

## Professional Ethics

Ethical behavior includes personal and professional conduct. For example, *personal ethics* may include such behaviors as complying with laws, complying with societal norms, telling the truth, and having concern for others' well-being and respect for their freedom and independence. *Professional ethics*, however, may include such behaviors as complying with company policies and procedures, telling the truth to the public and elected officials, protecting the confidentiality of coworkers and those one serves, and performing due diligence for the safety of the workers and the public.

Many professions are under the oversight of organizations that regulate the ethical behavior of their practitioners. For example, states may create boards made up of practicing professionals, such as doctors, lawyers, or engineers, to regulate a specific profession. As noted by philosopher and ethicist Sissela Bok in the book *Lying: Moral Choice in Public and Private Life*, these boards should expose any members of the profession who act illegally or unethically; however, this is rarely the case (Bok, 1999). A colloquial expression to describe this relationship is "the fox watching the hen house."

Although the FES profession does not have any organizations that specifically watch for unethical actions, a few fire organizations do have the potential to make judgments on professional conduct. These organizations include the National Fire Protection Association, the International Association of Fire Chiefs, the National Volunteer Fire Council, and the International Association of Fire Fighters. Some of these organizations, such as the Florida Fire Chiefs' Association (FFCA), use a suggested code of ethics to guide members' conduct. The FFCA's Code of Ethics cites the following principles (FFCA, 2012):

- Maintain the highest standards of personal integrity; be honest and straightforward in dealings with others; and avoid conflicts of interest.
- Place the public's safety and welfare and the safety of employees above all other concerns. Be supportive of training and education, which promote safer living and occupational conduct and habits.
- Ensure that the lifesaving services offered under the members' direction be provided fairly and equitably to all without regard to other considerations.
- Be mindful of the needs of peers and subordinates and assist them freely in developing their skills, abilities, and talents to the fullest extent; offer encouragement to those trying to better themselves and the fire service.
- Foster creativity and be open to consistent innovations that may better enable the performance of our duties and responsibilities.

Similar to this example, the National Association of Emergency Medical Technicians adopted a Code of Ethics in 1978. This Code acknowledges the EMS practitioners' obligation to make ethical patient-care decisions and to avoid opening their organization to liability. As stated within the Code of Ethics, "Professional status as an Emergency Medical Technician and Emergency Medical Technician-Paramedic is maintained and enriched by the willingness of the individual practitioner to accept and fulfill obligations to society, other medical professionals, and the profession of Emergency Medical Technician" (Gillespie, 1978). The Code also goes on to list a number of specific ethical behaviors to follow, including conserving life and alleviating pain, respecting the dignity of all patients, maintaining confidentiality, and upholding medical standards and laws.

EMS organizations, many of which are private companies, are expected to train to an ethical standard and to investigate incidents where unethical acts may have occurred. Most organizations are held accountable to local government oversight, which is also responsible for investigating any incident involving unethical behavior or a departure from city- or county-approved protocols or policies. If the oversight entity finds wrongdoing, the incident might then be referred to a state-level EMS authority that would perform its own investigation and decide whether or not to take disciplinary action, such as revoking an EMS provider's license.

In situations where the failure of an administrator, supervisor, or organization to comply with a safety regulation results in a death or serious injury, state or federal Occupational Safety and Health Administration agencies step in to assess unethical conduct. Although there have been some past examples of chiefs being dismissed after a firefighter death, generally in cases of noncompliance, the organization—not the individual—is found at fault.

## Why People Lie

Researchers report that extroverts and coworkers are some of the most likely culprits of lying (Feldman, 2005). The tendency of coworkers to lie to each other is an important phenomenon to understand in the FES field. For example, consider the fire chief who surrounds himself with staff who are "yes men." This practice prevents the discussion of controversial policy issues and hinders the delivery of honest feedback. Such lack of communication is unhealthy to the organization.

However, the most negative types of lies are not those that are classified as boasting or told in the name of discretion or politeness. The most harmful lies involve leaving out the truth or stating something misleading or absolutely false. These types of lies have the potential to destroy trust and intimacy in personal and professional relationships.

Therefore, it is important to understand why someone would lie in the first place. According to University of Massachusetts psychologist Robert Feldman, the act of lying is attached to a person's self-esteem: "We find that as soon as people feel that their self-esteem is threatened,

> **Chief Officer Tip**
>
> **Habits of Dishonesty**
> The article entitled "American teens lie, steal, cheat at 'alarming' rates" discusses the results of a 2008 study conducted by the Josephson Institute in which the ethical habits of nearly 30,000 high school students were examined. According to the article, the study revealed "entrenched habits of dishonesty for the workforce of the future," including the following alarming statistics (Josephson Institute, 2008):
> - 30% of students admitted to stealing from a store within the past year.
> - Between 78% and 83% of students admitted to lying to their parents about something significant.
> - 64% of students admitted to cheating on a test, with 38% saying they had done so more than once.
>
> Although these numbers may be alarming, what is even more surprising is the students' perspective on ethics. "Some 93 percent of students indicated satisfaction with their own character and ethics, with 77 percent saying that 'when it comes to doing what is right, I am better than most people I know'" (Josephson Institute, 2008). Keeping in mind that these same high school students may be future elected or appointed officials, members of the FES organization, and even officers, administrators must use a policy of no tolerance for lying for themselves and the members of their organization. This kind of behavior can be changed with perseverance and leading by example.

they immediately begin to lie at higher levels" (Feldman, 2005, p. 373). It is with this notion that lying can be viewed as a person's attempt to see himself or herself as he or she would like to be seen by others. "People are so engaged in managing how others perceive them that they are often unable to separate truth from fiction in their own minds" (Feldman, 2005, p. 375). For people to keep fact and fiction separate, they must consciously commit to telling the truth in every situation.

## Justification for Lying

Imagine a woman is in her kitchen at home when two very young children come running through the kitchen door. They tell her that a crazy person is chasing them with a knife. The woman quickly directs the children to hide in a closet. Almost immediately after the children close the closet door, a man bursts through the kitchen door with a large machete in his hand. He asks the woman in the kitchen if she has seen two children. The woman lies to him and tells him she has not seen any children. The intruder goes away.

Is this person justified in lying? The rational answer is yes. However, most situations are not so "black and white" **TABLE 11-1**. Trying to determine if a lie is justifiable is often very difficult, and opinions can vary drastically. The following sections address a few commonly cited justifications.

### "It's for the Public Good"

Administrators who see themselves superior by means of birth, wealth, training, or education often feel they have a right to make decisions for the public. Bok explains, "Convinced that they know the truth—whether in religion or politics—enthusiasts may regard lies for the sake of this truth as justifiable. They may perpetrate so-called pious frauds to convert the unbelieving or strengthen the conviction of the faithful. They see nothing wrong in telling untruths for what they regard as a much 'higher' truth" (Bok, 1999, p. 7). In practicing this belief, however, administrators are left with only their own consciences to judge whether their choices are morally right.

For example, suppose an incumbent mayor is running for reelection and is convinced that his continued leadership is in the best interest of the city. Most likely, all of his staff and his political party also support this notion. Under these circumstances, the mayor might hide an indiscretion, fail to be completely frank about his position on a controversial issue, or support a popular proposed law or allocation of funding even if he knows doing so would not be in the best interest of the city and most of the citizens.

One effective technique administrators could use is to look at issues through the eyes of the young and the old, minority members, men and women, members of the emergency service's agency, customers, and elected officials. They should then consider how the people in each of these roles would feel about choices being made for them. This can be difficult, because many members

**TABLE 11-1  Examples of Ethical Dilemmas**

- Should a paramedic lie to a patient who asks, "You won't let me die, will you?"
- Should professors and supervisors exaggerate the excellence of job applicants when giving references to give the applicants a better chance at being hired? Is this fair to the potential employer?
- Should journalists write false statements, or fail to tell the whole story, if doing so supports their personal views on a subject that they strongly believe is right?
- Should a fire chief exaggerate the consequences (e.g., more deaths and property loss) of not funding the budget properly?
- Should a police officer use fabricated evidence to help convict a known criminal?

of the FES industry have strong beliefs regarding what is needed for emergency operations and safety.

Consider the chief at an incident where a fire death has occurred. The chief is asked by a newspaper reporter, "What could have been done to prevent the death?" The chief could ignore several prevention factors (e.g., smoke alarms and sprinklers) and answer that having four firefighters on each engine would have averted the death. Although this answer may open up a conversation for additional staff funding, which would ultimately lead to better service by the department, it would not be entirely truthful. In the emergency services field, there is often more than one way to prevent a tragic incident and, to tell the whole truth, the chief officer should mention all solutions. Although beliefs regarding service needs may be well founded, members and officers still need to use truthful statements when justifying these beliefs to the public and elected officials FIGURE 11-1.

## "It's in Self-Defense"

There are situations in which lying seems justifiable when the alternative would be to use physical force. As considered by Bok, "If to use force in self-defense or in defending those at risk of murder is right, why then should a lie in such cases be ruled out?" (Bok, 1999, p. 41). Therefore, when physical force seems like the only means of self-defense, a person could also consider whether lying would accomplish the same objective.

However, even in instances of self-defense, it is important to consider the consequences of lying. For example, before the invasion of France during World War II, the Allies went to great efforts to deceive the Germans as to the location of the invasion to avoid an attack on the liberation forces. What the Allies did not consider was future discussions on surrender or peace agreements. How could Germany believe anything said to them if the Allies had always lied to them in the past? Even with enemies, it may ultimately be better to tell the truth in some cases.

**FIGURE 11-1** Members and officers need to use truthful statements when justifying their beliefs to the public and elected officials.
© Karin Hildebrand Lau/ShutterStock, Inc.

## "It's a Miscommunication"

Chief officers should be very careful that all communications are received accurately. A person may be accused of lying when instead something actually went wrong within the communication process. This is especially true with verbal communications. A person may have meant to say one thing, but inadvertently said something with a completely different meaning. If an officer conveys the circumstances of a fire and the service's response to it, but fails to do so clearly, the message may be misunderstood and people may believe that the chief has lied about the event or the department's response to it.

Another type of miscommunication might occur when what the listener has already assumed to be true differs from what is actually said. In this case, the listener might hear what he or she wants to hear. In the case of emergency incidents, prompt communications after a major event helps prevent the public from being informed by gossip and rumor, which might then color their perception of the official story.

For example, after a major fire in an apartment complex, occupants complain about the slow fire department response. Several occupants had been forced to jump from upper floors and sustained serious injuries after waiting a long period of time for the fire department to arrive. The fire department's on-scene public information officer immediately calls the dispatch office to inquire about the response times noted in the computer-aided dispatch system. After checking the system, it is determined that the response actually took place within 4 minutes, clearly an acceptable time range.

So what happened? This was an afternoon fire and a major incident, with heavy flames and smoke showing from numerous windows. After checking with several witnesses and asking if they called 911, it is determined that nobody actually called, because everyone thought that someone else would have already called. To correct the misinformation, more than the specific facts were needed. These facts had to be paired to a plausible explanation. This was undoubtedly a case where quick and accurate communications saved the fire department's reputation.

## Ethical Tests

The following is a list of methods to consider when evaluating an ethical dilemma:

- Assume the public is in the room and has adequate knowledge to review and discuss the issue (especially if the decision must be made secretly, a discussion using this assumption provides needed input).
- Follow the Golden Rule: treat others as you would want them to treat you.
- Consider the viewpoint of everyone who will be affected.
- Examine the validity of the "avoiding harm" and "greater truth" excuse.

- Investigate from top to bottom the "produces greater benefit" justification.
- Consider the fairness of the decision.

In addition to the considerations listed above, keep in mind the following quote from Mark Twain: "When in doubt, tell the truth. It will confound your enemies and astound your friends" (Twain, 1906).

## Consequences of Lies

### Denial of the Public's Free Choice

When we talk about taking away someone's freedom, most people think of a dictatorship or a prison where freedom is removed by physical force. But deceit and lying can also result in the loss of freedom. Furthermore, deceit does not have to be a direct communication of something that is untrue; it can also be the withholding of information that others have a right to know. For example, before the information became public knowledge, a well-known tire manufacturer kept many reports of vehicle rollover crashes involving its tires a secret from the public. Without knowing this information, the public was unable to choose the safest tires for their cars.

Now consider a common situation in the fire service. When the general public sees the fire engine responding with its red lights and siren operating, they are unable to see how many firefighters are on board; furthermore, they often have no knowledge of how many firefighters are even needed to do a safe and effective job. Even though the minimum number of firefighters on the truck has been established by a national standard (e.g., National Fire Protection Association 1710) and bolstered by the Occupational Safety and Health Administration's safety regulation, the FES administrator can easily mislead the public and elected officials into thinking that whatever level of service is being provided at the local level is appropriate. This may result in fire engines responding with only one, two, or three firefighters, which necessitates more than one fire company for a fire to be safely attacked within a structure. In many cases, this level of service (i.e., number of stations and staffing) is justified by such factors as cost avoidance, unique local situations, or tradition, none of which meets the test of truthfulness.

Similarly, members of the public often assume and expect that the response of an ambulance means that advanced life support care is provided. Although there is a distinct difference in the scope of practice between a basic life support emergency medical technician and an advanced life support paramedic, the uniform often looks very similar. Should the service be obligated to clarify if it offers only basic life support service? Just like the engine company that does not advertise it has only two crew members, the ambulance company may not attempt to change what the public perceives about their available level of service.

In cases such as these, it may be easy to get caught up in addressing the needs of employees or elected and appointed officials and fail to remember the primary goal of serving the public. A chief officer may not attempt a change because he or she knows the change would be resisted by specific groups or individuals, such as the union, volunteer members, the mayor, or the city council; however, the officer should continue to try, using appropriate preparation, leadership, and courage to gain acceptance.

### Damage to Professional Reputation

With the media's global reach, everyone has instant access to news about fraud, corruption, and cheating. It seems that the more immoral the act and the more famous the perpetrator, the bigger the story. It is now very difficult for public officials to cover up inappropriate behavior. If administrators have enemies or members in their organizations that simply do not approve of their leadership, impropriety can quickly be turned into ammunition.

"Paradoxically, once his word [the person lying] is no longer trusted, he will be left with greatly decreased power—even though a lie often does bring at least a short-term gain in power over those deceived" (Bok, 1999, p. 26). Some studies indicate that it may take up to 2 years for others to acknowledge any positive changes in behavior made by a person or an organization. As noted by social psychologist Roy Baumeister, "Bad impressions and bad stereotypes are quicker to form and more resistant to disconfirmation than good ones" (Baumeister, Bratslavsky, Finkenauer, & Vohs, 2001, p. 232). After an individual or an organization loses the confidence of the public and its own members, it becomes very difficult to ever regain that trust.

## Moral Obligations of Public Roles

There is no reason to think that individuals in public roles are exempt from traditional ethical and moral requirements, or that the end always justifies the means. The person who accepts the position of chief officer of an FES department or EMS organization has a special obligation to the public, and to the organization's members and elected officials. The FES or EMS chief officer is directly responsible for the service the public receives and for the safety of the members. Also, in many cases, the chief officer is among the few who have the technical expertise to judge the adequacy of the service provided. Right or wrong, the officer has the power to fool the public and politicians into believing the organization is providing great service when in fact the organization and its members may not be operating at 100% effort and efficiency.

### Duty to Obey

There are many times during a chief officer's career in government when the officer feels pulled in several directions by superiors, such as a mayor or city administrator. People entering authoritarian systems might no longer view themselves as acting for their own purposes, but acting rather as agents for executing the orders of another person. "[A]dministrators typically work in environments where the presumption of the obligation to obey is powerful—so powerful that, in an extreme case, the administrative environment may promote a willingness on the part of officials to give up all sense of personal responsibility" (Garvey, 1996, p. 328). Chief officers must be careful in these situations, because they can end up involved in a civil or criminal action.

Although obedience may be a deeply ingrained behavior for many public officials, it is important to remember that blind obedience is a powerful force that can override training in ethics, common sense, and moral conduct. For example, the chief officer receives a direct order from the mayor to eliminate all overtime. The officer knows that overtime is needed to maintain minimum staffing levels, but he obeys the order without a fight, possibly fearing loss of his job. A leader of an FES department or EMS organization must find a way to stay independent by nurturing a relationship with superiors that allows for open, honest communication. Although most of this frank communication must take place in private, the chief officer should be able to approach his or her superior in any setting without the fear of reprisals.

Unions, volunteer organizations, and business owners have all used their influence over the years to change or attempt to change fire- and EMS-related public policy. These groups were specifically created to look out for the best interest of their members. Administrators need to remember their obligation to the public because many of the groups attempting to influence the behavior or policies of the organization have their own personal agendas. Officers must always try to do what is ethically right, using some of the techniques and suggestions in this chapter as a guideline.

## Fairness Among Employees

Fairness is equated by many as being the "right" or ethical thing to do. This is not always achievable, but when possible, chief officers should treat all employees equally and fairly. Every person should have an equal opportunity to succeed in accordance with his or her abilities and hard work.

For example, in any FES organization, some members perform at a higher job performance level than their peers. A chief officer could easily start treating these few outstanding workers better than their peers. However, the officer should remember that it takes a team of players to win the game. All players contribute to the best of their ability if they are properly motivated. The job of a leader is to get all employees to operate near their maximum performance level, whatever may be that level.

Only in those cases where an employee's behavior is unacceptable should different treatment be administered. In these cases, the other employees are watching, and if they see an employee getting away with substandard performance, some will lower their own performance levels. Everything a chief officer says or does is a signal to all employees of what to expect in relation to job performance. If the officer allows a few self-motivated overachievers to carry the bulk of the work or does not discipline those not meeting minimum performance standards, that officer is not treating employees impartially or fairly.

## Cost-Benefit Analyses

In most cases, the presumption is that an existing or new program will not be funded if its costs outweigh its benefits. In private industry, the balance of cost to benefit is at the heart of any plan of action, and the financial bottom line is one of business owners' and stockholders' most important considerations. However, for environmental, safety, and health services and regulations, there may be many instances where a certain decision might be right even though the regulation's benefits do not outweigh its costs. There are good ethical reasons not to assign dollars to the value of life.

For example, if the chief proposed a new fire safety regulation that required the installation of sprinkler systems in all new one- and two-family dwellings, the opposition would probably be able to produce a cost-benefit

### Facts and Figures

**Misconduct in Government**

With employees at all levels of government witnessing a high incidence of ethical misconduct, and with many local and state entities failing to establish strong ethics programs, the public sector is at considerable risk of seeing major ethical scandals unfold. The Ethics Resource Center's National Government Ethics Survey points out some disturbing findings, especially for local government, which provides most FES. During a year's period, an estimated 34% of local government employees witnessed misconduct (e.g., putting one's own interests ahead of the organization, lying to employees, and behaving abusively) but did not report it.

The article goes on to make the case that governments should have more educational programs to discourage this misconduct. For example, whistle-blower programs were used by only 1% of government workers who saw misconduct. "Because government sets many rules to assure ethical practices in business, it is vital that government set a high standard of its own," says Dr. Patricia Harned, president of the Ethics Resource Council. "A world where almost one third of local government workers don't report ethics violations when they see them does not set a high standard" (Ethics Resource Center, 2008).

## Case Study

### An Example of a Tough Ethical Choice

During his first week as fire chief for the Prince George's County Maryland Fire Department, now retired Chief Steven Edwards faced a serious ethical dilemma that could have cost him his job. A citizen group in the county was holding a fundraising event for the county executive. The chief learned that the event, which was already advertised and had sold hundreds of tickets, was being held in a vacant store that had significant fire code violations. He contacted the event planners, but they were very upset and did not want to cancel the event.

Still, the chief believed he had no choice and contacted the county executive to explain that the event could not be held in the building. To his surprise, the county executive thanked the chief for presenting him with this information and wound up moving the event to a tent close to the original site.

In this situation, the chief *and* the county executive displayed courage and personal integrity, making a tough decision in the name of ethics.

*Source:* Edwards, S. T. (2010). *Fire service personnel management.* Upper Saddle River, NJ: Brady/Prentice Hall Health.

analysis showing that the costs (installation of the system) outweigh the benefits (savings in fire insurance, property damage, and life loss). Although the difference in these values has become almost insignificant as the cost of residential sprinkler systems continues to fall, the opposition would probably use the highest costs it could find to help support its argument that sprinklers would cost substantially more for the home buyer. In this case, the chief should point out the benefits of the sprinkler regulation that cannot be quantified, such as saved lives, prevented injuries, and preservation of the home and the irreplaceable personal items in it.

A classic problem in selling the need for adequate fire protection and EMS is the relatively infrequent need for these services by an individual. This aspect makes the job of an FES or EMS administrator very difficult, but not impossible. To be successful, the challenge must be met with knowledge, energy, courage, and planning.

## CHAPTER ACTIVITY #1: Automatic Sprinklers

A fire chief is scheduled to testify at a city supervisor's workshop regarding a proposed fire code amendment requiring installation of automatic sprinklers in all newly constructed parking garages.

An automobile fire in a parking garage several months ago prompted this proposed fire code amendment. Fire companies had a difficult time advancing hose lines to the area of the parking garage where the fire was and, subsequently, the fire spread to cars on each side of the original fire. After the fire was out, a reporter from a local newspaper interviewed the chief and asked why the fire companies seemed to have so much trouble putting out the fire. Without much thought, the chief pronounced that if a sprinkler system had been installed in the parking garage, the fire would have been contained with smaller property loss.

Sometime later, the chief received a report from a battalion chief indicating that the fire company had been very slow in advancing the hose line at the fire. The battalion chief then went on to recommend that the department should institute a "back to basics" program. The battalion chief also noted that the number of structural fires in the city was down, causing the proficiency of the companies to diminish.

*(continues)*

## CHAPTER ACTIVITY #1: (continued)

In addition, while researching his testimony for the city supervisor's workshop, the chief discovered that the model building codes did not require sprinklers in garages if there were appropriate openings on each level. The chief also discovered articles in fire publications based on real-life experience and controlled burns where significant justification could not be established.

The fire chief had been in his present job for several years and had a reputation as a very knowledgeable leader. In other presentations and appearances before public hearings, the chief had gained a lot of support for the department's opinions, goals, and ideas.

### Discussion Question

1. Make a list of options the chief might choose based on his situation and explain the ethical considerations that each option entails.

## CHAPTER ACTIVITY #2: Call Response

Your FES agency has just experienced a 45% increase in emergency responses after implementing a new policy that requires an engine company to respond to all calls for medical assistance, not just life-threatening calls.

### Discussion Questions

1. Based on a cost-benefit analysis, are the citizens getting more for their tax dollars under the new policy?

2. How would you defend the following situation: A first due engine company was slow to respond to a structural fire where a 5-year-old girl died. At that time, the first due engine was busy at a medical call scene where a 45-year-old construction worker had broken his arm. After providing first aid and splinting the arm, they were waiting for the arrival of a transport ambulance. Take into account all ethical considerations when answering.

3. With this new activity level, the officers of your engine companies are now reporting that they are not able to complete their in-service fire prevention inspections and physical fitness programs. How might this situation affect your organization's ability to meet its goal for quality service to the public?

4. Would you consider a change in response policy that required an engine company response only in those truly life-threatening emergencies (e.g., heart attacks, serious trauma, trouble breathing) to reduce the call volume?

## CHAPTER ACTIVITY #3: Right or Wrong

Charlie is a 54-year-old homeless man with coronary artery disease. He calls 911 almost every day, especially in the winter when he is looking for a warm meal and a dry bed. Every time the ambulance crew serving the primary service area in which Charlie lives transports him to the busy county hospital, they know that they will wait at least 20–30 minutes for a bed. The crew also knows that they are going to get grief from the emergency department staff when they arrive with Charlie.

On a particularly busy, rainy day, the crew finds Charlie at a telephone booth with the usual complaint of chest pain. Jack, the senior medic, tells Charlie this is a bad day to be fooling around and that the emergency department is full. Charlie says he is hungry and agrees to sign an Against Medical Advice form if Jack gives him $20 for food. George, the new intern at the hospital emergency department, questions whether they should have transported as the patient requested or at least done an assessment. Jack replies, "It's no different than any other day. The guy is just a system abuser."

Two hours later another crew responds to find Charlie in cardiac arrest. He is unable to be resuscitated. When the family finds out a crew had seen him earlier and did not treat or transport, they file a lawsuit.

### Discussion Questions

1. Was Jack's decision to talk Charlie out of transport ethical?
2. Should the intern have reported his discomfort to a supervisor?
3. Can Jack and his crew be liable for patient abandonment even with a patient signature?

## References

*American Heritage Dictionary of the English Language.* (4th ed.) (2005). Retrieved from http://education.yahoo.com/reference/dictionary/?s=ethics.

Baumeister, R. F., Bratslavsky, E., Finkenauer, C., & Vohs, K. (2001). Bad is stronger than good. *Review of General Psychology, 5*(4), 323–370.

Bok, S. (1999). *Lying: Moral choice in public and private life.* New York, NY: Vintage Books.

Edwards, S. T. (2010). *Fire service personnel management* (3rd ed.). Upper Saddle River, NJ: Brady/Prentice Hall Health.

Ethics Resource Center. (2008). *Governments at all levels show high rates of misconduct; next Enron could be in public sector, ERC survey finds.* Retrieved from www.ethics.org.

Feldman, R. (2005). Deflecting threat to one's image: Dissembling personal information as a self-presentation strategy. *Basic and Applied Social Psychology, 27*(4)

Florida Fire Chiefs Association. (2012). *Code of ethics.* Retrieved from http://www.ffca.org/i4a/pages/index.cfm?pageid=3287.

Garvey, G. (1996). *Public administration: The profession and the practice.* New York, NY: St. Martin's Press.

Gillespie, C. (1978). EMT oath and code of ethics. Retrieved from www.naemt.org/about_us/emtoath.aspx.

Josephson Institute. (2008). *The ethics of American youth: 2008.* Retrieved from http://charactercounts.org/programs/reportcard/2008/index.html

Robbins, S. P. (2011). *Fundamentals of management* (7th ed.). Upper Saddle River, NJ: Pearson Education.

Twain, M. (1906). Retrieved from http://www.twainquotes.com/freefind.html

# CHAPTER 12

# Public Policy Analysis

## Fire and Emergency Services Higher Education (FESHE) Course Objectives

### Module I: Leading and Managing Purposefully with a Community Approach

The students will:

1. Describe the role of the fire/emergency medical services department as a part of the community government and comprehensive plan. (pp 165–176)
3. Assess ways to develop a good working relationship with public officials and the community. (pp 169–172)

### Module II: Core Administrative Skills

The students will:

6. Recognize the formal and informal dynamics of public organizations and describe strategies to ensure success. (pp 166, 169–172, 175, 177)
8. Identify and discuss a practical agency evaluation process. (pp 165–169)

### Module III: Planning and Implementation

The students will:

1. Describe the process of consensus-building. (pp 169–170)
2. Describe the components of project planning. (pp 165–177)
3. Identify the steps of the planning cycle. (pp 165–175)
4. Discuss how an environmental assessment determines the strategic issues and direction of an organization. (pp 168–169)
5. Assess the interrelationship between budgeting, operational plans, and strategic plans. (p 166)
6. Analyze the importance of an organizational culture and mission in the development of a strategic plan. (pp 165–166, 177)

### Module V: CRM—A 21st Century FESA Responsibility

The students will:

2. Assess your organization's capabilities and needs based on risk analysis probabilities. (pp 167–169)
3. Describe the relationship between community risk analysis and strategic and operational planning. (pp 167–169)
4. Identify the major steps of a community risk assessment. (pp 167–169)
6. Analyze economic incentives that encourage and discourage fire prevention. (pp 174–176)

## Knowledge Objectives

After studying this chapter, the student will be able to:

1. Comprehend the different types of policy analysis. (pp 165–169, 174–175)
2. Examine the role of the administrator as a leader and facilitator in policy analysis. (pp 165–166, 169–172, 174–177)
3. Understand the impact and influence of policy analysis on the budget and on resource allocations. (pp 166–169)

4. Examine the necessity of consensus building and the political process used to implement policy changes. (pp 169–170)
5. Understand the importance of the public policy presentation and the different elements involved. (pp 170–172)
6. Understand the use of statistics in policy analysis. (pp 172–174)
7. Analyze and research case studies and other historical experiences. (p 174)

# Future Planning and Decision-Making

On the operational side of public service, emergency services personnel often take immediate action and complete emergency operations in minutes or hours. In the public policy arena, however, decisions may not be so immediate and a long-term implementation plan may take years to complete.

The concept of future planning in the fire and emergency services (FES) organization has been encouraged through such documents as the US Fire Administration's (USFA) *Urban Guide for Fire Prevention & Control Master Planning* (1977) and several courses at the National Fire Academy (executive planning and strategic analysis of community risk reduction). Many of the short- and long-term planning functions in an FES organization are relatively simple; advanced planning may allow one to anticipate needs and forestall problems for the future. The following are items for the chief officer to consider when planning for the future:

- If a project or policy in question is relatively simple, it may be possible to anticipate needs and forestall problems through advanced planning. For example, the basic measure of emergency service is response time. The determination of the appropriate response time can be complex, but once the maximum acceptable time is agreed upon, analysis to determine coverage areas for existing and proposed stations is fairly straightforward.
- If the policy area being studied makes it unwise to delegate broad discretionary authority to the frontline workers, the major policy decisions must be made at the department level so that each unit provides consistent, reliable emergency service to the public.
- If the implementation effort fits unambiguously within the responsibilities of an existing agency, the appropriate organization to be used for the job will not become the subject of discussion.
- In the case of FES operations, it is not a lack of knowledgeable experts at the company level that is a problem; it is a lack of adequate time and the big picture policy analysis that is needed to weigh all the pros and cons to make a valid decision. That is why standard operating procedures (SOPs) are extremely important in creating a safe working environment and attaining consistent quality service. The majority of emergency actions taken by first arriving companies need to be scripted to guarantee the same results each time. These decisions are best accomplished by a thoughtful unhurried analysis conducted through the task force process.
- If there is a high level of confidence in the causal theory underlying the policy, planners can predict the effects of executing changes even in the absence of additional study and research.

In many cases, a plan or policy can be implemented less painfully by incremental adjustment. This method works by making a small change and assessing the results. Incremental adjustment tends to be a safe way for policy makers to proceed with their decisions without the risk of making big changes that may have unforeseen consequences. This method spreads out a large financial impact over several years, allowing for small increases in tax revenues each year. Also, this can allow for adjustments to the plan if the outcomes are different than anticipated.

Caution should be exercised when implementing public policy; leading change is not a simple process. The notion that policy makers exercise, or ought to exercise, some kind of direct and determinant control over policy implementation might be considered an erroneous belief of conventional public administration and policy analysis. However, once a new policy is identified, techniques discussed in the chapter, *Leading Change*, will help implement change and overcome resistance. One successful method for determining future plans is to use public policy analysis—also known as strategic planning.

# Strategic Planning

Strategic planning is normally conducted at the department level. This type of planning identifies the department's vision, mission, and strategies. The USFA document *America Burning* (1973) contains helpful guidelines for the vision and mission of a modern FES organization. Generally, this type of planning is conducted by an appointed task force of FES officials and senior members. In some cases, an official master plan is produced with input from the public, other agencies, and appointed or elected officials of the municipality. Strategic planning can help identify where the organization is at the present time and where a new vision can take the organization in the future.

To implement the strategic plan, tactical plans should be prepared and monitored on a monthly basis. For example, suppose a department has just received additional funding for a new station and part of the justification for the funding was to reduce response time to a specific area of the jurisdiction. The strategic plan lists a specific response time as one of its goals; the tactical plan lists the details. In this case, a new fire station is needed. After the tactical plan is implemented, a monthly check of response times to emergency incidents should be maintained and a formal report to brief the executive and elected officials on the results. At the appropriate

> **Chief Officer Tip**
>
> **Rational-comprehensive Decision-making**
> In the book *The Policy-Making Process* (Lindblom & Woodhouse, 1993), the concept of rational-comprehensive decision-making is explained. This method is based on logical reasoning, accepting significant change, and favoring the solution that completely resolves the issue rather than temporarily fixes it. The FES administrator can use this decision-making process to:
> 1. Identify the problem
> 2. Rank goals
> 3. Identify all possible alternatives that solve the problem
> 4. Perform a cost-benefit analysis (if possible)
> 5. Perform a comprehensive analysis of various ways of solving the problem
> 6. Select alternatives that accomplish goals

time, a public summary of this new program should be prepared to inform citizens and the press of the impact of the new spending of tax dollars.

There are also operational items that need to be accomplished on a day-to-day basis. For example, it is common to have minimum staffing levels that must be maintained daily. There may also be guidelines for the number of people who may be on annual leave for any particular day. If staffing levels fall below the minimum level, personnel must be called back on overtime or transferred from other stations. It is very important to document the exact reasons for unexpected expenses, such as overtime. At the tactical planning level, this information can be used to suggest changes in departmental policies. For example, if the timesheets show an increase in overtime for personnel using sick leave, an evaluation of leave policy and decreasing the number of personnel authorized to be off on annual leave may be appropriate.

## The Role of Budgeting Within Public Policy Analysis

The budget and planning processes are a continuous and inseparable cycle. Plans should dictate the items placed in the budget; the budget is the process of funding the plans. When the planning process is completed, the administration should determine the financial resources needed to achieve the goals. Creative strategies for funding may be available and should be pursued (refer to the chapter, *Financial Management*, for examples of some of these strategies). Proactivity is always a good way to gain respect from elected officials.

If the numbers in the budget become the focus of the policy analysis, administrators may find they are simply adding an incremental percentage to each line item in the proposed budget each year. If a strategic plan is followed, however, the chief may decide to move funds from one part of the budget to another to fund a new goal, priority, or objective, or they may find a justifiable need for a significant increase in the budget.

For medium to large FES departments and emergency medical services (EMS) organizations, budget preparations should start at the lowest feasible level of the organization. For example, the chief may direct the fire prevention division to prepare its own budget based on an incremental increase over last year's budget. In addition, division chiefs may be asked to submit justifiable enhancements as separate budget items or a budget based on a percentage decrease, including identification of programs that could be eliminated. There might be as much competition for the budget dollars internally as there is externally. Internal conflicts can be avoided or at least reduced by including division chiefs in the preparation, justification, and prioritization of the budget; this helps them to see the big picture (strategic plan) and better support the unified budget.

In addition, including the division chiefs in the process gives the administration many options to fine-tune the budget. Using groups of people to study and make recommendations about the organization's goals and objectives can help to ensure the administration is not blindsided by the lack of appropriate information and expert advice when analyzing a management or functional problem. The chief may be able to eliminate funding for some items while at the same time add funds for new programs. Also, after the decision has been agreed to by consensus, the FES administrator can expect to receive support from all the managers.

After the division chiefs have prepared their budgets, the chief administrator may then ask them to prioritize between adding back the proposed percentage cuts and funding the enhancements that have been requested. After this step is complete, the chief administrator must combine the proposed budgets from all divisions and have the entire organization prioritize any decreases and enhancements. After this prioritization, it is up to the FES administrator to balance funding by choosing decreases that are justified, adding funding for those enhancements that warrant support, and requesting additional funds for other enhancements that the administration believes are justified.

## Measuring Outcomes

To carry out a public policy analysis, outcomes must be measured. An outcome is a measured benefit to stakeholders, created as a direct result of dollars spent by the organization. However, many of the measurements that are kept by the traditional FES organization are measures only of workload. For example, retired fire marshal Mike Love notes, "[I]f you have robust public education and an aggressive fire sprinkler ordinance, you may be able to reduce the risk of fire or control it before it reaches flashover. So this measure can help you see how well you are doing" (Love, 2008). In this example, the measure is based on the preventive actions, which cannot be measured directly, rather than any specific outcome.

Although the FES organization may keep accurate statistics on the number of emergency incidents to which they respond, that does not show the actual outcome. This concept is further explained by research analyst Jennifer Flynn (2009, p. 3):

> Performance measurement relies on the evaluation of achieved outcomes, compared to desired outcomes. For the fire service, the desired fire protection outcomes—which are not easily measured—include fires prevented or suppressed, and ultimately the human life and property preserved… Performance measures are the quantitative or numerical representation of activities and resources that help evaluate whether the goal is met.

Consider the department that responds to a small trashcan fire in a high-rise building and extinguishes the fire quickly. Could the department then claim to have saved the entire dollar value of the high-rise building? What about if a complete floor was on fire and the department was successful at extinguishing the fire before it extended to any other floors? As illustrated, assessing the dollar value of properties saved is very difficult, if not impossible.

It is also rare to be able to document saving a life. If firefighters were to search, find, and remove an unconscious person from a building on fire, they would be able to claim an unquestioned save. In the fire prevention arena, however, lives are more often saved from built-in fire protection equipment, such as smoke detectors that allow occupants to escape without injury. Many of these types of saves are never reported to the fire department. Therefore, outcomes for a FES organization must be measured using other more easily and verifiable items.

Because some outcomes are based on relative or subjective items, the accuracy of the outcome can often be evaluated by looking at efficiency, responsiveness, or equity. For example, a newly purchased computer-aided dispatch system might reduce dispatch time by 50%. This measured outcome is an indication of greater efficiency, greater responsiveness, and equity because it equally serves everyone who calls for emergency assistance. In general, many outcomes of response time improvements cannot be measured directly (e.g., three lives saved last week); however, relative improvement can be measured. A better outcome is the result of arriving sooner, starting emergency operations sooner, and using professionally trained and well-equipped responders.

If it can be demonstrated that response times can be reduced by changing existing procedures (e.g., streamlining the call-taking process) or adding new resources (e.g., increasing the budget for new staff or new facilities), the argument can be made that improved service to the public has been accomplished.

## Staffing Levels

Personnel make up over 80% of the career FES budget; this section provides a detailed discussion of the critical need for this line item. In addition, the minimum number of personnel needed for safe emergency operations is addressed in the Occupational Safety and Health Administration's Respiratory Protection Standard 29 CFR 1910.134 (two in/two out) and NFPA 1500, *Standard on Fire Department Occupational Safety and Health Program*. The Occupational Safety and Health Administration and the National Fire Protection Association (NFPA) agree that a minimum of four firefighters must be at the scene of a structural fire before an interior attack can be initiated. Therefore, one measure of an adequate level of emergency service has to be the time needed for four trained firefighters to arrive on the scene. It is believed that anything less than this number could cause delays in attacking the fire or unsafe actions by firefighters.

Several studies have been conducted relating efficiency—measured by the time required to complete fireground operations—to staffing levels. The most recent of these studies, conducted by the National Institute of Science and Technology (NIST), investigated "the effect of varying crew size, first apparatus arrival time, and response time on firefighter safety, overall task completion, and interior residential tenability using realistic residential fires" (National Institute of Science and Technology, 2010, p. 9). Of the 22 critical fire ground tasks performed in this study, the four-person crew was the clear winner by 25–30%, completing tasks 5.1 minutes faster than the three-person crew **FIGURE 12-1**.

Providing on-duty crews is the key to a fast, coordinated response to an emergency incident. In many cases this means career personnel, but volunteer departments have been able to provide on-duty crews by assigning a duty time to each of their members. This has worked successfully in a small number of volunteer departments, and the trend seems to be spreading. In areas where public funding is not available for career firefighters and volunteers are not able to provide complete coverage of the station with on-duty members, staffing must rely on alerting volunteer members by outside sirens or pagers to respond to the emergency incident. Generally, these volunteer departments are in areas that serve 2,500 or fewer residents.

As a general guideline for planning purposes, if the fire company or EMS unit responds to more than 2,500–3,000 calls per year, planners should look very closely at the need for a second unit, either fire or EMS. Actual response data should be used, including the time out of service spent either responding to or working at emergency incidents. If time out of service totals more than 10% of the available time during the busiest 8-hour shift, it might be time to plan for a second unit. In some cases, these high call numbers could be the result of a dispatch policy that sends too many unnecessary units to each call. For example, a department may send a full commercial structural response to an automatic alarm and then find that during weekday business hours it is rare to need

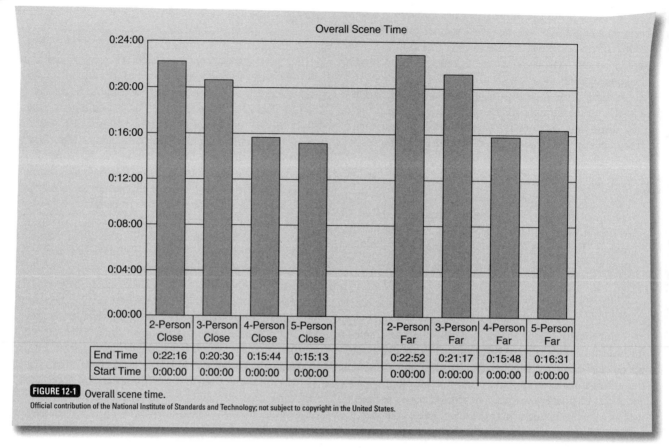

**FIGURE 12-1** Overall scene time.
Official contribution of the National Institute of Standards and Technology; not subject to copyright in the United States.

more than one engine. One other technique that is more commonly seen for EMS transport units is to staff additional units only during the busiest hours. The New York City Fire Department also used this technique by staffing part time ladder companies during busy times.

## Response Times

The number of fire stations serving an area has the greatest impact on response time for an FES department. In addition, the more stations needed, the greater number of firefighters needed. Some jurisdictions use their existing fire station spacing as the local benchmark. Paid firefighters are used to staff units that respond immediately, keeping response times to a minimum. Some volunteer-staffed companies also may respond immediately if an appropriate number of members are in the station. A few volunteer organizations require members to stand by in the station to provide an immediate response.

NFPA 1710, *Standard for the Organization and Deployment of Fire Suppression Operations, Emergency Medical Operations, and Special Operations to the Public by Career Fire Departments*, was adopted in 2001 and has since set a benchmark for response times. According to the standard, the first due company has 80 seconds to start its response after dispatch, and 4 minutes for travel time to 90% of emergency incidents.

The NIST report mentioned in the section on staffing levels does not validate the response time benchmark as clearly. In this study, NIST examined only two response times: 3 minutes and 5 minutes. A substantial difference in the effectiveness of the fire companies was not found. The data (four-person staffing) extrapolated from the study indicates that the 3-minute response time would result in a 1-minute improvement over the 4-minute response time. The same 1-minute difference, this time longer for the 5-minute response time, would be realized (NIST, 2010, p. 48). Because a 4-minute vs. 5-minute response time takes substantially more fire stations to cover the same area (36% more for the 4-minute time), it could be deduced that a cost-benefit analysis would conclude that the municipality would have to fund 36% more stations and staffing for a 10% quicker victim rescue (i.e., one minute).

> ### Words of Wisdom
>
> "Personnel requirements are not merely a matter of numerical strength, but are also based on the establishment of a well-trained and coordinated team necessary to utilize complicated and specialized equipment under the stress of emergency conditions. Attempting to operate more fire companies than can be effectively staffed, even if some response distances must be somewhat increased, is less desirable than fewer but appropriately staffed companies."
>
> — NFPA
>
> *Source:* National Fire Protection Association. (2003). *Fire protection handbook* (19th ed.). Quincy, MA: National Fire Protection Association, p. 7–37.

Some consultants and planners use response times for locating stations but do not consider the staffing of the companies. As previously mentioned, the number of firefighters that should be on the scene before an interior attack on a hostile fire is four. Therefore, the response times should be reported only when four firefighters arrive. NFPA 1710 specifies that each company be made up of a minimum of four firefighters, and this is justified by the safety and efficiency criteria validated by the NIST fireground studies.

The NFPA standard also dictates that a minimum of two people trained at the emergency medical technician and paramedic level and two people trained at the emergency medical technician level should arrive within the established time of 8 minutes on all advanced life support emergencies. NFPA 1720, *Standard for the Organization and Deployment of Fire Suppression Operations, Emergency Medical Operations, and Special Operations to the Public by Volunteer Fire Departments*, also addresses response times for volunteer fire departments, allowing 9 minutes for "urban" areas (greater than 1,000 people per square mile), 10 minutes for "suburban" areas (500–1,000 people per square mile), and 14 minutes for "rural" areas (<500 people per square mile) 80% of the time. There is no limit on turnout time.

The headline of a 2011 news report from San Diego, California, read "Study Suggests Smaller Fire Crews to Help Shorten Response Times." The study made the argument for several two-person crews to reduce response times in areas of the city. However, as noted in the previous section on staffing levels, most credible sources state that four firefighters should ultimately respond to the scene. Response time should be measured only after these four firefighters arrive on the scene. It may seem understandable for a department with understaffed fire companies to report a response time when the first fire engine arrives if it is also planning and budgeting to increase the staffing to four. However, if there is no plan to increase staffing, response time should be reported when four firefighters arrive, which would be almost twice as long as the response time of a four-person crew, because at least two fire companies must be on the scene.

If a four-person response time is used for reporting, it can also be used to justify staff increases. For example, if the department has minimum staffing of two or three firefighters, the response time is the arrival at the scene of the second engine, which in many cases is twice as long as the response time of a fully staffed company. The maximum response time with understaffed companies is actually at the two stations. For properly staffed companies it is exactly the opposite. Remember, the standard is a maximum response time. This is a difficult concept to comprehend, but critical to justifying adequate safe staffing.

## Professional Competency

Professional competency is an additional area that relates to better service. EMS responders have nationally recognized certification for training and education to ensure equal competency across the United States; however, individual departments and states require different levels of training for firefighters. The NFPA report *Fire Service Performance Measures* states, "This measure [the competency of firefighter training] is a proxy for quality of service provided. It is assumed that a high percentage of responders with completed training and certification are providing high quality service when responding to calls" (Flynn, 2009, p. 23).

## Consensus Building

To be successful at maintaining the department's budget and adding enhancements when needed, it is very important to understand the approval process. In general, if the administration can make arguments that are backed up by common sense and solid research, there is a good chance of receiving approval for the budget request. If a national consensus standard can be used—such as those of the NFPA—to support the budget request, it is easier for elected officials to justify their vote.

For example, a chief may propose a new program to certify all officers to NFPA 1021, *Standard for Fire Officer Professional Qualifications*. Certification might require a training program, incentive pay, and overtime pay to cover classroom attendance. The selling point for this proposal is that, when finished, the city or county has a more professional fire department. This is the kind of proposal that the municipal administration and elected officials can clearly understand.

Still, there has been some controversy over the use of national consensus standards in the fire service. In the EMS field, responders are now certified to national standards for emergency medical technician and paramedic; in the fire service arena, however, national standards for training and safety have been available for a relatively short period of time. Although these national standards can provide an opportunity to justify the existing budget or gain approval for a budget increase, local FES officials often want the freedom to provide the level of public protection that is determined necessary by the local community. With the use of national standards, there is the understanding that a group of experts derived consensus to determine the level of protection needed.

At the local level, consensus building must start with the municipal administrator or elected officials. A chief should seek their full advice and consent. Influential private citizens and any administrators who give trusted advice to the chief administrative officer, such as a budget director or public safety director, should also be included in the consensus-building analysis. The task is to build support among these influential people. The chief officer should begin by preparing a list of all those people and groups who could influence the final decision, such as firefighters' unions or volunteer fire associations **TABLE 12-1**.

### TABLE 12-1 Consensus Building Example

| Person or Group | Potential Influence (high, medium, or low) | For, Against, or Neutral |
|---|---|---|
| Chief administrative officer | | |
| Mayor | | |
| Power elite member (*your choice*) | | |
| Firefighters' union | | |
| Local newspaper | | |
| Police chief | | |
| Budget director | | |
| The public | | |
| City attorney | | |

The chief should rate each individual or group's potential impact on the final decision. For example, in a mayor-council form of government, the mayor and the most influential private citizen members should be rated as having a very strong potential influence. Next, each person or group's support of the proposal should be assessed. Again, if the proposal were to hire additional firefighters, the union president would be rated as strongly supportive.

After this list is completed, the chief should seek out those people who are very influential and will more than likely support the request. In private one-on-one discussions, these likely supporters should be briefed on the proposed request and asked for their input before their support is requested. These discussions should also take place with those individuals who are deemed neutral on the issue. The object is to bring these people to a point where they mildly support or do not oppose the proposals.

Finally, the people who will most likely oppose the proposal need to be approached. The object is to present the best case for support of the proposal. The hope is that these people will become neutral on the issue. Discussing the issue with these individuals also keeps them from feeling like they have been defeated by an organized conspiracy by their opponents, without a fair chance to influence the decision. Each person or organization should be personally contacted. If they do not feel that they have been fairly treated, these individuals or groups may become a permanent opponent.

This detailed process should be necessary only for potentially controversial and expensive enhancement items in the proposed budget. For every new tax dollar the department receives in its budget, an equal amount must be taken from another agency's budget or generated from increases in taxes.

### Chief Officer Tip

**Political Action Plan**

A document from the International Association of Fire Chiefs (IAFC), entitled *How to Develop a Political Action Plan* (2012), is available for members of the IAFC at the organization's website (www.iafc.org). Many of the suggestions in this document can be used in the everyday political process to facilitate approval of a new policy or spending initiative. The document also has some good ideas for approval processes that involve a vote by the local town or city. A document from the International Association of Fire Fighters (IAFF) advises that political goals should be as follows:

- Explaining the value of the department or organization and of the services provided
- Discussing the value of the changes sought
- Creating public pressure on political leaders
- Producing any necessary votes by the city/council or commission

### Words of Wisdom

"Planners who neglect to consider the desires and special problems of interested individuals in the locale of intended effect are apt also to forget that affected individuals can organize coalitions to resist programs that they judge to be harmful or intrusive."
– Gerald Garvey, political scientist

*Source: Garvey, G. (1996). Public administration: The profession and the practice. New York: St. Martin's Press, p. 466.*

## Public Policy Presentation

The next step would be for the chief administrator to prepare a presentation to be delivered in public for the elected and appointed municipal officials. These officials act as the arbitrator between the competing agencies' requests for public funds.

The administrator's presentation should be easy to understand and include common sense arguments. For example, if a department is requesting funding for a new fire station and firefighters to staff the station to reduce response time in one section of the city, the administration should determine the response time in the other surrounding areas FIGURE 12-2. It can then be argued that all residents in the city deserve equal fire and rescue protection (similar response times) and that the proposed funding increase would work to increase the percentage of city residents who fall into the adequate response time area.

The following sections discuss suggestions for handling public policy presentations as professionally and successfully as possible.

## Consider the Audience

What is the audience's real knowledge of FES operations? Many appointed and elected officials have no background in FES. Do these officials understand jargon or acronyms that are typically used?

Before scheduling a policy or budget hearing appearance, consider arranging a tour and orientation for the officials at one of the department's stations. Adapt an educational program used for children, having the officials try on the protective clothing and the self-contained breathing apparatus. Explain the environment that the fire and EMS personnel operate in when performing emergency operations. Use worst-case—but plausible—examples to illustrate those incidents when the department is pressed to its maximum effort.

Chief officers who are unsure of elected and appointed officials' understanding of the budget request and its impact on service efficiency should explain the details in their presentations along with emphasizing the improvements. Never assume anything. If in doubt, ask questions.

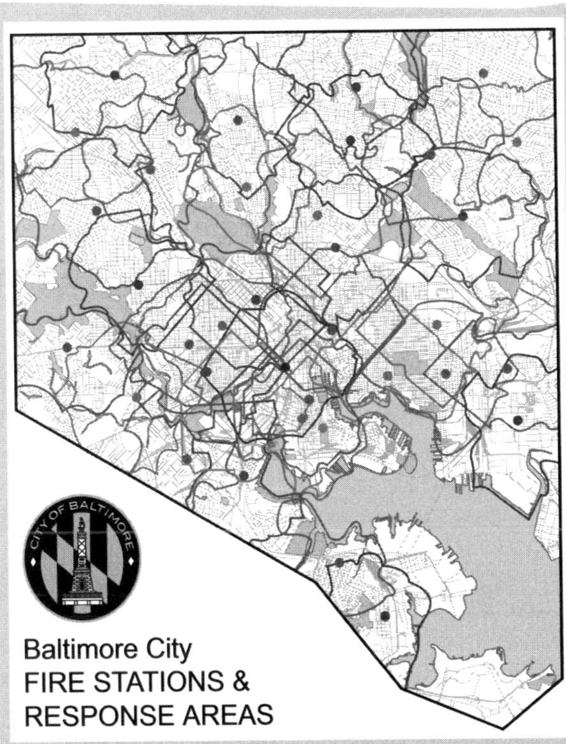

## Practice

At the first level of preparation for the presentation, the chief should become familiar with the entire budget and justifications. One technique that has merit is to ask the staff to review the budget and write down questions that the appointed and elected officials may ask. For questions that require in-depth knowledge to answer, an appropriate staff member should prepare a briefing document for review. This person should also be present at the budget hearing to fill in details or correct any misunderstandings.

If a chief lacks presentation experience, he or she may want to practice the presentation in front of a friend or staff member or videotape the simulated presentation and review the recordings. There are two different types of situations to practice: the prepared presentation and the response to the questions at the public hearing. Practice helps the chief be confident, relaxed, and persuasive. As with any presentation, the chief should be very knowledgeable about the subject.

## Be Positive and Cheerful

Tell the audience, in simple terms, the positive aspects of the proposals. From practice, the chief should be able to maintain a positive, confident appearance that reflects professional ability and credibility. State the justification in positive terms. For example, point out that the department will be able to provide better service with the approval of the enhanced budget requests.

## Maintain Eye Contact

Reading is distracting and prevents the speaker from maintaining eye contact. The presentation has more impact and is more believable if the speaker gives the presentation extemporaneously from notes, rather than reading through pages and pages of material. Maintain as much eye contact with the elected officials as possible; many people relate eye contact to honesty and sincerity. Scan the officials without staring, which usually involves eye contact of about 6–8 seconds. A glance at their eyes and facial expressions may give an indication of their reaction to the presentation, and the presentation can then be adjusted accordingly.

**FIGURE 12-2** Four-minute response contours.
Courtesy of City of Baltimore, Office of the Mayor

### Use Down-to-Earth Ideas

Whenever possible, use an analogy or a real-life story, perhaps including a humorous anecdote, to help explain a theoretical or abstract proposal. The meaning behind the old saying "a picture is worth a thousand words" also applies to the use of analogies.

### Use Visual Aids

Include visual aids such as handouts, charts, slides, and videos. However, make sure they do not distract from the main presentation. If using films or videos, be careful not to use too much of the time allocated for the presentation. If necessary, review the film or video and limit it to no more than 5–10 minutes. Selective use of video aids can increase the understanding and attention of the elected officials.

### Involve the Audience

Appropriate techniques for involving the audience may vary, so it is helpful to have a good understanding of procedural rules and the respect that is expected by the elected officials. At a public hearing, elected officials operate for the public and media who may be present. A direct confrontation or disrespect can cause a loss of support.

Research the backgrounds of each elected official. A thorough understanding of the elected officials and their past experiences increases the chance of involving them positively in the discussion. In addition, the chief should have personal one-on-one meetings to get to know all the officials before the presentation.

At the end of the presentation, ask for questions. It is at this point that the chief officer really has to think on his or her feet. Remember, the elected officials may have prepared questions for the hearing. Answer truthfully and as simply as possible. Keep to the facts and professional opinion. If the answer does not come to mind, say so and offer to locate the answer after the hearing.

## Using Statistics

Statistics can also be useful for creating and justifying policy analysis and recommendations. National FES statistics from 2003 to 2007 reflect the following trend: fire incidents per million population decreased 2.7%, civilian deaths per million population decreased 19.7%, and civilian injuries decreased 8.8%. In addition, firefighter deaths per 100,000 fires took a downward trend in 2009, an improvement from the upward trend seen from 2005 to 2008 (USFA, 2009). The chief officer should be aware, however, of trends that might affect these statistics in the future. For example, consider the effect the following trends might have on these national fire statistics:

- Smoke alarms have been installed in 96% of US homes; however, approximately 63% of fire deaths occur in homes where the smoke alarm was not installed or was installed improperly, was dysfunctional, or was missing the battery (Ahrens, 2011). To further add to the complexity of statistics regarding these devices, homes protected with working smoke alarms might have a fire that is detected early and extinguished by occupants, therefore never resulting in a call to the fire department.
- The greatest percentage (21%) of fatal fires is started by misuse of smoking materials. In 1965, 42.4% of the adult population smoked; in 2004, that figure had dropped to 20.9%. This decrease in smokers most likely had a large impact on fire deaths over the past years and will continue to do so in the future.
- Mandatory installation of residential sprinklers continues to raise strong opposition from homebuilders and elected officials. As a result, fire sprinklers protect few single-family homes; however, NFPA 101, *Life Safety Code* and the *ICC Residential Code*, now require residential sprinklers for new one- and two-family dwellings, which will increase the adoption of sprinkler requirements over time.
- The addition of first-responder medical service by fire departments is very common. In most cases, citizens are receiving emergency medical treatment a lot faster because the response times of fire companies are shorter than for the typical EMS unit. However, this trend may also cause an increase in response times for fire calls if the closest fire company is working at an EMS incident when a fire is reported in its first due area.

As with the concept of outcomes, statistics in the FES can be misused and misunderstood. For example, consider the structural fire data shown in **FIGURE 12-3**. The data show that an increase in the average dollar loss per incident equates with an increase in the number of firefighters who respond to the incident. In considering the situation, this makes sense; a large fire resulting in extensive property damage requires a large number of firefighters on scene, and a fire that results in minimal property loss requires only a few firefighters to extinguish the fire.

A city manager using this data and graph to determine how to reduce fire loss might notice that reduced property loss is associated with fewer numbers of responding firefighters. The city manager may then conclude that sending fewer firefighters to fires results in smaller losses. Although the data is correct, the analysis lacks a sensible approach to interpret the results.

Another example of a misused statistic is that because most fires are extinguished by using small amounts of water, the department needs only one or two firefighters to put out fires. However, when looking at these statistics more closely, the analysis concludes that a small percentage of fires causes the greatest number of deaths and the highest amount of property

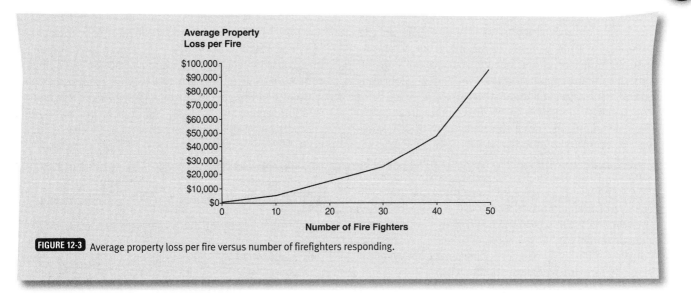

**FIGURE 12-3** Average property loss per fire versus number of firefighters responding.

damage. From a quality service perspective, the department should be prepared to handle these challenging fires and should give 100% effort at all times.

As exemplified, statistics from surveys and studies can be interpreted and analyzed in many different ways. If administrators or their staffs do not have a good understanding of statistics, they can be unintentionally misled by some of these reports. When it comes to surveys, there are several items the chief officer can look for in determining reliability. The first is that the survey was truly random; most surveys sample a small number of people and then generalize the results for a large population. One method of selecting participants is to use a random number generator (on a computer) to select telephone numbers from all participants in the area being surveyed. Furthermore, a sufficient number of respondents are needed to ensure that the standard of error is small, which is expressed as a plus or minus percentage error. Although not everyone in the population can be surveyed, the more people included, the better chance of receiving an accurate prediction of the outcome for the larger group. At least 250 observations are required for a low standard of error. Many surveys use greater numbers primarily to guarantee a random selection and gain the confidence of the general public.

## Facts and Figures

**Total Quality Management**
What does it mean to the public if a business or agency gives less than 100% effort at all times? In the book *Root Cause Analysis: A Tool for Total Quality Management*, this question is examined. The following calculations are based on the assumption that businesses lowered their goals by only 0.1%, to 99.9% effort (Wilson, Dell, & Anderson, 1993):
- 16,000 pieces of mail would be lost per hour.
- 20,000 incorrect drug prescriptions would be prescribed each year.
- 500 incorrect surgical operations would be performed each week.
- 50 newborn babies would be dropped at birth by doctors each day.
- 22,000 checks would be deducted from the wrong account each hour.

## Case Study

**Firefighter Heart/Lung Disease Study**
Studies of firefighters and the incidence of heart/lung illnesses conducted by the Medical Research Council (London, England) concluded that firefighters had a much higher incidence of these illnesses than the general public (Douglas, 1985). These studies were used to justify most of the heart/lung presumption retirement and health benefits that are common throughout the country. Later, that same data was reanalyzed and normalized for cigarette smokers. The new studies found that firefighters who do not smoke had a statistical chance of heart/lung disease slightly less than the general public, most likely as a result of hiring fit employees. However, the firefighters who smoke had an incidence rate much higher.

The results pointed to a synergistic outcome from the combination of cigarette smoking and firefighting, rather than the initial conclusion that connected the illness to all firefighters. As this example points out, if more than one variable can cause the outcome, all potential variables should be studied.

It is important that the source providing the statistics is credible and reliable. This does not guarantee that a mistake will not be made in the study, but it can give a strong indicator of whether the statistics are correctly reported. Some organizations or individuals collecting statistics may have a particular interest in a study being conducted. Although many reported studies are accurate, the chief should be aware that some may be structured with a narrow focus toward a preferred outcome.

Each valid statistical conclusion from a random sampling study should have a confidence level reported. For example: "There is a 95% certainty that the average age of firefighters in this country is between 36 and 39 years old." Or, a statistic may report that there is a plus or minus variance of some percentage. In any case, consider each study and its conclusion using common sense before incorporating it into your public policy analysis. Statistics and other mathematical descriptions of problems in the public sector can be very helpful for understanding a problem, analyzing possible solutions, and finalizing the plan to solve the problem.

## Case Studies

In addition to statistics, researching case studies can be a good approach to studying, reviewing, and analyzing administrative and policy problems. Administrators can use case studies to learn from others' experiences. For example, when dealing with problems relating to humans, administrators should also consider subtle factors, such as how people feel about an issue or how they will behave. In many cases, without being able to prove these factors scientifically, as in a controlled statistical study, administrators must treat these human behavior traits as facts. For example, in the case of smoke alarms, a human behavior problem is that many smoke alarms that have been installed have been found to be nonfunctional because of missing or dead batteries; this is very troublesome to fire protection professionals. Case studies are anecdotal evidence and may not meet the test of reality for larger groups; however, they can also be very powerful at persuading others that the policy analysis is valid.

## Program Budgeting

Senior public administrators and politicians demand analytical, quantitative justifications to support budget requests and policy decisions. Formal policy analysis can provide the justification needed when the leader of a fire department or EMS organization is faced with scarce resources.

Program budgeting—sometimes referred to as the Planning, Programming, and Budgeting System (PPBS)—is not a formal part of the budget process but rather a separate policy analysis. This budgeting system describes measurable outcomes and gives detailed costs of every activity or program that is to be carried out in a budget. If justifications for the budget requests are supported by irrefutable quantitative outcomes or expected results, the FES administration has an advantage to bringing increased funding to the agency; however, with politics there are no guarantees. In addition, because of its time-consuming effort, it often results in opposition from managers and, therefore, is not widely used. The administration needs to make its best effort to get the resources needed to accomplish the organization's goals, and program budgeting is one of the tools that can be used to accomplish this task.

## Cost-Benefit Analysis

Because it is often difficult to measure benefits or direct outcomes from FES programs, city or county administrators often do not rely on cost-benefit analysis to provide guidance in allocating public funds and developing public policy.

If an administrator were to survey the public about the value of FES, there would probably be a great variance in the valuation from person to person, situation to situation, and group to group. What makes cost-benefit analysis even more difficult for FES organizations is the realization that the outcome—saving lives and property—is in most cases the result of a complex set of circumstances. The following is a hypothetical set of circumstances that could lead to a fire death in a house fire:

1. Smoke alarms were not required by the local fire prevention regulations to be installed in existing dwellings.
2. Home fire inspections along with fire safety education were not provided by the local FES organization.
3. The occupants of the dwelling did not see any value in installing smoke alarms.
4. The occupants of the dwelling did not recognize the hazard of careless smoking in the home.
5. The first arriving FES unit was understaffed or personnel were not adequately trained.

### Chief Officer Tip

**Consultants**

There are numerous FES consultants who can perform professional policy analysis and master planning. The following is a list of items the chief should consider before deciding to obtain professional consultant services:

- Does the department have members with the skills, knowledge, and talent to do this type of planning in-house?
- Check out the consultant's past work and references.
- Cooperate fully with the consultant and be as accessible as possible.

6. When the victims were eventually rescued and brought to the front yard, the FES organization was not equipped or trained to provide EMS.

It could be argued that if any of these six elements were different, lives could have been saved. This makes it almost impossible to construct a valid cost-benefit analysis. Although some of these items are clearly more expensive than others to correct, it would be good to implement programs to correct all the deficiencies indicated by each item. Anything less than 100% effort in all areas leaves open opportunities for preventable life and property loss in this example.

## Program Analysis

One method that can be used in place of cost-benefit analysis is program analysis. This systematic process examines alternative means, methods, or policies to accomplish the goals and objectives of the fire department or EMS organization. This type of analysis accepts the use and reporting of intuition and judgment, and statistical studies and sensitivity analysis. This method should be used only for those problems that have a big impact on the services the department delivers.

Program analysis works best when a chief appoints a task force representing all members of the department. This technique is used to make sure that all possible solutions are discussed and considered. There may be cases when citizens, special interest groups, and elected officials are also requested to participate. This participation can be used as part of a strategy to gain support for a new program.

The FES or EMS administrator leads the task force and defines the problem to study. For example, a report on response times in a city notes that one geographic area has substantially longer response times to emergency calls than the averages in other parts of the city. A task group is called together, the problem is outlined, and the group is asked to complete an analysis and recommend several solutions. This group is given the following guidelines:

- Brainstorm. Allow all ideas to be listed and considered. Identifying the problem can be the most critical point in this process. Many times a symptom is confused with the real cause. All solutions should be analyzed. Have the group consider all traditional and inventive ways to solve the problem.
- Evaluate. Remember to test each solution for political sensitivity, human behavior problems, and common sense. Review any public preference surveys that have been conducted (see the chapter, *Customer Service*). Surveys can be a strong justification to support the plan.
- Implement. If the selected solution requires additional funding, consider all options, such as increasing taxes, consolidating or relocating existing stations, charging fees for service, or establishing mutual aid with an adjoining jurisdiction. In addition, a plan that contains phases is very useful to the decision maker. The administrator can then focus on obtaining resources for part of the plan each year, rather than dealing with the entire financial impact in the first year. For example, the plan may call for the construction and staffing of three new fire rescue stations. The administration may be able to use its skills in negotiation, compromise, and political sensitivity to gain approval to construct one new fire station every 2 years, with full implementation of the plan in 6 years.

If the FES or EMS administration is not successful in gaining approval or financing for the plan or change initiative, the plan should be stored for use in the future if an opportunity surfaces.

## Policy Analysis Reference Sources

One of the most comprehensive outlines of public FES is contained in NFPA 1201, *Standard for Providing Emergency Services to the Public*. This document, which was created in 2010, references the following sources:

- NFPA 1500, *Standard on Fire Department Occupational Safety and Health Program*
- NFPA 1710, *Standard for the Organization and Deployment of Fire Suppression Operations, Emergency Medical Operations, and Special Operations to the Public by Career Fire Departments*
- NFPA 1720, *Standard for the Organization and Deployment of Fire Suppression Operations, Emergency Medical Operations, and Special Operations to the Public by Volunteer Fire Departments*
- Insurance Service Office (ISO), Public Protection Classification Service; Fire Suppression Rating Schedule (contact the ISO at ISO Customer Service Division, 545 Washington Boulevard, Jersey City, NJ 07310–1686)
- Center for Public Safety Excellence; Commission on Fire Accreditation International (CFAI); http://www.publicsafetyexcellence.org/agency-accreditation/about-accreditation-cfai.aspx

These sources should all be reviewed and incorporated into policy analysis. In addition, for the emergency management function, NFPA 1600, *Standard on Disaster/Emergency Management and Business Continuity Programs*, should be consulted. Note that each NFPA standard may contain references to other standards and reference sources that need to be referred to while developing a policy analysis. The following sections further discuss NFPA 1710 and the ISO documents.

## NFPA 1710

NFPA 1710, *Standard for the Organization and Deployment of Fire Suppression Operations, Emergency Medical Operations, and Special Operations to the Public by Career Fire*

*Departments*, was first issued by the NFPA in 2001. There was opposition to the adoption from many special interest groups. For example, the International City/County Management Association stated in an article that several managers believed the standard was "being specifically cited in their communities as a means to leverage more resources for only one aspect of the fire service: staffing and deployment" (International City/County Management Association, 2002).

The IAFC and the IAFF originally published separate documents in 2001 explaining their interpretation of this standard. There were some differences. However, in a 2002 joint analysis of the standard, these two groups came to an agreement and jointly published a new document titled *NFPA 1710: Implementation Guide* (2002). This guide states the following (IAFC & IAFF, 2002, p. 1–9):

> The NFPA 1710 standard could be found highly relevant to the question of whether a jurisdiction has negligently failed to provide adequate fire or emergency medical protection to an individual harmed in a fire or medical emergency. To prevail in such a claim, the individual would have to show that the jurisdiction failed to provide the level of service required by the standard, and that this failure was a cause of his or her injury.

The guide is an excellent source for any analysis on staffing and deployment of fire and rescue services.

## ISO Fire Suppression Grading Schedule

ISO gathered 5 years' worth of data, reviewing the cost of fire claims around the country for homeowners and commercial property insurance. The data showed that "the communities with better classifications [Classes 1–4] experienced noticeably lower fire losses than communities with poorer classifications" (ISO, 2001, p. 3). Because better ratings are very sensitive to reduced response distances, adequate staffing, training, and water supply, it is not surprising that departments that have these resources are able to more effectively extinguish fires.

The NFPA *Fire Protection Handbook* points out one very important factor to keep in mind when reviewing a recommendation from the ISO for public policy analysis: "The purpose [of ISO grading] is to aid in the calculation of fire insurance rates and is not for property loss prevention or life safety purposes" (NFPA, 2003, p. 7–40). Therefore, if the department is looking for recommendations to improve emergency services to the public, use the ISO recommendations as part of the analysis, but not exclusively.

ISO measures fire company coverage by using a response distance calculation. If the area served by the station is within 1.5 miles of the station, the department gets full credit for this item. This is a numerical evaluation and partial credit is allowed if the station is over 1.5 miles from the protected area. However, using only distance can be misleading and results in big differences in response time from area to area. For example, a fire station that is located in an urban area that has heavy traffic or a volunteer station that relies on members traveling to the station before responding will have substantial increases in response times.

There are two advantages to using ISO ratings. The first is that the evaluation is free to the municipality, although there are costs if the municipality attempts to improve its rating. The other advantage is that a chief can characterize any development that would improve the public fire protection rating as being able to reduce insurance costs to the owners of property in the city. However, only owners of certain structures receive substantially reduced insurance premiums. Properties protected by automatic sprinklers—such as large industrial complexes—already have lower rates. Single-family dwellings are also rated in standard policies that do not see lower premiums unless the rating is dropped several classes. For example, in Florida, the insurance for single-family dwellings and commercial structures protected by automatic sprinklers is essentially the same for Classes 1–7. In addition, several national jurisdictions with populations of over 200,000 have their insurance rates based on loss experience, with the ISO rating having no effect.

The problem with using ISO for policy analysis, especially in cost-benefit calculations, becomes evident when the financial pluses and minuses are considered. Any improvements for the department are funded by tax dollars. Any benefit to the citizens and property owners comes from lower insurance premiums. The two may not seem connected or noticeable; however, a business owner may pay less insurance but have to pay higher property taxes to fund improvements in public fire protection.

It is incumbent on the department to educate the business community and elected officials about the advantages of a more effective FES department. Many jurisdictions have used lower insurance premiums to attract new commercial development; however, as can be seen from the ISO fire loss data, complying with ISO recommendations does make fire departments more effective.

**TABLE 12-2** further compares the requirements of NFPA 1710 to the requirements in ISO (to gain the greatest credit in the rating calculations). In general, a department using NFPA 1710 for deployment and staffing has a lower budget than a Class 1 ISO department because fewer stations are needed to meet the response times (distances) and fewer firefighters are needed per company. The calculations for costs using ISO are very complex, so it is hard to generalize for other protection classifications, although some fire experts have commented that they think the break-even point in a cost-benefit analysis for an ISO rating is somewhere around a Class 4 rating.

**TABLE 12-2  NFPA 1710 vs. ISO Class 1 Requirements**

|  | NFPA 1710 | ISO |
|---|---|---|
| Engine company travel distance | 2–2.5 miles* | 1.5 miles |
| Staffing | 4 firefighters | 6 firefighters |

*Note: Because NFPA 1710 uses response times, the distance will vary depending on items such as road layout and traffic conditions.

## Changing Social Perspective

Administrators who use a group to make recommendations on policy should be aware of the attitudes and values of members of the organization when conducting policy analysis. Because of such management techniques as total quality management and Six Sigma, many new members are under the impression that they will be a part of every decision-making and planning process. Formal task groups in many organizations are slowly restructuring how policies and decisions are made.

Members' attitudes and traits can affect policy analysis; therefore, it is important to know the characteristics of the different generations of members. For example, baby boomers, born between 1946 and 1965, typically reflect values of a strong work ethic, believe in the integrity of family, are more trusting of government, and show loyalty to their employers. Those born after 1970 may accept divorce more readily, tend to distrust government, and change jobs more frequently. Generation Y or the Millennial Generation—born between 1980 and the early 2000s—is characterized by egotism, vanity, pride, or selfishness.

Members and their support groups today may feel entitled to benefits of the job and may not be open to discussions about sacrifices that are necessary to implement change. In the past, most FES members lived in the community they served. This is no longer true; many members—volunteer and paid—live miles from the districts they serve. These commuters may not be as committed to the municipality they serve because they do not live there. Several conflicts of values can be seen in these groups; for example, group commitments versus individual needs, respect for authority and obedience versus participation and democratic rule, a melting pot of cultures and races versus diversity and appreciation for differences, and materialism versus value gained through new approaches and experimentation.

Therefore, the administrator must be prepared to guide any policy task group toward the strategic goal, which for simplicity would be the best emergency service for the tax dollars spent. There may be conflicts between the member's wants and the citizen's needs; the administrator must make it clear that the public comes first.

## Empowering Employees

Empowerment has been praised by many management gurus as the way to put policy analysis at the worker's level; however, the chief officer should not adopt and implement these or any business management methods without first conducting a detailed study. Empowerment could cause havoc if used by front-line emergency service workers at the company level. Should paramedics choose the medical protocol on an ad hoc basis for each call? Should the members of each engine company be allowed to change the location and setup of equipment on their engine? With no consistency or adherence to generally accepted practices and procedures, each shift, crew, station, or battalion would do its own thing. This encourages "freelancing," which invariably ends up as a safety issue.

If there is a need for policy analysis or change in operations for emergency service, empowerment is not a good idea. However, if, for example, a chief assigns an officer the responsibility to supervise, plan, and create a new collapsed building and confined space rescue training area, this officer may be empowered with the authority to do whatever is necessary to complete the project. The officer may reach out to the private sector for donations of equipment, materials, and even help with the construction. The officer might then be empowered to mount a plaque in a prominent place at the training area, thanking the donor for the help.

## CHAPTER ACTIVITY #1: The Quint Concept

One mature metropolitan fire department has adopted a new strategy for equipping and staffing its fire companies. The quint concept is named after a fire apparatus that is constructed to replace an engine and ladder company with one piece of equipment containing hose, water, a pump, and an aerial ladder. Some experienced fire officers have observed that this type of apparatus should have a crew of six or seven to fully use the quint's potential; however, this department uses only four firefighters for each company. The senior officials of the department are quick to point out several tactical advantages that mainly focus on their ability to direct many elevated streams quickly on major defensive strategy fires. However, the original pressure for the new policy may have come from other sources. This city was faced with a substantial shortfall of tax revenues, and the elected officials were pressuring the fire department for reductions in personnel. The fire department had also not been able to purchase any new fire apparatus for many years.

Generally, many older central cities in this country are protected by fire departments that have a long tradition and history. These older departments typically have some stations that were constructed very close to each other at a time when horses pulled the steam pumpers, but this spacing is no longer necessary with motorized apparatus.

In addition, most of these cities have a greater number of ladder companies than typically found in newer cities or suburban areas. In the past, there was justification for these ladder companies to rescue trapped occupants from windows in taller buildings, but the enforcement of modern building and fire codes, including automatic sprinklers and fire-resistive exit stairs, has eliminated many of these firetraps.

Before changing to quints, the department operated 36 engines and 16 ladders out of 30 stations, totaling 52 companies. After the conversion, the department ended up with 30 quints and 4 ladders totaling 34 companies. Slightly over one-third of its fire companies were disbanded. Even with this major cutback in personnel and companies, the department managed to keep all fire stations open by placing one quint engine in each of the 30 stations. It has been argued that the original pressure for this change was not a better way to extinguish building fires, but to comply with a major cutback in the fire department's budget.

### Discussion Questions

1. How would you research the facts in this case to assess their accuracy?

2. If the original justification was budget cutting, is this new system actually giving better or equivalent fire protection?

3. What harm would there be from not informing the public of all the circumstances behind the new policy decision?

4. Do customers believe FES members are their protectors and trustworthy, and therefore blindly trust policy decisions?

5. How does reduced revenue for the city affect the customers' demand for FES services?

## CHAPTER ACTIVITY #2: Smoking Ban

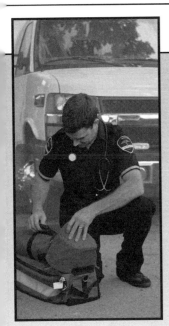

Many fire departments provide automatic retirement benefits for firefighters developing any heart/lung illness. One major cause of these illnesses is smoking cigarettes. An article describing the results of studies by the Medical Research Council (London, England) indicated that cigarette smoking has greater harmful effect than fire fumes on measures of pulmonary function. "The prevalence of respiratory symptoms was higher in smokers than non-smokers and increased with the number of cigarettes smoked" (Douglas, Douglas, Oakes, & Scott 1985). In addition, a 1973 study by H. D. Peabody discovered that cigarette smoking and exposure to carbon monoxide interacted in a synergistic fashion to substantially lower pulmonary function, reducing the ability of the firefighter to perform heavy work for sustained periods of time (Peabody, 1973).

In March 1978, the city of Alexandria, Virginia, stopped hiring smokers as firefighters because smokers accounted for a disproportionate number of early retirees from the fire department. Although the argument to support banning cigarette smoking by all firefighters seems strong, there are some practical and legal questions that need to be resolved when implementing this type of policy. For example, many unions or individuals would argue that this is an infringement on personal rights.

### Discussion Questions

1. Is it reasonable to adopt the policy of not hiring smokers and requiring all new firefighters to sign a condition of employment that they will not smoke while employed as a firefighter? Annual medical examinations could test for nicotine as part of the enforcement process.

2. Should the city fund a recognized program for smokers to quit and encourage firefighters to attend?

3. Should a policy that prohibits smoking in the fire stations be adopted?

4. Would this new policy save the city or county money by reducing costs for medical insurance, reducing sick leave, and in general providing a more physically fit firefighter? Provide references.

## CHAPTER ACTIVITY #3: General Discussion Questions

Choose a major program in an FES services agency and assume the executive official or the elected legislature of your city has cut the department's proposed budget for this program. As the FES administrator of the department, answer each of the following questions with consideration to justifying the budget.

### Discussion Questions

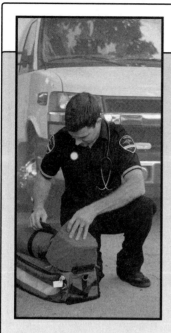

1. Should you mount a campaign to support your proposed budget using family, friends, and the firefighters' union?

2. Should you work quietly within the system to determine the political feasibility of fighting publicly?

3. Should you fight just enough to show your employees that you care, but not enough to do political damage to yourself? (How would you determine what is "just enough"?)

4. Should you leak information to the media about the disastrous consequences of the budget cut?

5. Should you accept the cut and prepare to be creative about how you implement the cut?

6. Should you be prepared to point out problems with the budget cut when the next catastrophic fire or emergency incident occurs?

7. How would the old adage, "You don't want to win the battle and lose the war," affect your actions?

8. What influence would knowing that your boss has the authority to fire you have on your actions?

## CHAPTER ACTIVITY #4: Problematic Patients

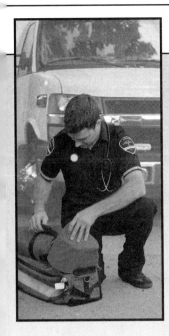

An 84-year-old woman named Hazel Henry is known by the agency as a "frequent flyer." She has limited income and lives alone in an old house that is badly in need of repair. She lost her husband a year ago and has no other family close by. Her daughter has tried to convince her to move into a senior home, but she refuses.

Hazel has mobility issues and a number of medical problems including diabetes, high blood pressure, and a heart condition. She does not drive and her doctor is a considerable distance away. She falls frequently, is often hypoglycemic, and has the occasional hypertensive crisis and chest pain. She calls your fire-based EMS resources at least once a week. She loves your crews and sometimes calls just because she wants the company.

### Discussion Questions

1. Because your local and regional policies require response and transport for any patient that requests it, patients like Hazel tax your minimal resources. Is there anything you can do to prevent this?

2. What kind of programs could help create efficiency for your service as well as benefit the patient?

3. How will you know if this type of program would be a viable option in your community?

## References

10News. (2011). *Chief Reacts to Report on City Fire Coverage, Study Suggests Smaller Fire Crews to Help Shorten Response Times*. Retrieved from www.10news.com-news-chief-reacts-to-report-on-city-fire-coverage.

Ahrens, M. (2011). NFPA's "Smoke Alarms in U.S. Home Fires." Retrieved from http://www.nfpa.org/itemDetail.asp?categoryID=2467&itemID=55728&URL=Research/Statistical%20reports/Fire%20protection%20systems/Smoke%20Alarms%20in%20U.S.%20Home%20Fires

Douglas, D. B., Douglas, R. D., Oakes, D., & Scott, G. (1985). Pulmonary function of London firemen. *British Journal of Industrial Medicine, 42*(1), 55–58.

Flynn, J. D. (2009). *Fire service performance measures*. Quincy, MA: National Fire Protection Association.

Garvey, G. (1996). *Public administration: The profession and the practice*. New York, NY: St. Martin's Press.

Insurance Service Office. (2001). *ISO's PPC program: Helping to build effective fire-protection services*. Retrieved from http://www.iso.com/Research-and-Analyses/Studies-and-Whitepapers/ISO-s-PPC-Program.html

International Association of Fire Chiefs. (2012). *How to develop a political action plan*. Retrieved from www.iafc.org.

International Association of Fire Chiefs, & International Association of Fire Fighters. (2002). *NFPA 1710: Implementation guide*. Fairfax, VA: International Association of Fire Chiefs.

International City/County Management Association. (2002). *NFPA 1710: Your community may be next*. Retrieved from http://www.icma.org

Lindblom, C. E., & Woodhouse, E. J. (1993). *The policy-making process* (3rd ed.). Upper Saddle River, NJ: Prentice Hall.

Love, M. (2008). On the way to a performance measure. *Fire Chief*. Retrieved from http://firechief.com/blog/way-performance-measure

National Fire Protection Association. (2003). *Fire protection handbook* (19th ed.). Quincy, MA: National Fire Protection Association.

National Institute of Science and Technology. (2010). *Report on residential fireground field experiments*. Retrieved from http://www.nist.gov/customcf/get_pdf.cfm?pub_id=904607

Peabody, H. D. (1973). San Diego Fire Department Health Survey. In: Survival in the Fire Fighting Profession: Second Symposium on Occupational Health and Hazards of the Fire Service. University of Notre Dame, Indiana. John P Redmond Memorial Fund, International Association of Fire Fighters.

US Fire Administration. (1977). *Urban guide for fire prevention & control master planning*. Washington, DC: US

Deptartment of Commerce, National Fire Prevention & Control Administration, National Fire Safety & Research Office.

US Fire Administration. (1989). *America burning*. Retrieved from www.usfa.fema.gov/downloads/pdf/publications/fa-264.pdf

US Fire Administration. (2009). *Fire in the United States 2003-2007* (15th ed.). Retrieved from http://www.usfa.dhs.gov/downloads/pdf/statistics/fa_325.pdf

Wilson, P. F., Dell, L. D., & Anderson, G. F. (1993). *Root cause analysis: A tool for total quality management*. Milwaukee, WI: ASQC Quality Press.

# CHAPTER 13

# The Future

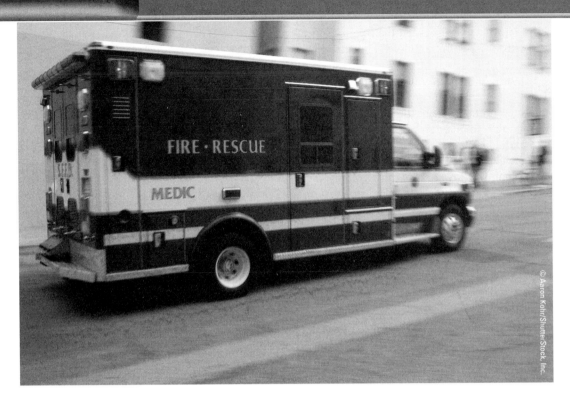

## Fire and Emergency Services Higher Education (FESHE) Objectives

### Module III: Planning and Implementation
The students will:
8. Explain how a fire and emergency service administrator creates a vision of the future for his or her organization. (pp 183–188)

### Module IV: Leading Change
The students will:
1. Describe the importance of accepting and managing change within the fire and emergency service department. (pp 183–187)
5. Describe how an organization can respond to current or emerging events or trends. (p 183–187)

### Module V: CRM – A 21st Century FESA Responsibility
The students will:
1. Assess the importance of integrating fire and emergency services into a community's comprehensive plan. (pp 185, 186–187)

## Knowledge Objectives

After studying the chapter, the student will be able to:
1. Comprehend the creation, effect, and importance of progressive change. (pp 183–184)
2. Understand the impact and influence that the future will have on fire and emergency services (FES) administration. (pp 184–187)
3. Examine and comprehend the effect that outside influences have on the future. (pp 184–187)
4. Identify the responsibility and accountability of administration to constantly strive for improved service to the public. (pp 184–187)
5. Comprehend and understand the ability to predict future trends in the public FES. (pp 184–187)

## Persistent Change in the Past and Future

Taking into consideration that the US fire service is more than 350 years old, many changes in FES are relatively new. For example, the federal fire focus of the US Fire Administration and the National Fire Academy (NFA) is only slightly more than 20 years old. The first National Fire Protection Association (NFPA) professional qualification standard, NFPA 1001, *Standard for Fire Fighter Professional Qualifications*, was adopted in 1974, and many more recently adopted standards are still finding their way into the training and operations of the fire service.

In the free enterprise business community, where turning a profit is the number one goal, the urgency for change is of a high intensity. There is always the possibility that the company might go out of business if it does not keep up with the latest and greatest consumer products and services. Although there are also some outside pressures for change in the FES organization (e.g., regulations and standards for operations and safety), most of the urgency for change must come from within, leaving the future in the hands of the chief officer. In many cases, this requires competent leadership and personnel who are comfortable with change.

In the FES industry, the typical employee and organization itself receive little feedback on their performance. This lack of feedback is not the result of poor management but rather the result of a lack of performance measurements. To measure future performance, valid outcomes directly resulting from input are needed. Leadership expert John Kotter further explains this notion: "The combination of valid data from a number of external sources, broad communications of that information inside an organization, and a willingness to deal honestly with the feedback will go a long way toward squashing complacency. An increased sense of urgency, in turn, will help organizations change more easily and better deal with a rapidly changing environment [future]" (Kotter, 1996, p. 163).

One recent example of a document addressing this issue of measurable outcomes is NFPA 1710, *Standard for the Organization and Deployment of Fire Suppression Operations, Emergency Medical Operations, and Special Operations to the Public by Career Fire Departments*. Adopted in 2001, this standard addresses the issues of response times, deployment capabilities, and staffing in FES. It also contains requirements for training, operations, and safety through the reference of other NFPA standards. This standard now provides the FES community with legitimate measurements of service delivery that have been validated by consensus of subject matter experts using scientific research and extensive experience, allowing FES administrators to set goals for the future.

New positions and promotions should go to those who believe in these future goals and are comfortable with change. Individuals should be compatible with the intimate family atmosphere of the typical FES company. This characteristic can be difficult to assess, but it is not impossible. Psychological testing is becoming more and more common in the public safety professions. Professional consultants can help devise tests to screen applicants for personality traits that are desirable in a new member. Incumbents who aspire for promotions can also be evaluated periodically based on a set of known qualities.

In the past, departments have typically attracted and accepted risk-takers who were physically strong and preferred to interact agreeably with others who had similar personalities. Although these traits are still valued, compassion, open-mindedness, and being a team player are also now essential. Moving forward, departments should also look for individuals who enjoy and feel comfortable interacting with the public. These individuals should be tolerant of others and want to help the public. They should also be able to deal with the rapidly changing world and the effects these changes have on the FES.

## Technology and Research

Competency at the emergency scene is the result of three separate functions: (1) competent personnel; (2) equipment; and (3) command (standard operating procedures). To have FES organizations equipped with the newest and best technology is only one part of providing quality emergency services to the public; additional training and increases in physical fitness also dramatically improve performance at an emergency incident.

Still, technology seems to be a never-ending source of change, some good, some bad, and some unnecessary. Examples of new technology in FES are 1.75- and 2-inch hoses, large diameter supply hoses, positive pressure ventilation, automatic defibrillators, the National Incident Management System (NIMS), compressed air foam, infrared cameras, and GPS navigation **FIGURE 13-1**. Major research participants, such as the National Fire Academy and the US National Institute of Science and Technology, are studying respiratory protection, dermal protection, thermal imaging cameras, and other personal protective equipment technologies. Other studies focus on firefighter safety and health, relating to such issues as cardiovascular disease, hearing loss, and fatigue caused by shift work.

In addition, with the ability to collect better data and the use of quality improvement programs to evaluate the data, long-standing medical protocols are now being called into question. "Evidence based medicine," which includes accurate, definitive studies of the effectiveness of a product, drug, or treatment protocol, drives decisions on diagnostic and therapeutic equipment, ultimately improving the quality of patient care.

Although many of these new technologies may be very helpful for the FES organization, the process for research, experimentation, and acceptance tends to be sporadic and decentralized. The US Fire Service is made up of more than 30,000 departments that characteristically serve small numbers of people and have

**FIGURE 13-1** Firefighter using an infrared camera.

minimal resources, especially for research. There is no large, well-funded, independent FES research agency to investigate and study new technologies, methods, or equipment. As stated by former Federal Emergency Management Agency administrator Dave Paulison, "It's just a start [the present level of fire research]. We must continue funding this type of research if we are truly committed to 'Everyone goes home'" (Grant, 2009). Although most of this current research has been funded through the Department of Homeland Security, pressure to reduce the national deficit may threaten this funding in the future.

Furthermore, although many organizations label themselves as progressive, state-of-the-art agencies, there still tends to be reluctance in the industry to accept all that is "new" and innovative, particularly in terms of equipment. As noted by NFPA research director Casey Grant, this issue seems to stem from the fact that the fire service is deeply rooted in tradition. "While this brings a certain stability to the profession, it can also mean that the fire service is sometimes slow to embrace improvements" (Grant, 2009).

However, just because equipment or a technique is old does not automatically mean it is bad. Each new tool, idea, or technique needs to be judged on its own merits through the use of research techniques that are based on standards and broad-based beta testing. This research is often best conducted independently. It is important to question information, observations, and studies cited by manufacturers. It is the manufacturer's job to tell the customer the best reasons to buy its product, and there is the chance that the study process may not have been completely unbiased.

The future is always difficult if not impossible to predict. In many cases, the final results of a product are not even apparent until many years after purchase. Although it is not realistic to be 100% certain before making a decision or implementing a change, it is necessary to be cautious and take the time necessary to gain as much information as is needed. Flexibility and a willingness to adapt are key.

## Economic Impacts

Local governments primarily fund FES and are required to have a balanced budget. Unlike the federal government, local governments do not have the luxury of borrowing huge sums of money in times of economic downturn. In addition, elected officials seem less and less likely to try to raise taxes. Therefore, when tax revenues fall, services must be cut.

When this occurs, departments that are required to cut their budgets generally target nonpersonnel purchases first. Eventually, however, staffing also needs to be reduced. Often, this reduction begins with nonoperations staff, such as training and prevention personnel, but may also include fire-suppression personnel. For safety and efficiency issues, it is better to close stations than run companies with inadequate and unsafe crews. Closing fire stations may be a more difficult decision for elected officials. When closed, these empty stations send a reminder to the public that their local government is in an economic struggle; reducing staffing levels, however, is rarely even noticed by the public.

One idea to justify and acquire new funding can be found in the emergency medical services (EMS) field. Recently, community paramedicine has been replacing traditional in-hospital service in some regions. Community paramedicine is an effort to provide primary and preventive health care by having paramedics make house calls, especially in rural communities where access to health care is limited. These paramedics perform limited physical assessments, confirm prescription compliance, and evaluate an older person's home for fall hazards. Fees can be used to cover these services.

Another way to battle the issue of reduced funding is to consolidate fire departments. Consolidation has some substantial cost savings, such as reductions in mid- and upper-level management, staff functions, stations, and special services. In many cases, consolidation involves a county fire department absorbing a city fire department; however, there are also cases of a city absorbing a county department, such as the case of Jacksonville, Florida, consolidating with Duval County.

Even though a majority of fire departments are experiencing economic difficulties, consolidation continues to be looked upon with apprehension; but this was not the case for Palm Beach County in 2009. The department, which was originally created in 1984 as a consolidation of several departments in the unincorporated area of the county, consolidated again with the smaller city of Lake Worth, absorbing the city's 56 fire employees, two fire engines, and two rescue trucks. The result was that property owners in the city began paying $2.95 per $100,000 taxable value to the county for fire and rescue service versus $4.70 per $100,000 they were paying for the city fire service (Howard, 2009). The city was then able to use the savings toward paying the $12 million fire pension obligation and replacing $1.1 million in revenue lost by the community redevelopment agency.

As can be seen from this case, consolidation usually results in a savings for the city and its residents. In addition, these larger fire departments tend to be more efficient and effective because of their greater resources and funding levels. In fact, the largest fire department in the country is a consolidation of five counties: Kings, Richmond, Queens, Bronx, and New York Counties form the New York City Fire Department. This example seems to suggest that anything is possible. Even though there is the potential for many pitfalls and very detailed planning must precede these efforts, consolidation is an alternative that should not be overlooked for departments that are required to cut their budgets or that have shortages of volunteer members.

## Higher Education and Training

The results of an informal survey conducted by Firehouse.com in 2001 indicated that more than 27% of fire chiefs at the time had a bachelor's degree or higher and 32.7% had no college background (Firehouse.com, 2001). It is now common for municipalities advertising for a vacant fire chief's position to list a minimum college degree requirement. At the national level, the NFA Executive Fire Officer program requires a minimum education level of a bachelor's degree; however, the NFPA Fire Officer professional standards still do not require any college education to comply with their criteria, which are commonly used for certification in many areas. This discrepancy may send a mixed message to future officers.

One study shows that fire service college degrees for fire officers are readily available. Specifically, the study reported that there are 222 associate-level programs and 26 bachelor fire service degree programs in the United States. Many serve only the geographic areas near their campuses, however, there is a growing trend to offer courses at some higher education institutions via the Internet. This report went on to comment that "the most important reported challenges facing degree programs included means and methods to increase enrollment, updating curriculum, finding quality instructors, lack of funding, and lack of incentives for earning an advanced degree" (Sturtevant, 2001, p. v).

Many of the structural parts for a professional FES training and education system already exist. For example, state training agencies and other regional training centers are in place to provide firefighter and officer training. The problem is that most have been developed independently and many have different requirements. Dr. Denis Onieal, Superintendent of the NFA, summarizes the issue of training and education in the present fire service (2004, p. 1):

> It varies from place to place, depending upon the organization, the structure of the department and the governing agency. The process isn't the same wherever you go; frequently it is a slow and uneven process, or one solely based on popularity....Right now, there is no one universally recognized and reciprocal system to acquire the knowledge and skills required in the Fire and Emergency Services.

Although most training programs use the NFPA professional qualifications, there are differences in the length of these training programs and the testing used for certification. In addition, some states maintain separate educational systems that provide officer and other specialized fire training. For example, a community college may offer a fire science program, but that does not mean the courses are recognized by the state training and certification agency. To become certified, applicants would then have to take the courses offered by the state training agency, even when the fire science program covers the same educational material.

In contrast, the EMS field has a model education, training, and certification system in place. The National Registry of Emergency Medical Technicians is a national certification system that was created in 1970 by the National Highway Traffic Safety Administration. This system is used by 90% of all states, with the other 10% still using the same training and skills but doing their own testing.

To be on the safe side, members who strive to become future leaders in the emergency services profession should, at a minimum, pursue a bachelor's degree. Fire services degrees that can be obtained by distance education, such as the University of Florida FES bachelor's degree program or the NFA Degrees at a Distance program, are often a good choice for students. However, if the student is fortunate enough to be in driving distance from a college that offers these degrees, face-to-face interaction with the instructor and other FES members is an advantage.

This is an area of emergency services that needs work and some type of consensus in the future. The key to leadership and change management is higher education. Without change, there cannot be progress. Education and training can give the chief officer the confidence and knowledge to deal with elected officials and the public, including union presidents, and to successfully lead the organization through change.

## Risk Management

"America at Risk," a study that was concluded in 2002 by the Federal Emergency Management Agency, contained the first discussion of total risk management at the federal level. According to the report, the future of emergency management involves more than just the risk of fire and may include other disasters such as hurricanes, floods, earthquakes, etc. "The establishment of FEMA, the growth of the emergency management community as a profession, the increase in disaster losses in America, and other factors, had dictated a different context for the fire service [in the future]" (Federal Emergency Management Agency, 2002, p. 6).

The following is a list of the major topics discussed in this report:

- Implementation of loss prevention strategies (multihazard)
- The application and use of sprinkler technology
- Loss prevention education for the public
- The acquisition and analysis of data
- Improvements through research
- Codes and standards for fire loss reduction in the built environment
- Public education and awareness
- National accrediting and certification
- Firefighter health and safety
- Emergency medical services
- Diversity

"America at Risk" changed the expectations of elected officials and the public, including the fire services

**FIGURE 13-2** Terrorist attacks of September 11, 2001.
© Todd Hollis/AP Photos

perception of itself. FES agencies should focus on these topics when formulating goals to manage risk in the future.

Another theme that is present in this report is the emphasis on the effectiveness of prevention efforts. The United States has not experienced a major terrorist event since September 11, 2001; this could be the result of extensive prevention efforts by federal agencies such as the CIA, FBI, and US military agencies **FIGURE 13-2**. For national security reasons, the federal government is not commenting at this time on any prevention successes—they just confirm that security is being pursued through prevention. While it is also difficult to document in the FES field, prevention should continue to be a goal for the future.

## Homeland Security

The war on terrorism seems to be driving contemporary changes in FES administration and operations. On March 1, 2004, the US Department of Homeland Security adopted NIMS. Previously, the emergency services had several incident command systems to choose from, each a little different from the others. The NIMS "provides a consistent nationwide template to enable Federal, State, local, and tribal governments and private-sector and non-governmental organizations to work together effectively and efficiently to prepare for, prevent, respond to, and recover from domestic incidents, regardless of cause, size, or complexity, including acts of catastrophic terrorism" (US Department of Homeland Security, 2004, p. ix).

The International Association of Fire Chiefs (2001, p. 3) has offered the following suggestions for NIMS-compliance:

- Every responding agency (local, state, national) must use a standardized incident command system.
- Continual coordination between local, state, and national responders is essential for success in protecting our personnel and the citizens we serve.
- National agencies need to involve local first responders in every level of discussions for resource allocation, especially training and equipment.

Many of the details of the NIMS are scheduled for study and adoption at a later date. These are very complex subjects, especially for the fire service, which has traditionally developed standards of service at the local level. As this project proceeds, expect to see efforts to define and implement the following for responses to a federal declared emergency:

- Mutual or automatic aid agreements
- Credentialing emergency responders for different competencies
- Resource typing
- Training for NIMS
- Integrated communications systems

For a look into the future, the federal system for responses to wildfires is a good model. This system is able to provide direction and coordination for hundreds of resources and thousands of firefighters operating at major wildfires. Wildfire resources and firefighters are brought in from hundreds and even thousands of miles away and from numerous individual departments and are able to work smoothly side by side. This type of effort to provide efficient emergency services will be needed at the next major catastrophic hurricane, earthquake, wildfire, building fire, or terrorist event.

## Advice for the Future

*Investor's Business Daily* has spent years analyzing leaders and other successful individuals. Most of these leaders have 10 traits that, when combined, can turn dreams and high goals into reality. The FES chief officer can refer to these traits when leading the organization toward change in the future (*Investor's Business Daily*, 2004):

1. Be optimistic and place yourself in a positive environment with positive people.
2. Decide on your dreams and goals, and develop a plan to reach them.
3. Act on your goals. Now is a good time to start.
4. Continue to learn and acquire new skills.
5. Work hard and do not give up on your goals. Perserverance can overcome most obstacles.
6. Be analytical. Get the facts and learn from mistakes.
7. Stay focused. Do not let others distract you.
8. Do not be afraid to be different and innovative.

> **Words of Wisdom**
>
> "There are those people in this world who like to be challenged. They enjoy doing the jobs that no one else will do. These people are motivated by more than just money. They have an inner motivation that gravitates them towards their current job. Ask any one of them why they do it and they'll tell you they can't imagine doing anything else. They are firefighters."
>
> – Scott Paulson, WDVE Pittsburgh DJ
>
> *Source:* Paulson, S. "They don't make many." WDVE, Pittsburgh. March 15, 2004. Radio.

9. Learn to effectively interact and communicate with others.
10. Be honest, dependable, and responsible.

## References

Federal Emergency Management Agency. (2002). *America at risk, America burning recommissioned*. Retrieved from http://www.usfa.fema.gov/applications/publications

Firehouse.com. (2001). *What is the formal educational level of your fire chief?* Retrieved from http://www.firehouse.com/polls/2001/

Grant, C. (2009, March/April). The future of fire: Research renaissance? *NFPA Journal*.

Howard, W. (2009). City OKs fire merger and water, sewer rate increase. *Palm Beach Post*. Retrieved from: http://articles.sun-sentinel.com/2009-03-24/news/0903230141_1_sewer-rates-city-fire-increase

International Association of Fire Chiefs. (2001). *Leading the way: Homeland security in your community*. Retrieved from http:// www.iafc.org/downloads/chckweb.pdf

*Investor's Business Daily*. (2004). IBD's 10 secrets to success. *Leaders and Success*. Retrieved from http://www.stuartlevine.com/wp-content/uploads/2010/12/Investors-Business-Daily.pdf

Kotter, J. P. (1996). *Leading change*. Boston, MA: Harvard Business School Press.

Onieal, D. (2004). *Professional status: The future of fire service training and education*. Retrieved from http://www.usfa.fema.gov/downloads/pdf/nfa/higher-ed/ProfStatusArticle.pdf

Paulson, S. (2004, March 15). "They don't make many." WDVE, Pittsburgh. March 15, 2004. Radio.

Sturtevant, T. B. (2001). *A study of undergraduate fire service degree programs in the United States*. Retrieved from http://www.bookpump.com/dps/pdf-b/112130Xb.pdf

US Department of Homeland Security. (2004). *National Incident Management System*. Retrieved from http://www.dhs.gov/interweb/assetlibrary/NIMS-90-web.pdf

# APPENDIX A

## Fire and Emergency Service Higher Education (FESHE) Correlation Grid for Fire and Emergency Services Administration Course (Bachelor's Core)

| FESHE Objective | Chapter | Page |
|---|---|---|
| **Module I: Leading and Managing Purposefully with a Community Approach** | | |
| 1. Describe the role of the fire/emergency medical services department as a part of the community government and comprehensive plan. | 1, 4, 7, 12 | pp 1–10, 47, 49–51, 102–110, 165–176 |
| 2. Explain the importance of a good working relationship with public officials and the community as a whole. | 2, 7, 10 | pp 19, 21–22, 23–24, 102–105, 107–110, 144, 145–146, 152 |
| 3. Assess ways to develop a good working relationship with public officials and the community. | 2, 7, 12 | pp 21–22, 23–24, 102–110, 169–172 |
| 4. Identify local, state, and national organizations that will be beneficial to your department. | 1, 6 | pp 2–7, 95–96 |
| 5. Describe how to take a proactive role in local, state, and national organizations. | 6 | pp 95–97 |
| 6. Identify effective skills for developing a cooperative relationship with fire and emergency services personnel as well as public officials and the general public. | 2, 6, 7 | pp 19–24, 90–92, 102–110 |
| **Module II: Core Administrative Skills** | | |
| 1. Identify the core skills essential to administrative success. | 2 | pp 12–14 |
| 2. Describe the integrated management of financial, human, facilities, and equipment and information resources. | 3, 5, 6 | pp 29–40, 57–74, 79–99 |
| 3. Explain the importance of public access to government operations. | 7 | pp 102–103, 107 |
| 4. Describe the key elements of successful communication. | 3, 10 | pp 30–32, 144 |
| 5. Recognize the basic management theory in use in your agency. | 3 | pp 37–39 |

| FESHE Objective | Chapter | Page |
|---|---|---|
| 6. Recognize the formal and informal dynamics of public organizations and describe strategies to ensure success. | 2, 6, 12 | pp 19–20, 95–97, 166, 169–172, 175, 177 |
| 7. Discuss the components and styles of leadership. | 2 | p 13 |
| 8. Identify and discuss a practical agency evaluation process. | 3, 12 | pp 35–37, 165–169 |
| **Module III: Planning and Implementation** | | |
| 1. Describe the process of consensus-building. | 3, 12 | pp 30–32, 169–170 |
| 2. Describe the components of project planning. | 12 | pp 165–177 |
| 3. Identify the steps of the planning cycle. | 12 | pp 165–175 |
| 4. Discuss how an environmental assessment determines the strategic issues and direction of an organization. | 12 | pp 168–169 |
| 5. Assess the interrelationship between budgeting, operational plans, and strategic plans. | 12 | p 166 |
| 6. Analyze the importance of an organizational culture and mission in the development of a strategic plan. | 3, 12 | pp 30–32, 165–166, 177 |
| 7. Describe the purpose, function, and current and future security concerns of working document publication, storage, and integrity. | 10 | pp 143–144 |
| 8. Explain how a fire and emergency service administrator creates a vision of the future for his or her organization. | 13 | pp 183–188 |
| **Module IV: Leading Change** | | |
| 1. Describe the importance of accepting and managing change within the fire and emergency service department. | 1, 3, 4, 8, 9, 13 | pp 1–10, 29–40, 45–46, 114–124, 126–127, 129–138, 183–187 |

# Appendix A

| FESHE Objective | Chapter | Page |
|---|---|---|
| 2. Identify models of change commonly used in organizations. | 4 | p 53 |
| 3. Summarize the steps of the change management process. | 3, 4 | pp 29–37, 46–53 |
| 4. Assess ways to create a positive climate for change and introduce new ideas within the organization. | 3, 4, 8 | pp 30–32, 34–35, 37–39, 49–50, 115, 116–120, 121–124 |
| 5. Describe how an organization can respond to current or emerging events or trends. | 4, 5, 8, 9, 13 | pp 46–53, 73–74, 116–117, 118–120, 131–138, 183–187 |
| 6. Explain the benefits of employee involvement in departmental decisions. | 3, 4, 9 | pp 30, 48–49, 133–134 |
| 7. Demonstrate innovative ways to address traditional problems within the organization. | 4, 10 | pp 50–52, 143–146, 151–152 |
| 8. Describe ways to increase and reward professional development efforts. | 3, 4, 8, 9 | pp 34–35, 51–52, 121–124, 132–134 |

### Module V: CRM—A 21st Century FESA Responsibility

| FESHE Objective | Chapter | Page |
|---|---|---|
| 1. Assess the importance of integrating fire and emergency services into a community's comprehensive plan. | 7, 13 | pp 102–110, 185, 186–187 |
| 2. Assess your organization's capabilities and needs based on risk analysis probabilities. | 12 | pp 167–169 |
| 3. Describe the relationship between community risk analysis and strategic and operational planning. | 12 | pp 167–169 |
| 4. Identify the major steps of a community risk assessment. | 12 | pp 167–169 |
| 5. Identify direct and indirect costs associated with fire. | 1, 5, 7 | pp 2–3, 8, 9, 66–69, 102, 103–104, 107–108 |
| 6. Analyze economic incentives that encourage and discourage fire prevention. | 1, 5, 7, 12 | pp 2–3, 61–66, 107–108, 174–176 |
| 7. Describe the role of fire and emergency services in the economic development and neighborhood preservation programs of the community. | 5, 7 | pp 60–61, 102–103, 107–110 |

| FESHE Outcome | Chapter |
|---|---|
| 1. Define and discuss the elements of effective departmental organization. | 2, 3, 5, 6, 7, 11, 12, 13 |
| 2. Classify what training and skills are needed to establish departmental organization. | 2, 3, 5, 6, 7, 8, 11, 12, 13 |
| 3. Analyze the value of a community-related approach to risk reduction. | 2, 4, 5, 7, 9, 11, 12 |
| 4. Outline the priorities of a budget planning document while anticipating the diverse needs of a community. | 5, 11, 12 |
| 5. Assess the importance of positively influencing community leaders by demonstrating effective leadership. | 2, 3, 4, 5, 7, 9, 10, 11, 12, 13 |
| 6. Analyze the concept of change and the need to be aware of future trends in fire management. | 1, 2, 3, 4, 6, 8, 9, 10, 12, 13 |
| 7. Report on the importance of communications technology, fire service networks, and the Internet, when conducting problem-solving analysis and managing trends. | 3, 8, 10, 11, 12, 13 |
| 8. Develop a clear understanding of the national assessment models and their respective approaches to certification. | 2, 6, 8, 9, 10, 12, 13 |

# Index

## A

"Accidental Death and Disability: The Neglected Disease of Modern Society" (paper), 3
accountability, 69
ADA. *See* Americans with Disabilities Act
adaptive policies, 70–71
administrative power, 24
administrative rule making, 152
administrators
   appointment of, 15–16
   change effected by, 19–25
      peer feedback in, 20
      staff member feedback, 19–20
   courage of, 22
   direct supervision and standardization, 18
   education, 13
   election of, 15
   experience, 13
   leadership skills of, 13
   management skills of, 13
   negotiation, 25
   power sources, 25
      administrative, 24
      community, 23–24
      political, 21–22
      tradition, 20–21
      unions, 22–23
   professional qualifications for, 13–15
   responsibilities, 16
   rules and regulations, 16–17
   selection of, 15–16
   standard operating procedures, 18
   tips for, 14
   training, 13
advanced EMT, 94
advanced life support (ALS), 104
affirmative action cases, 81
Age Discrimination in Employment Act, 89
AIDS, 53
alcohol testing, 87
ALS. *See* advanced life support
ambulance
   civilian services, 2
   crash prevention, 130–132
   private companies, 23, 24, 96
"America at Risk" (study), 186
"America Burning Report", 2, 3, 13, *165*
American firefighters, early, 2
Americans with Disabilities Act (ADA), 93, 146
   hiring practices and, 85–86
Anderson, G. F., 173
annual meetings, 20
appointment of administrators, 15–16
Army Corps of Engineers, 6
attitude. *See also* customer service
   loyalty, 15
   motivation, 34–35
   positivity in, 171
attorneys, 151–152
audience, public policy presentation, 171, 172
auditing, 69
automatic sprinklers, 161
awards, 35

## B

baby boomers, 21
Baltimore, 2
banding, 82
bargaining units, 96
barriers, overcoming, 50–52
   tradition as, 51
basic firefighter training, 115–116
battalion chiefs (BC), 54
BCCRS. *See* Bethesda-Chevy Chase Rescue Squad
benefits, 66
benefit of doubt, 40
*Benshoff v. City of Virginia Beach*, 88
Bethesda-Chevy Chase Rescue Squad (BCCRS), 98
Bok, Sissela, 156
bonds, 63
borrowing, 64
Boston Fire Department, 81
Boy Scout syndrome, 70
brainstorming, 175
   in group discussions, 31
broad-based empowerment, 39
Brown, Chief, 5
brownouts, 70
budgets
   adaptive policies, 70–71
   adjustments, 75
   approval, 58–59
   bureaucrats' role in, 60
   cost-benefit analysis, 174–175

budgets (*cont.*)
  cuts, 57, 70
    common areas for, 71
  cycle, 58
  expenditure reports, 67
  financial manager's role in, 59–60
  increases, 74
  key players in, 59–61
  line-item, 66
  management, 59
  performance, 57
  planning, 58
  program, 57, 174–175
  protecting, 71
  in public policy analysis, 166
  public role in, 60–61
  review, 58
    stages of, 59
  submission, 58
  training program, 57
  types of, 57
  zero-based, 57
building and fire codes, 144–145
Bureau of Indian Affairs, 7
Bureau of Land Management, 7
bureaucrats, budgeting and, 60

## C

CAFS. *See* Compressed Air Foam System
call response, 162
cancer, 136
Candidate Physical Ability Test (CPAT), 134
capital improvement project, 63
Carter, Harry, 9, 132
case studies, 174
Census Bureau, 60
Center for Public Safety Excellence, 175
Central City, 75–76
certification, 114–117
Certification Standards for Training, Experience, and Credentialing Regulation, 98
CFAI. *See* Commission on Fire Accreditation International
change
  at all levels, 54
  bringing about, 19–25, 46
  embracing, 45–46
  feedback in
    peers, 20
    staff member, 19–20
  guiding coalition for, 48–49
  institutionalizing, 52–54
  member resistance to, 51–52
  overcoming barriers to, 50–52
  public services and, 46
  public support of, 51
  resistance to, 46
  short-term goals, 52–53

SOPs and, 46
supervisors opposed to, 51
urgency and, 46–48
vision communication, 49–50
vision development for, 49
Charleston, 3
chief officer tips, 14, 30, 34, 38, 64, 166
cigarette smokers, 173, 179
Cincinnati, 2
CISM. *See* Critical Incident Stress Management
citizen groups, 21
citizen protection, 142
civil rights, 85
Civil Rights Act, 81
Civil War, 2
civilian ambulance services, 2
Coalition for Fire-Safe Cigarettes, 9
Coast Guard, 7
Cochran, Kelvin, 105
Code of Ethics
  FFCA, 156
  National Association of EMS Technicians, 156
Coleman, Ron, 19
collaborative training, 115
combination departments, 98–99
Commission on Accreditation of Ambulance Service, 5
Commission on Fire Accreditation International (CFAI), 3, 175
communication. *See also* feedback; management
  miscommunications, 158
  with unions, 95–96
  of vision, 49–50, 55
community demographic change, 103
community powers, 23–24
complaints, silencing, 90
compliance audit, 69
Compressed Air Foam System (CAFS), 65
Computer Aided Dispatch, 35
conflict
  avoiding, 33
  grievances, 96–97
  insubordination, 90–92
  member resistance, 51–52
  with power elite, 23
  with supervisors, 51
consensus building, 169
  example, 170
consistency, 118
consolidation, 185
constructive discharge, 92
consultants, 175
Contra Costa, 115
contracts, union, 145–146
controlling, 29, 42
  corrective action as, 36–37
  management and, 35–37
  performance measuring, 36
  process, 37

cooperative purchasing, 68
corrective action, 37
    feedback in, 36
costs. *See also* expenditures
    of equipment, 63
    facilities, 80
    of injuries on the job, 86, 138
    of lawsuits, 89
    staffing, 80
    total cost purchasing, 68
    of training, 95, 117
cost-benefit analysis, 64
    in budgets, 174–175
    of lying, 160–161
courage, 22
court system, 151
CPAT. *See* Candidate Physical Ability Test
critical incident stress, 139
Critical Incident Stress Management (CISM), 137
Croom, Norris W., 105
Cuba, 25
customer service
    considerations, 107
    duties, 106–110
    at emergencies, 110
    expanded
        EMS, 104–105
        fire suppression, 103–104
    fire prevention, 107
    fire safety codes enforcement, 109–110
    fire safety inspections, 108–109
    new models of, 111

**D**
Dalberg-Acton, John Emerich Edward, 34
dating policies, 88
debt ceiling, 63
decision-making
    direct hierarchical, 39
    group, 30–33
    of management, 29–33
    in public policy analysis, 165–166
    rational-comprehensive, 166
    SOPs and, 29
delayed implementation, 32
Dell, L. D., 173
demographic change, community, 103
Department of Agriculture, 6
Department of Education, 121
Department of Homeland Security (DHS), 7, 10, 65, 185, 187
Department of Transportation (DOT), 4
    training standards, 60
devil's advocacy, 32
Dewar, Jeff, 38
DHS. *See* Department of Homeland Security
diet, 133
direct hierarchical decision-making, 39

direct supervision and standardization, 18
directing
    rules and regulations, 34
    staff motivation, 34–35
disaster planning, 77
disaster purchasing plans, 69
discipline, 85
    fair, 90–91
    public sector, 90
discussion questions, 180
dishonesty, habits of, 157
Disney, 13
disparity of duties, 104–105
dispatch policy, 150
diversity, 80
    selection in practice, 81–82
    sensitivity training, 82–83
documentation, 167
DOT. *See* Department of Transportation
doubt, benefit of, 40
down-to-earth ideas, 172
Drennan, Vina, 107
drive, 38
driving, danger in, 38
drug and alcohol testing, 87
duties, disparity of, 104–105

**E**
early American firefighters, 2
economy
    global, 73–74
    impacts on, 185
    regionalized, 74
education. *See also* training programs
    in America, 74
    fire chief, 122
    higher, 121–123, 186
    officer, 122
Edwards, Steven, 80, 95, 151
EEOC. *See* Equal Employment Opportunity Commission
Eisenhower, Dwight D., 24
election of administrators, 15
elite, power, 23
*Emergency!* (television), 4
emergency incident research, 30
emergency medical response, 94
    history of, 2
emergency medical services (EMS), 46
    certification, 116
    challenges, 105–106
    contemporary, 9–10
    development of, 3–4
    disparity of duties, 104–105
    expanded services, 104–105
    financial challenges, 9
    fire service melding with, 5–6
    healthcare integration, 106

emergency medical services (EMS)(cont.)
　　healthcare reform issues, 106
　　injury and fatality prevention, 130
　　medical call volume, 105–106
　　NHTSA on, 106
　　personnel safety, 127
　　preventive medicine, 110
　　problems facing, 8–9
　　resource overcommitment, 105
　　standards, 5
　　system overloading, 106
　　training, 116
emergency medical technician (EMT), 46
　　advanced, 94
　　job classification, 94
emergency response, unified federal, 6–8
*Emergency Response Operations for Releases of, or Substantial Threats of Releases of, Hazardous Substances Without Regard to the Location of the Hazard*, 66
employees
　　empowerment, 18, 177
　　fairness among, 160
　　noncompliance, 25
empowerment
　　broad-based, 39
　　of employees, 18, 177
EMS. *See* emergency medical services
EMS Agenda for the Future, 5
EMT. *See* emergency medical technician
Equal Employment Opportunity Commission (EEOC), 85, 88, 146
Equal Protection Clause, 81
equipment
　　costs of, 63
　　fire apparatus crash prevention, 130–132
　　personal protective, 136
　　replacement plans, 63
　　technology, 184–185
　　total cost purchasing, 68
ethical dilemmas, 157
ethics
　　behavior, 155–156
　　personal, 156
　　professional, 156
　　tests of, 158–159
evaluation, 175
evidence based medicine, 184
example, leadership by, 38–39, 54
Executive Fire Officer program, 186
exercise, 133. *See also* physical fitness
expanded services
　　EMS, 104–105
　　fire suppression, 103–104
expenditures. *See also* costs
　　budget expenditure report, 67
　　legal considerations, 68–69
　　tracking, 66–68

extraversion, 38
eye contact, 171

**F**
facilitating, 51
facilities, costs of, 80
fair discipline, 90–91
Fair Labor Standards Act (FLSA), 88–89, 146
fairness, 160
Family and Medical Leave Act (FMLA), 87
fatal fires, 172
fatality prevention
　　EMS, 130
　　firefighter, 129–130
fear, 47
Federal Emergency Management Agency (FEMA), 5, 122, 185
　　risk management, 186–187
federal excise taxes, 72
federal government. *See* government
*Federal Register*, 143
federal regulations, 143
　　input on, 144
　　safety, 128
feedback
　　on change
　　　　from peers, 20
　　　　staff member, 19–20
　　in corrective action, 36
fees, 62
Feldman, Robert, 156
FEMA. *See* Federal Emergency Management Agency
FES. *See* fire and emergency services
FFCA. *See* Florida Fire Chief's Association
Field Training Officer, 86
finance. *See also* budgets; economy; expenditures; funding; taxes
　　challenges, 9
　　crisis of, 63
　　manager, 59
Fire Administration, U.S. (USFA), 5, 165
fire and emergency services (FES), 30. *See also* emergency medical services
　　budget expenditure report, 67
　　customer service duty, 106–110
　　as free public services, 64
　　funding of, 47
　　future of, 1, 187–188
　　legal aspects of, 152
　　persistent change in, 183–184
fire and rescue department, 33
fire apparatus crash prevention, 130–132
Fire Brigades, 148
fire chief. *See also* leadership; management
　　career, 13, 20
　　education of, 122
　　obeying, 24
　　power of, 20–25

*Fire Chief* (magazine), 3, 17, 30
*Fire Engineering Magazine*, 30
fire flow formula, 62
fire inspections, home, 109, 174
fire prevention, 36
   customer service, 107
Fire Prevention Code, 152
Fire Research and Safety Act, 3
fire safety
   codes, 109–110
   education, 109
   inspections, 108–109, 174
fire service
   consolidation of, 185
   contemporary, 9–10
   EMS melding with, 5–6
   financial challenges, 9
   history of, 2
   problems facing, 8–9
   tax, 62
fire sprinklers. *See* sprinklers
fire standards, 4–5
fire suppression, 103–104
Fire Suppression Grading Schedule, ISO, 176
firefighters
   early American, 2
   hiring issues, 84
   injury and fatality protection, 129–130
   job requirements, 93
   recruitment of, 83
   relationships of, 83
   safety, 126–127
Firefighter Close Calls, 30
Firefighter Fatalities, 89
Firefighters Grant Program, 65
firefighting training, 114. *See also* education; training programs
   basic, 115–116
   collaborative, 115
   goals, 120–121
   live-fire, 115
   recertification, 117
   seminars, 117
   sessions, 117
   SOPs, 117–118
*Firehouse Magazine*, 30, 70
FIREPAC, 22
First Amendment, 81, 85
first-responder medical service, 172
Fish and Wildlife, US, 7
Fisher, Robert, 23, 97
fitness. *See* exercise; physical fitness
flat taxes, 72
Florida Building Code Commission, 152
Florida Fire Chief's Association (FFCA), 156
FLSA. *See* Fair Labor Standards Act
Flynn, Jennifer, 167
FMLA. *See* Family and Medical Leave Act

Forest Service, US, 6, 7
forward thinking, 32, 33
four-minute response contours, 171
four-person companies, 149
four-person response time, 169
Fourteenth Amendment, 81
14th Amendment, 85
free choice, 159
free enterprise, 183–184
free public services, 64
freedom of speech, 85
freelancing, 177
funding. *See also* budgets; expenditures
   increasing, 64–66
   reduced, 185
   revenues, 61–64
   source definition, 65–66
*Funding Alternatives for Emergency Medical Fire Services*, 65
future of fire and emergency services, 1, 187–188
future planning, 165–166

**G**

*Garcetti v. Ceballos*, 85
Garvey, Gerald, 152, 170
*Getting to Yes, Negotiating Agreement Without Giving In* (Fisher & Ury), 23, 97
Gladwell, Malcolm, 30
global economy, 73–74
goals. *See also* change; vision
   political, 170
   short-term, 52–53
   training, 120–121
Good, Neil, 97
good of public, lying and, 157–158
government
   bonds, 62–63
   debt of, 63
   financial crisis of, 63
   intervention, 102
   local representatives, 96
   misconduct in, 160
   regulation, 141–143
      citizen protection and, 142
Grading Schedule of Municipal Fires, ISO, 2, 150
Grant, Casey, 185
grant programs, 65
grievances, 96–97
groups. *See also* teams
   decision-making, 30–33
      disadvantages of, 33
      selecting members, 30
      size of group in, 30
   discussions
      brainstorming in, 31
      delayed implementation, 32
      discussion in, 31–32
      forward thinking in, 32, 33

# Index

groups (*cont.*)
    information gathering, 31, 32
    limitations of, 30–31
    national consensus standards, 31
    techniques aiding, 30–32
    voting in, 32
  effectiveness of, 41
  member selection, 30
  size, 30
guiding coalition, 48–49

## H

harassment, sexual, 87–88
hazardous materials, training for, 118–119
Hazardous Waste Operations and Emergency Response, 147
HazMat districts map, 119
health and safety. *See also* Occupational Safety and Health Administration
  committee meetings, 18
  EMS, 127
  federal regulations, 128
  firefighter, 126–127
  NFPA standard, 127–128
  standards, 60
healthcare integration, 106
healthcare reform, 106
heart attacks, 132
Heightman, A. J., 105
higher education, 121–123, 186
Highway Safety Act of 1966, 4
hiring issues, 84
  ADA and, 85–86
  local, 92–94
  regulations, 146
  state, 92–94
history
  of emergency medical response, 2
  of fire service, 2
  of ISO, 2–3
Holmes, Oliver Wendell, Jr., 71
home fire inspections, 109, 174
home fire sprinkler systems, 107–108
homeland security. *See* Department of Homeland Security
honesty, 20, 38. *See also* lying
house fire, 5
*How to Develop a Political Action Plan* (document), 170
human resources (HR). *See also* staff
  function of, 80
  legal issues, 83–89
hurricane damage, 6
Hurricane Katrina, 47, 106

## I

IAFC. *See* International Association of Fire Chiefs
IAFF. *See* International Association of Fire Fighters
imperfect information, 19

implementation, 175
Incident Command System, 24, 119
  NIMS-compliant, 29
income taxes, 61
information gathering, 31, 32
infrared cameras, 184
infrequently used skills, 125
injuries on the job (IOJ), 86, 138
injury prevention
  EMS, 130
  firefighter, 129–130
inspections
  fire safety, 108–109
  home fire, 109, 174
institutionalizing change, 52–54
insubordination, 90–92
Insurance Service Office (ISO), 103, 175
  class 1 requirement, 176
  Fire Suppression Grading Schedule, 176
  history of, 2–3
  Municipal Fire Grading Schedule, 2, 150
integrity, 38
intelligence, 38
International Association of Fire Chiefs (IAFC), 3, 105, 122
  Candidate Physical Ability Test, 3
  on NIMS compliance, 187
  political action plans, 170
  recommendations of, 3
  on standards, 176
International Association of Fire Fighters (IAFF), 2, 84
  Candidate Physical Ability Test, 83, 134
  on physical fitness, 134
  on political goals, 170
  on standards, 176
International City/County Management Association, 3
International Code Council, 144
International Society of Fire Service Instructors, 107
internet sales, 62
investments, 63–64
*Investor's Business Daily*, 187–188
IOJ. *See* injuries on the job
ISO. *See* Insurance Service Office

## J

Javitch, David, 35
Jennings, Charles, 3
job actions, 96
job classification, 94
job requirements, firefighters, 93
job-relevant knowledge, 38
Johnson, Robert, 142
Joint Labor-Management Wellness and Fitness Initiative, 134–135
*Journal of Emergency Medical Services*, 106
Jury Verdict Research, 89

## K

Kennedy, John F., 25
*Kimel et al v. Florida Board of Regents*, 89
Klinger, Keith E., 74
Korda, Michael, 50
Kotter, John, 52, 54, 184

## L

Lancaster, Patricia, 145
law enforcement, 83
lawsuits, financial impact of, 89
leadership
   of administrators, 13
   by example, 38–39, 54
   management combined with, 12–13
   qualities of, 46
lease-purchase contracts, 64
legal considerations, 68–69
   advice on, 153
   FES, 152
   HR, 83–89
   lawsuit costs, 89
   in regulations, 150–152
life saving, 167
light duty, 86
limitations, group discussion, 30–31
Lindblom, C. E., 166
line distinctions, 14–15
line functions, 15–16
line-item budget, 57, 66
live-fire training, 116
local government representatives, 96
local hiring laws, 92–94
Love, Mike, 166
loyalty, 15
Ludwig, Gary, 106
lying
   consequences of, 159
   cost-benefit analysis of, 160–161
   justifications for, 157–159
   miscommunications and, 158
   public good and, 157–158
   reasons for, 156–157
   reputation and, 159
   as self-defense, 158
*Lying: Moral Choice in Public and Private Life* (Bok), 156

## M

management
   of administrators, 13
   of budgets, 59
   controlling, 35–37
   decision-making, 29
      group, 30–33
      defined, 29
   directing, 34–35
   leadership combined with, 12–13
   organization, 33–34
   performance assessment, 39–40
   tactics
      broad-based empowerment, 39
      chief officer tips, 38
      leading by example, 38–39
      technology-based programs, 37–38
      total quality management, 38
      walking around, 39
management by walking around (MBWA), 20, 41
   as tactic, 39
market failures, 142–143
Maryland Fire and Rescue Institute, 80
MBWA. *See* management by walking around
McGlothlen, Condon, 85
measuring outcomes, 166–169
   professional competency, 169
   response time, 168–169
   staffing levels, 167–168
Medicaid, 62
medical call volume, 105–106
Medical Research Council, 173
Medicare, 23, 62
mentors, 40
Metro Fire Chiefs, 20
miscommunications, 158
misconduct, 160
Mississippi River, 7
Mobile Intensive Care Unit, 4
   nurse, 24
modern fire service, 2
Molitor, Chris, 84
monopolies, 143
Montecalvo, Michael, 131
moral obligations, 159–161
motivating staff, 34–35
motivation, 97–99
multitasking, 40
municipalities, 2

## N

Napoleon, 2
National Association of EMS Directors, 5
National Association of EMS Physicians, 5
National Association of EMS Technicians, 156
National Association of Home Builders, 107
National Association of State Emergency Medical Services, 131
National Association of State EMS Directors, 5
National Board of Fire Underwriters, 2
National Commission on Fire Prevention and Control, 3, 13
national consensus standards, 17–18
   in group discussions, 31
   reviewing, 31
national development efforts, 2–4
National Fallen Firefighters Foundation, 115, 129

National Fire Academy (NFA), 7, 20, 165, 183–184
    Degrees at a Distance program, 186
    Executive Fire Officer program, 186
National Fire Protection Association (NFPA), 3, 16, 89, 103, 167, 183
    building codes, 144–145
    compliance, 153
    on response times, 149–150, 168–169
    standards, 148–150
        NFPA 1001 *Standard for Firefighter Professional Qualifications*, 92, 115, 169
        NFPA 1201 *Standard for Providing Emergency Services to the Public*, 175
        NFPA 1410 *Standard on Training for Initial Emergency Scene Operations*, 120
        NFPA 1500 *Standard for the Fire Department Occupational Safety and Health Program*, 31, 64, 116, 148–149, 167, 175
        NFPA 1582 *Standard on Comprehensive Occupational Medical Program for Fire Departments*, 89
        NFPA 1710 *Standard for the Organization and Deployment of Fire Suppression Operations, Emergency Medical Operations, and Special Operations to the Public by Career Fire Departments*, 4–5, 17, 60, 103, 149–150, 168–169, 175–177, 184
        NFPA 1720 *Standard for the Organization and Deployment of Fire Suppression Operations, Emergency Medical Operations, and Special Operations to the Public by Volunteer Fire Departments*, 175
        NFPA 1917 *Ambulance Vehicle Safety*, 131
    safety and health, 127–128
    training, 60
National Firefighter Burn Study, 127
National Highway Traffic Safety Administration (NHTSA), 4, 64, 116, 186
    on EMS, 106
    establishment of, 150
National Incident Management System (NIMS), 6, 7, 10, 34, 184
    IAFC on compliance with, 187
    Incident Command System of, 29
    training for, 119–120
National Institute for Occupational Safety and Health (NIOSH), 131, 136
National Institute of Standards and Technology (NIST), 147, 167
    on staffing levels, 168
National Labor Relations Act, 96
National Park Service, 7
National Professional Development Model, 123–124
National Registry of EMTs, 5, 60
National Research Council, 3
National Response Framework (NRF), 119
National Standard Curriculum, 5

national standards, 60
    evidence from, 64–65
    results compared to, 36
National Volunteer Fire Council, 134
negotiation process, 25
New Amsterdam, 2
New Haven, 81
new officers, 99
New York City, 2
NFA. *See* National Fire Academy
NFPA. *See* National Fire Protection Association
NFPA 1001 *Standard for Firefighter Professional Qualifications*, 92, 115, 169
NFPA 1201 *Standard for Providing Emergency Services to the Public*, 175
NFPA 1410 *Standard on Training for Initial Emergency Scene Operations*, 120
NFPA 1500 *Standard for the Fire Department Occupational Safety and Health Program*, 31, 116, 167, 175
    adoption of, 17
    bureaucracy and, 17
    evidence from, 64
    overview of, 148–149
NFPA 1582 *Standard on Comprehensive Occupational Medical Program for Fire Departments*, 89
NFPA 1710 *Standard for the Organization and Deployment of Fire Suppression Operations, Emergency Medical Operations, and Special Operations to the Public by Career Fire Departments*, 4–5, 103, 177, 184
    bureaucracy and, 60
    four person company compliance, 149
    in policy analysis, 175–176
    response time requirements in, 149–150, 168–169
NFPA 1720 *Standard for the Organization and Deployment of Fire Suppression Operations, Emergency Medical Operations, and Special Operations to the Public by Volunteer Fire Departments*, 175
NFPA 1917 *Ambulance Vehicle Safety*, 131
NFPA *Fire Protection Handbook*, 176
NHTSA. *See* National Highway Traffic Safety Administration
NIMS. *See* National Incident Management System
NIOSH. *See* National Institute for Occupational Safety and Health
NIST. *See* National Institute of Standards and Technology
Nixon, Richard, 143
noncompliance, employee, 25
NRF. *See* National Response Framework

## O

obeying, 159–160
Occupational Exposure to Bloodborne Pathogens, 147
occupational hazards, 136–138

Occupational Safety and Health Administration
  (OSHA), 5, 53, 65, 167
    general duty clause, 148
    regulations, 128, 146
        29 CFR 1910.120, 147
        29 CFR 1910.134, 147
        29 CFR 1910.146, 147
        29 CFR 1910.156, 148
        29 CFR 1910.1030, 147
    state plan jurisdictions, 146
*Officer Development Handbook*, 122
officer education and training, 122
Oklahoma City, 6
on-duty crews, 167
Onieal, Denis, 186
Ontario Association of Fire Chiefs, 89
organization, 33–34
OSHA. *See* Occupational Safety and Health Administration
outcome measuring, 166
    response time, 168–169
    staffing levels, 167–168
*Outliers: The Story of Success* (Gladwell), 30
overloading of system, 106
overtime, 87

# P
Page, J., 34
paramedics, 4, 105
    certification, 16
    job classification, 94
Parow, Jack, 114
Paulison, Dave, 185
Paulson, Scott, 188
Peabody, H. D., 179
peers, 20
performance budgets, 57
performance measuring
    controlling in, 36
    Flynn on, 167
Permit-required Confined Spaces, 147
personal ethics, 156
personal protective equipment, 136
personnel requirements, 168
Pfiefer, Joseph, 132
Philadelphia, 2
physical fitness
    fairness, 135–136
    IAFF on, 134
    need for, 132–134
    programs, 134–135
planning, 29
    budget, 58
    disaster, 77
    future, 165–166
    strategic, 165–166
Planning, Programming, and Budgeting System
    (PPBS), 174

podcasts, 117–118
policy. *See also* public policy
    adaptive, 70–71
    dating, 88
    dispatch, 150
    purchasing, 68
*The Policy-Making Process* (Lindblom & Woodhouse), 166
political action plan, 170
political power, 21
    opposition, 22
positivity, 171
posttraumatic stress disorder (PTSD), 137
Powell, Colin, 19, 22, 48
power base, building, 40, 43
power elite, 23
power sources, 25
    administrative, 24
    community, 23–24
    political, 21–22
    tradition as, 20–21
    unions as, 22–23
PPBS. *See* Planning, Programming, and Budgeting System
pregnancy issues, 86–87
preventive medicine, 110
private ambulance companies, 23, 24, 96
private sector, 102–103
    unions, 96
probationary period, 91
procedural audits, 69
professional ethics, 156
professionalism
    administrator, 13–15
    competency, 169
    universal standards, 4–5
program analysis, 175
program budgets, 57, 174–175
progressive labor relations, 97
progressive taxes, 72
Project 51, 4
property loss, 173
property taxes, 61, 72
PTSD. *See* posttraumatic stress disorder
public
    in budgeting process, 60–61
    free choice of, 159
    hearings, 152
    lying and good of, 157–158
    moral obligations to, 159–161
    relations, 35
    unions and, 95
public policy
    budgeting within, 166
    consensus building, 169–170
    decision making in, 165–166
    future planning in, 165–166

Index

public policy (cont.)
  outcome measuring, 166
    professional competency, 169
    response time, 168–169
    staffing levels, 167–168
  presentation, 170–172
    audience in, 171, 172
    down-to-earth ideas, 172
    eye contact, 171
    positivity in, 171
    practicing, 171
    visual aids, 172
  program, 175
  reference sources, 175–176
  statistics in, 172–173
    case studies, 174
  strategic planning, 165–166
public sector, 102–103
  discipline, 90
  unions, 23, 95–96
public services
  change and, 46
  expanded, 103–105
  free, 64
public support and opinion, 47
  for change, 51
  consensus building, 169, 170
  public preference surveys, 102–103
purchasing policies, 68

Q

quint concept, 178

R

race. *See* diversity
radio communication towers, 47
rational-comprehensive decision-making, 166
recertification, 117
recruitment, 94
  selection process, 95
reductions in force (RIF), 91
re-election, 21
reference checks, 84–85
regressive taxes, 72
regulations. *See* rules and regulations
reliability, 118
reputation, damage to, 159
research, 184–185
residential sprinklers, 172
resource overcommitment, 105
Respiratory Protection, 147
response time
  four-minute, 171
  four-person, 169
  NFPA standards for, 149–150, 168–169
  outcome measuring, 168–169
  reduction of, 167
  requirements, 149–150

revenues, 61–64
RIF. *See* reductions in force
right-to-work states, 22, 95–96
risk management, 186–187
Risk Watch programs, 16
risk-taking, 184
*Root Cause Analysis: A Tool for Total Quality Management* (Wilson, Dell & Anderson), 173
rules and regulations, 34. *See also* standards; *specific types*
  administrative, 16–17
  directing, 34
  federal, 143
    input on, 144
    safety, 128
  of government, 141–143
    citizen protection and, 142
  hiring, 146
  OSHA, 128, 146–148
  state, 144–145
  taxes, 146
  zoning, 145

S

safety
  committee meetings, 18
  education, 109
  EMS, 127
  federal regulations, 128
  fire codes, 109–110
  firefighter, 126–127
  inspections, 108–109
  NFPA standard, 127–128
  personnel, 127
  standards, 60
Safety to Life, 4
Saint Joseph's Hospital, 132, 133
salaries, 66
sales tax, 61, 62
Salka, John J., Jr., 134
Santayana, George, 1
scene time, overall, 168
*Schwarzbek v. City of Wauseon*, 130
self-defense, 158
seminars, 117
sensitivity, diversity, 82–83
September 11th, 2001, 5, 47, 187
sexual harassment, 87–88
Shaw, Eyre, 113
short-term borrowing, 62–63
short-term goals, 52–53
skills, infrequently used, 125
Slater, Jim, 73
sleep deprivation, 137–138
*Sleep Medicine* (journal), 106
smoke alarms, 172, 174
smoking ban, 179

social perspective, 177
Social Security, 21, 23
SOPs. *See* standard operating procedures
special interest groups, 24, 51
sprinklers
    automatic, 161
    home systems, 107–108
    residential, 172
staff, 14. *See also* education; human resources; training programs
    costs, 80
    division resolution, 15
    feedback from, 19–20
    hiring issues, 84
        ADA and, 85–86
        local, 92–94
        regulations, 146
        state, 92–94
    levels, 167–168
        NIST on, 168
    motivation of, 34–35
    personnel requirements, 168
    recruitment, 94
        selection process, 95
    terminations, 91–92
    training, 121
standards. *See also specific standards*
    EMS, 5
    fire, 4–5
    IAFC, 176
    IAFF, 176
    national, 60
        evidence from, 64–65
        results compared to, 36
    national consensus, 17–18
        in group discussions, 31
        reviewing, 31
    NFPA, 148–150
        safety and health, 127–128
        training, 60
    safety, 60
    universal professional, 4–5
standard operating procedures (SOPs), 18, 90, 128, 147
    change and, 46
    compliance with, 100
    consistency and, 118
    creating, 50
    critique of, 118
    decision-making and, 29
    reliability and, 118
    in training, 117–118
    variances in, 118
    vehicle speeds, 131
standardization, 18
State Farm, 3
state hiring laws, 92–94
state regulations, 144–145

statistics
    case studies, 174
    in public policy analysis, 172–174
stock markets, 73
strategic planning, 165–166
stress, 137
    critical incident, 139
strikes, 96
subzone rating factors, 3
supervisors, change opposed by, 51
Supreme Court, 81, 89, 151
surveys, public preference, 102–103
*Sutton et al v. United Air Lines*, 89

## T
tactics, management
    broad-based empowerment, 39
    chief officer tips, 38
    leading by example, 38–39
    technology-based programs, 37–38
    total quality management, 38
    walking around, 39
task force, 19–20
taxes
    avoidance, 73
    considerations, 71–73
        interstate, 73
    federal excise, 72
    fire service, 62
    flat, 72
    incidence analysis, 72–73
    income, 61
    progressive, 72
    property, 61, 72
    reduction, 73
    regressive, 72
    regulation, 146
    sales, 62
    use, 62
teams, 184. *See also* groups
    guiding coalition, 48–49
    union member relations, 95–96
technology, 184–185
technology-based programs, management, 37–38
technology-based training, 116–117
terminations, 91–92
terrorism, 7, 187
13th Amendment, 85
total cost purchasing, 68
total quality management, 38, 173
tracking expenditures, 66–68
tradition
    as barrier, 51
    power of, 20–21
tragedy, 110
training programs, 113–114, 186. *See also* education
    budget cut, 57
    costs of, 95, 117

training programs (cont.)
  efficacy of, 124
  firefighting, 114
    basic, 115–116
    collaborative, 115
    goals, 120–121
    live-fire, 115
    recertification, 117
    seminars, 117
    sessions, 117
    SOPs, 117–118
  hazardous materials, 118–119
  NIMS, 119–120
  officer, 122
  seminars, 117
  staffing and, 121
Trauma Care Systems Planning and Development Act, 5
Truman, Harry S., 24
trust, 43

## U

Unified Command, 24
unified federal emergency response, 6–8
unions, 15, 95–97
  bargaining, 23
  change resisted by, 51
  communication with, 95–96
  contracts, 145–146
  power of, 22–23
  private sector, 96
  public representatives and, 95
  public sector, 23, 95–96
United Air Lines, 89
United States, financial problem of, 73
universal professional standards, 4–5
University of Michigan, 81
*Urban Guide for Fire Prevention & Control Master Planning* (document), 165
urgency, change and, 46–48
Ury, William, 23, 97
use tax, 62
USFA. *See* Fire Administration, U.S.

## V

validation, 92–94
vehicle crashes, 150
vehicle speeds, 131
Vietnam War, 4
violence, 136–137
vision
  communication, 49–50, 55
  developing, 49
visual aids, 172
*The Voice* (publication), 107
voluntary termination, 91
volunteer departments, 97–98
voting, 32

## W

walking around, management by, 20, 41
  as tactic, 39
*Wall Street Journal*, 142
Washington, D.C., 6
wildfires, 69, 187
Wilson, P. F., 173
Wisconsin, 23
Woodhouse, E. J., 166
World War II, 24

## Y

*Your 21st Century Firefighter* (Page), 34

## Z

zero-based budgets, 57
zoning regulations, 145